Lecture Notes in Artificial Intelligence 4828

Edited by J. G. Carbonell and J. Siekmann

Subseries of Lecture Notes in Computer Science

T0241069

Marcus Randall   Hussein A. Abbass
Janet Wiles (Eds.)

Progress in
Artificial Life

Third Australian Conference, ACAL 2007
Gold Coast, Australia, December 4-6, 2007
Proceedings

 Springer

Series Editors

Jaime G. Carbonell, Carnegie Mellon University, Pittsburgh, PA, USA
Jörg Siekmann, University of Saarland, Saarbrücken, Germany

Volume Editors

Marcus Randall
Bond University
School of Information Technology
Robina, QLD 4229, Australia
E-mail: mrandall@bond.edu.au

Hussein A. Abbass
University of New South Wales
Australian Defence Force Academy
School of Information Technology & Electrical Engineering
Canberra, ACT 2600, Australia
E-mail: h.abbass@adfa.edu.au

Janet Wiles
The University of Queensland
School of Information Technology & Electrical Engineering
Division of Complex & Intelligent Systems
Brisbane, QLD 4072, Australia
E-mail: janetw@itee.uq.edu.au

Library of Congress Control Number: 2007939513

CR Subject Classification (1998): I.2, J.3, F.1.1-2, G.2, H.5, I.5, J.4, J.6

LNCS Sublibrary: SL 7 – Artificial Intelligence

ISSN   0302-9743
ISBN-10   3-540-76930-7 Springer Berlin Heidelberg New York
ISBN-13   978-3-540-76930-9 Springer Berlin Heidelberg New York

Springer is a part of Springer Science+Business Media

springer.com

© Springer-Verlag Berlin Heidelberg 2007
Printed in Germany

Typesetting: Camera-ready by author, data conversion by Scientific Publishing Services, Chennai, India
Printed on acid-free paper   SPIN: 12195459   06/3180   5 4 3 2 1 0

Preface

The field of artificial life (Alife) is a rapidly emerging area that draws on expertise from computer science, biology, psychology, to name a few. In essence it is the study of systems related to life, its processes and evolution. These systems commonly use computer model simulations. The past decade has seen an increasing stream of scientific articles devoted to the exploration of Alife.

The Australian Conference on Artificial Life (ACAL) series is a testament to the above. It is a biannual event that originated in 2001 as the "Inaugral Workshop on Artificial Life" as part of the 14th Joint Conference on Artificial Intelligence. ACAL 2007 received 70 quality submissions of which 34 were accepted for oral presentation in the conference. Each paper was peer reviewed by two or three members of the Program Committee. Apart from Australian researchers, the conference attracted participants from a number of countries across Europe, America, Asia-Pacific and Africa.

ACAL 2007 was fortunate to have four distinguished speakers in Alife to address the conference. They were David Abramson (Monash University), Kenneth A. De Jong (George Mason University), K.C. Tan (National University of Singapore) and Rodney Walker (Queensland University of Technology).

The organizers wish to thank a number of people and institutions for their support of this event and publication. Importantly we would like to acknowledge the effort and contributions of the Program Committee members and advisory board. Our sponsors were: The Australian Computer Society, the ARC Complex Open Systems Research Network, Bond University, The University of New South Wales (Australian Defence Force Academy), University of Canberra, Australian National University and the Gold Coast City Council. Their financial and in-kind support ensured the costs were minimized for attendees. Finally, the editors must pay tribute to the team at Springer.

We hope to repeat the success of ACAL 2007 with ACAL 2009. The venue of this event will be announced in 2008.

December 2007

Marcus Randall
Hussein A. Abbass
Janet Wiles

Organization

ACAL 2007 was organized by the School of Information Technology, Bond University in association with the University of New South Wales (Australian Defence Force Academy) and the University of Queensland.

Chairs

General Chair   Marcus Randall (Bond University, Australia)
Co-chairs   Hussein A. Abbass (University of New South
                         Wales, Australia)
                       Janet Wiles (University of Queensland,
                         Australia)

Program Committee

Mark Bedau (Reed College, USA)
Alan Blair (UNSW, Australia)
Eric Bonabeau (Icosystem, USA)
Juergen Branke (University of
  Karlsruhe, Germany)
Angelo Cangelosi (Plymouth
  University, UK)
Stephan Chalup (Newcastle University,
  Australia)
Tan Kay Chen (NUS, Singapore)
Vic Ciesielski (RMIT, Australia)
Oscar Cordon (University of Granada,
  Spain)
David Cornforth (UNSW@ADFA,
  Australia)
Marco Dorigo (ULB, Belgium)
Alan Dorin (Monash University,
  Australia)
Daryl Essam (UNSW@ADFA,
  Australia)
David Green (Monash University,
  Australia)
Tim Hendtlass (Swinburne University,
  Australia)

Takashi Ikegami (Tokyo University,
  Japan)
Christian Jacob (Calgary University,
  Canada)
Ray Jarvis (Monash University,
  Australia)
Graham Kendall (Nottingham
  University, UK)
Kevin Korb (Monash University,
  Australia)
Xiaodong Li (RMIT, Australia)
Frederic Maire (QUT, Australia)
Bob McKay (Seoul National
  University, Korea)
James Montgomery (Swinburne
  University, Australia)
Chrystopher Nehaniv (University of
  Hertfordshire, UK)
David Newth (CSIRO,
  Australia)
Stefano Nolfi (CNR-ISTC, Italy)
Marcus Randall (Bond University,
  Australia)
Alex Ryan (DSTO, Australia)

Russell Standish (UNSW, Australia)
Charles Taylor (UCLA, USA)
Jason Teo (Universiti Malaysia Sabah, Malaysia)
Athanasios Vasilakos (University of Western Macedonia, Greece)

Peter Wills (University of Auckland, NZ)
Janet Wiles (University of Queensland, Australia)

Additional Reviewers

Lucas Hope (Monash University)
Antony Iorio (RMIT, Australia)
Irene Moser (Swinburne University, Australia)
Andy Song (RMIT, Australia)
Gayan Wijesinghe (RMIT, Australia)
Owen Woodberry (Monash University)
Qinying Xu (RMIT, Australia)

International Advisory Committee

Mark Bedau (Reed College, USA)
Eric Bonabeau (Icosystem, USA)
David Fogel (Natural Selection, USA)
Peter Stadler (Leipzig, Germany)
Masanori Sugisaka (Oita, Japan)

Table of Contents

Complex Systems II

Heuristics III

Biological Systems II

Alternative Solution Representations for the Job Shop Scheduling Problem in Ant Colony Optimisation

James Montgomery

Complex Intelligent Systems Laboratory
Centre for Information Technology Research
Faculty of Information & Communication Technologies
Swinburne University of Technology
Melbourne, Australia
jmontgomery@ict.swin.edu.au

Abstract. Ant colony optimisation (ACO), a constructive metaheuristic inspired by the foraging behaviour of ants, has frequently been applied to shop scheduling problems such as the job shop, in which a collection of operations (grouped into jobs) must be scheduled for processing on different machines. In typical ACO applications solutions are generated by constructing a permutation of the operations, from which a deterministic algorithm can generate the actual schedule. An alternative approach is to assign each machine one of a number of alternative dispatching rules to determine its individual processing order. This representation creates a substantially smaller search space biased towards good solutions. A previous study compared the two alternatives applied to a complex real-world instance and found that the new approach produced better solutions more quickly than the original. This paper considers its application to a wider set of standard benchmark job shop instances. More detailed analysis of the resultant search space reveals that, while it focuses on a smaller region of good solutions, it also excludes the optimal solution. Nevertheless, comparison of the performance of ACO algorithms using the different solution representations shows that, using this solution space, ACO can find better solutions than with the typical representation. Hence, it may offer a promising alternative for quickly generating good solutions to seed a local search procedure which can take those solutions to optimality.

Keywords: Ant colony optimisation, job shop scheduling, solution representation.

1 Introduction

Ant colony optimisation (ACO) is a constructive metaheuristic, inspired by the foraging behaviour of ant colonies, that produces a number of solutions over successive iterations of solution construction. During each iteration, a number of artificial ants build solutions by probabilistically selecting from problem-specific

M. Randall, H.A. Abbass, and J. Wiles (Eds.): ACAL 2007, LNAI 4828, pp. 1–12, 2007.

solution components, influenced by a parameterised model of solutions (called a pheromone model in reference to ant trail pheromones). The parameters of this model are updated at the end of each iteration using the solutions produced so that, over time, the algorithm learns which solution components should be combined to produce the best solutions. When adapting ACO to suit a problem an algorithm designer must first decide how solutions are to be represented and built (i.e., what base *components* are to be combined to form solutions) and then what characteristics of the chosen representation are to be modelled.

Shop scheduling problems consist of a number of jobs, made up of a set of operations, each of which must be scheduled for processing on one of a number of machines. Precedence constraints are imposed on the operations of each job. The majority of ACO algorithms for these problems represent solutions as permutations of the operations to be scheduled (operations are the base components of solutions), which determines the relative order of operations that require the same machine (see, e.g., [1,2,3,4]). A deterministic algorithm can then produce the best possible schedule given the precedence constraints established by the permutation. This approach is more generally referred to as the *list scheduler algorithm* [2].

An alternative approach is to assign different heuristics to each machine which determine the relative processing order of operations, thereby searching the reduced space of schedules that can be produced by different combinations of the heuristics. Building solutions in this manner may offer an advantage by concentrating the search on heuristically good solutions. A previous study compared these two solution representations in ACO algorithms for a real-world job shop scheduling problem (JSP) with staggered release and due dates modelled using fuzzy sets [5]. Applied to that single real-world instance the alternative approach performed extremely well, finding better solutions than the list scheduler ACO in considerably less time. An open question was whether the same relative performance would be observed on other, benchmark JSP instances.

This paper examines, in greater detail than in [5], the search space produced by the alternative solution representation when applied to a number of commonly used benchmark JSP instances (Section 4). An empirical comparison is subsequently made of ACO algorithms using the typical and alternative solution construction approaches (Sections 5–6). Section 7 describes the implications of the results for the future application of ACO to such problems. A formal description of the JSP and further details of the typical solution construction approach are given first.

2   Job Shop Scheduling

The JSP examined in this study is of the $n \times m$ form, with a set of n jobs $J_1, \ldots, J_n$ and m machines $M_1, \ldots, M_m$. Each job consists of a predetermined sequence of m operations, each of which requires one of the m machines. Only one operation from a job may be processed at any given time, only one operation may use a machine at any given time and operations may not be pre-empted.

Table 1. JSP instances used in this study

Instance	Best known	n	m
abz5	1234	10	10
abz6	943	10	10
abz7	656	20	15
abz8	669	20	15
abz9	679	20	15
ft10	930	10	10
ft20	1165	20	5
la21	1046	15	10
la24	935	15	10
la25	977	15	10
la27	1235	20	10
la29	1152	20	10
la38	1196	15	15
la40	1222	15	15
orb08	899	10	10
orb09	934	10	10

The objective is to schedule operations for processing on machines such that the total time to complete all jobs, the *makespan*, is minimised. The makespan of a solution s is denoted $C_{max}(s)$.

Table 1 describes the instances used in this study to compare the alternative solution representations. They are commonly used benchmarks in the ACO and wider operations research literature and are all available from the OR-Library [6].

3   Typical Solution Construction for the JSP

To generate a solution to the JSP it is sufficient to determine the relative processing order of operations that require the same machine. A deterministic algorithm can then produce the best possible schedule given those constraints. Indeed, it is common in ACO applications for the JSP and other related scheduling problems to generate a permutation of the operations, which implicitly determines this relative order (e.g., [1,2,3,4,7]). These algorithms are restricted to creating permutations that respect the required processing order of operations within each job, which can consequently be called *feasible permutations*.

Different approaches to constructing solutions produce different search spaces. The space of feasible permutations of operations for a JSP is very large (a weak upper bound is $O(k!)$, where $k = n \cdot m$ is the number of operations) and is certainly much larger than the space of feasible schedules [8]. This space also has a slight bias towards good solutions, which can be exploited by some pheromone models and proves disastrous for others. Another notable feature of this search space is that while all solutions can be reached, solutions (schedules) are represented by differing numbers of permutations. These issues are discussed in some detail by Montgomery, Randall and Hendtlass [8,9].

4   Search Space Created by Dispatching Rules

An alternative approach to building solutions is to assign different *dispatching rules* (i.e., ordering heuristics) to each machine, which subsequently build the actual schedule. The search space then becomes the space of all possible combinations of rules assigned to machines, which is $O(|D|^m)$ where D is the set of rules and m the number of machines. Given a small number of dispatching rules this search space will correspond to a subset of the space of all feasible schedules. Further, given that dispatching rules are chosen with the aim of minimising the makespan or number of tardy jobs, this is probably the case even for large sets of rules. However, if the dispatching rules individually perform well it is expected that this reduced space largely consists of good quality schedules.

Clearly, such an approach is inappropriate for single machine scheduling problems or problems in which too few criteria are available to heuristically determine the processing order of competing operations, as in either situation the search space is reduced by too great an amount. It is, however, entirely appropriate for problems with multiple machines and various criteria upon which to judge competing operations. This study examines its application to a number of common benchmark JSPs using four dispatching rules. The remainder of this section examines whether, for these instances using these four rules, the approach is appropriate.

The four rules used in this study are Earliest Starting Time (EST), Shortest Processing Time (SPT), Longest Processing Time (LPT) and Longest Remaining Processing Time (LRPT). SPT and LPT relate to an individual operation's processing time while LRPT refers to the remaining processing time of a candidate operation's containing job. EST is perhaps the simplest heuristic, choosing the operation that can start the soonest, with ties broken randomly. Note that the three other rules are not followed blindly: the earliest available operation is always chosen except when there are two or more such operations, in which case the rule determines which is given preference.

For small instances and a set of four rules it is possible to completely enumerate the set of assignment solutions.[1] This was performed for the test instances with up to 200 operations to discover the distribution of the cost of schedules described. The distributions for the larger instances were estimated by sampling 4×10^6 randomly generated solutions. Note that as the EST rule breaks ties randomly, there is some degree of error in the lower and upper bounds presented, although it is likely the distributions described here are good approximations of the true distributions. Fig. 1 presents box-plots of the distributions discovered, expressed in terms of the relative percentage deviation (RPD) from the best known cost, defined as

---

[1] Although complete enumeration of the search space obviates the need for a metaheuristic, on any moderate-sized instance or as the number of rules grows it quickly becomes impractical.

$$RPD = \frac{C_{max}(s) - C_{max}(s^*)}{C_{max}(s^*)} \cdot 100 \tag{1}$$

where s is a solution and s^* is the best known solution.

The most striking feature of the distributions is that they *do not include the optimum.* Additionally, tests with a smaller number of rules found that many unique assignment solutions generate the same schedule, as was anticipated.[2] Nevertheless, it is still possible that the assignment approach does focus on a good region of the space of schedules, and thus may present a good starting point for the subsequent application of a local search algorithm. As the worst cost is not known for these instances it cannot be proved that these distributions are biased towards good solutions. However, examination of the cost distribution of schedules produced by randomly generated feasible *permutations* lends some support to that conjecture. Fig. 2 presents box-plots for the cost distributions for 4×10^6 randomly generated feasible permutations. Notably, the minima of those distributions are in most cases above the median of those for assignment solutions while the body of those distributions typically lies above the maximum of that for assignment solutions. Of course, sample distributions for the permutation approach do not represent the full space of solutions that can be represented by permutations and indeed an ACO algorithm constructing permutations can improve on the minima of those randomly generated samples (see Section 6 for such results).

Table 2 summarises the characteristics of the search spaces created by the alternative construction approaches. With respect to search space size, the space of assignments of rules to machines (for four rules) for the instances studied is hundreds of orders of magnitude smaller than the upper bound on the space of feasible permutations.

Clearly, the two alternative approaches offer a mixture of advantages and disadvantages to any heuristic that uses them. The likelihood that, across a wider range of instances, the dispatching rules approach excludes the optimal certainly impacts on its utility. However, a previous comparative study of ACO algorithms using both approaches applied to a large, complex JSP instance found that the approach outperformed an ACO algorithm that constructs permutations in terms of both solution quality and computation time [5].[3]

Nevertheless, in a practical application of the approach, a local search component is required if the schedules described by dispatching rules are to be fully optimised. Furthermore, the local search cannot operate on the assignments directly, as that space does not contain the optimum. The next section compares ACO algorithms using both solution construction approaches. To avoid the confounding effects of an integrated local search procedure, local search has not been included in the algorithms compared in this paper.

---

[2] Determining the number of *distinct* solutions was impractical with four rules.

[3] The number of construction steps per solution in ACO for the JSP is $n \cdot m$ when constructing permutations but only m when assigning dispatching rules.

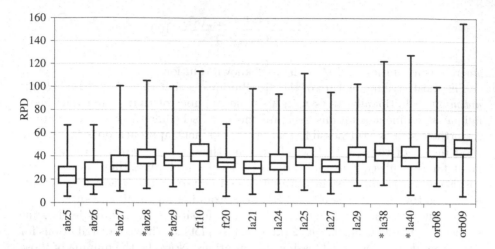

Fig. 1. Cost distributions (expressed as relative percentage deviation (RPD) from the best known) for solutions obtainable using EST, SPT, LPT and LRPT rules. Distributions marked with * are approximations based on 4×10^6 sampled solutions.

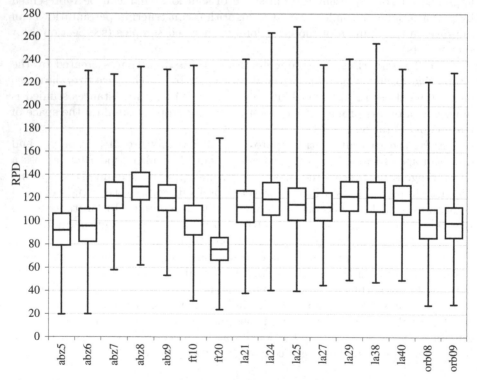

Fig. 2. Cost distributions (expressed as relative percentage deviation (RPD) from the best known) for 4×10^6 randomly generated permutation solutions. While the search space includes the optimum, it is unlikely to be found using random search.

Table 2. Comparison of permutation and dispatching rules search spaces. k is number of operations, D is set of dispatching rules, m is number of machines. Typically $|D| < m < k$. Notes: [#] this result is true for the rules and instances used in this study.

| | Solution approach | |
Search space feature	permutation	dispatching rules
Size	$\ll O(k!)$	$O(|D|^m)$
Includes optimal solution	yes	no[#]
Solution representation bias	yes	yes
Biased towards good solutions	yes, but not practically so [8,9]	yes

5 Comparison ACO Algorithm Details

Two ACO algorithms were developed based on the $\mathcal{MAX} - \mathcal{MIN}$ Ant System ($\mathcal{MMAS}$), which has been found to work well in practice [10]. The first of these, denoted $\mathcal{MMAS}$-P, constructs solutions as permutations of the operations, while the second, denoted $\mathcal{MMAS}$-R, assigns dispatching rules to machines. The set of dispatching rules D consists of the four rules described in Section 4. Although local search is considered an integral part of state-of-the-art ACO applications [11,12], in order to observe the differences between the two approaches, local search is not incorporated into either.

The two solution representations require different pheromone models. The models chosen have been found to produce the best performance for their respective solution representations [9]. For $\mathcal{MMAS}$-P, a pheromone value, denoted $\tau(o_i, o_j)$,[4] exists for each directed pair of operations that use the same machine, and represents the learned utility of operation o_i preceding operation o_j [13]. There may be several such precedence relations affected by the selection of a single operation. During solution construction, the set of unscheduled operations that require the same machine as a candidate operation o is denoted by O_o^{rel}. Blum and Sampels [13] recommend taking the minimum of the relevant pheromone values. This approach, like many ACO algorithms, benefits from the incorporation of heuristic information in the construction decision, by convention denoted η. While any dispatching rule could conceivably be used for this purpose, Blum and Sampels [2] have found that the EST rule works well on a range of instances. Accordingly,

$$\eta(o) = \frac{1}{t_{es}(o, s^p)} \tag{2}$$

where $t_{es}(o, s^p)$ is the earliest time operation o could start given the current partial solution s^p. Combining this measure with the pheromone information, at

---

[4] τ is historically used in ACO due to the pheromone model's inspiration in ant *trail* pheromones.

each step of solution construction, the probability of selecting an operation o to add to the partial permutation p is given by

$$
P(o,p) = \begin{cases} \dfrac{\left(\min_{o_r \in O_o^{rel}} \tau(o,o_r)\right) \cdot \eta(o)}{\sum_{o' \notin p} \left(\min_{o_r \in O_{o'}^{rel}} \tau(o',o_r)\right) \cdot \eta(o')} & \text{if } o \notin p \text{ and } |O_o^{rel}| > 0 \\[2ex] 1 & \text{if } o \notin p \text{ and } |O_o^{rel}| = 0 \\ 0 & \text{otherwise.} \end{cases} \tag{3}
$$

Note that the second branch is required so that the last operation on each machine is scheduled immediately, as there is no meaningful pheromone value that can be used.

For $\mathcal{MMAS}$-R, a pheromone value $\tau(M_k, d)$ is associated with each combination of machine and dispatching rule $(M_k, d) \in M \times D$, where M is the set of machines. At each step of solution construction, a machine is assigned a dispatching rule. Although the order in which assignments are made is significant in problems where certain items may only be assigned a limited number of times (e.g., in the generalised assignment problem [14]), here there is no limit to the number of times a rule can be used, so the assignment order is immaterial [5]. The probability of assigning a dispatching rule $d \in D$ to machine M_k is given by

$$
P(M_k, d) = \frac{\tau(M_k, d)}{\sum_{d' \in D \setminus \{d\}} \tau(M_k, d')}. \tag{4}
$$

Pheromone values are updated the same way in both algorithms, with each value τ (corresponding to some value from either model) updated according to

$$
\tau \leftarrow (\rho - 1)\tau + \rho \cdot \Delta\tau \tag{5}
$$

where ρ is the pheromone evaporation rate and $\Delta\tau$ is the amount of reinforcement given to a particular pheromone value determined by

$$
\Delta\tau = \begin{cases} \dfrac{1}{C_{max}(s)} & \text{if } \tau \text{ is part of iteration best solution} \\ 0 & \text{otherwise} \end{cases} \tag{6}
$$

where $C_{max}(s)$ is the makespan of the solution s. Pheromone values are bounded by $[\tau_{min}, \tau_{max}]$, the values of which are controlled using the value of the current best solution and size of the pheromone update in accordance with the rules defined by Stützle and Hoos [10].

6   Computational Results

The performance of the algorithms was compared on the benchmark instances described in Table 1. The algorithms were implemented in the C language and executed under Linux on a 3.2GHz Xeon processor. Each run used 100 ants

Table 3. Minimum, median, maximum and interquartile range (IQR) of solution cost (in RPD) for $\mathcal{MMAS}$-P and $\mathcal{MMAS}$-R. The last column shows the *estimated* best possible RPD in the space of dispatching rules used in this study. Bold values indicate the smaller value for that measure and that instance between $\mathcal{MMAS}$-P and $\mathcal{MMAS}$-R. *M-W test* indicates the direction of the difference between the distributions of RPD scores if the difference is statistically significant for $\alpha \leq 0.05$.

Instance	$\mathcal{MMAS}$-P				M-W test	$\mathcal{MMAS}$-R				lower bound
	min	med	max	IQR		min	med	max	IQR	
abz5	**2.6**	**4.3**	**6.3**	1.2	<	5.3	5.3	**5.3**	**0.0**	5.3
abz6	**0.4**	**2.4**	**4.0**	1.8	<	7.1	7.1	7.3	**0.0**	7.1
ft10	**8.6**	**13.5**	**14.8**	2.5	<	11.7	15.6	15.6	**0.4**	11.7
ft20	12.8	17.5	24.8	8.3	>	**5.9**	**7.1**	**8.2**	**0.6**	5.8
orb08	**9.7**	19.6	21.9	6.5		14.9	**18.0**	**18.4**	**0.3**	14.9
orb09	**1.5**	**6.3**	**9.3**	3.6	<	6.1	9.2	12.0	**3.5**	6.1
la21	**7.0**	**9.2**	11.6	2.3		7.8	9.3	**10.7**	**1.6**	7.6
la24	**7.6**	10.0	12.7	2.8		9.5	**9.5**	**9.5**	**0.0**	9.5
la25	**8.2**	**12.3**	13.8	4.5		12.5	13.1	**13.3**	**0.4**	11.2
la27	11.3	14.0	18.0	3.4	>	**8.3**	**10.1**	**10.9**	**1.5**	8.3
la29	**15.5**	**16.8**	**20.0**	**1.0**	>	15.6	16.1	16.2	**0.4**	15.1
la38	**12.7**	**14.7**	**17.1**	**2.3**	<	16.6	18.4	19.7	2.6	15.7
la40	**6.5**	**8.1**	**10.1**	**1.7**	<	7.4	9.0	10.4	2.0	7.4
abz7	12.2	**14.1**	19.1	2.7	>	**10.1**	10.9	**11.7**	**0.9**	9.9
abz8	14.1	16.2	19.1	3.4	>	**12.1**	**12.6**	**15.2**	**1.4**	12.1
abz9	18.3	20.3	27.5	2.0	>	**13.8**	**15.5**	**16.9**	**1.9**	13.8

and executed 500 iterations of solution construction. The $\mathcal{MMAS}$ pheromone decay control parameter $\rho = 0.1$. These settings were found to produce the best performance in both algorithms. Each algorithm and instance combination was executed across 10 random seeds.

6.1 Makespan

Table 3 describes, for each instance, the distributions of best solution cost (expressed in RPD) for $\mathcal{MMAS}$-P and $\mathcal{MMAS}$-R found across multiple runs of each algorithm. The instances appear in non-decreasing order of number of operations. Bold values indicate the smaller result within that instance and measure (min, median, max or interquartile range (IQR)) between the alternative algorithms. Although smaller values for IQR are not necessarily an indicator of better performance, they do indicate more consistent performance. To give an indication of the performance of $\mathcal{MMAS}$-R in exploring the space of assignments of dispatching rules, the last column gives the estimated lower bound on solution cost for each instance. Mann-Whitney tests were used to compare the distributions within each instance. Where those tests indicated a statistically significant result (at or below the 5% level), the central column indicates the direction of the difference (i.e., < means $\mathcal{MMAS}$-P outperformed $\mathcal{MMAS}$-R while > indicates the opposite).

Table 4. Median CPU time in seconds used to complete 500 iterations and until best solution found, and iteration when best solution found, for $\mathcal{MMAS}$-P and $\mathcal{MMAS}$-R

| Instance | Mean CPU time (s) | | | | Iteration when | |
| | total | | best solution | | best found | |
	$\mathcal{MMAS}$-P	$\mathcal{MMAS}$-R	$\mathcal{MMAS}$-P	$\mathcal{MMAS}$-R	$\mathcal{MMAS}$-P	$\mathcal{MMAS}$-R
abz5	23.7	3.1	2.7	0.1	58	11
abz6	23.7	2.9	2.5	0.7	52	124
ft10	23.5	2.9	5.3	0.1	113	17
ft20	46.4	3.7	17.5	1.7	189	238
orb08	22.9	2.9	4.6	0.6	100	102
orb09	23.6	3.1	4.7	2.0	100	324
la21	63.6	5.3	28.6	0.6	225	53
la24	63.3	5.2	19.6	0.3	155	31
la25	63.6	5.0	19.0	0.3	150	25
la27	130.6	8.1	58.9	1.9	226	114
la29	130.8	7.7	54.0	0.5	206	30
la38	118.3	8.4	32.2	0.6	136	35
la40	117.9	8.8	40.0	0.9	170	49
abz7	247.2	12.8	71.1	1.5	144	59
abz8	247.6	12.9	71.1	2.0	144	78
abz9	246.5	12.8	118.1	1.2	240	49

Based on these results, neither algorithm is clearly better than the other across all instances studied. The apparently aberrant statistical result for the la29 instance is because, even though $\mathcal{MMAS}$-P found a better solution on one of its runs, $\mathcal{MMAS}$-R produced solutions of similar cost more consistently. Considering just those instances where statistically significant differences were found there is an apparent trend showing better performance from $\mathcal{MMAS}$-R on larger instances, although this may be an effect of the actual instances used. In several cases $\mathcal{MMAS}$-R was able to locate assignment solutions at the estimated (for large instances) lower bound for the space it searches. Notably, it appears that, in the absence of a local search procedure, the traditional construction approach is unable to find the optimal solution even though it exists in the space of solutions it searches. Thus both algorithms require local search in order to find optimal solutions.

6.2 CPU Time

Table 4 summarises the median computation time required to complete 500 iterations and until the best solution was found, as well as the iteration in which the best solution was found. As predicted, $\mathcal{MMAS}$-R is significantly faster than $\mathcal{MMAS}$-P due to the difference in the number of required construction steps each iteration—as the number of operations grows the ratio between $\mathcal{MMAS}$-P's and $\mathcal{MMAS}$-R's runtimes approaches the number of jobs n. $\mathcal{MMAS}$-R also frequently locates its best solution after fewer iterations than $\mathcal{MMAS}$-P. The faster execution of $\mathcal{MMAS}$-R commends it as a good alternative for integration with a potentially computationally intensive local search, and would also allow for a greater number of separate runs of the algorithm to be performed than $\mathcal{MMAS}$-P given the same amount of time.

7 Conclusions

Typical ACO algorithms for shop scheduling problems such as the JSP build solutions as permutations of the operations to be scheduled, from which actual schedules are generated deterministically. An alternative approach when the problem has multiple machines and various criteria upon which to judge the urgency of competing operations is to assign different dispatching rules to each machine. The chosen dispatching rules are then responsible for determining the relative processing order of operations on each machine.

This paper examined the solution space produced by the space of dispatching rule assignments on a number of commonly studied benchmark JSP instances. Crucially, when using the four dispatching rules examined in this paper, that space does not contain the optimal solution. Given that dispatching rules are themselves simple heuristics, it is plausible that even with a vastly expanded range of rules the optimal solution may still be out of reach. Consequently, any real-world application employing this solution representation not only requires a local search component, but that local search must work directly on the schedules described by the dispatching rules and not the pattern of assignments.

Despite this severe drawback to the alternative solution representation, it does appear to concentrate the search on promising areas of the solution space and, in a constructive algorithm such as ACO, leads to a dramatic reduction in required computation. A comparison of ACO algorithms employing both the traditional solution representation and the alternative show a mixture of results, with neither algorithm clearly outperforming the other across the test instances. However, a slight trend for better performance from the new approach on the larger instances, coupled with its reduced computation times, suggest that it is a good candidate for seeding a local search procedure. As there is an unavoidable interaction between ACO and the local search procedure it uses (as the locally optimised solutions are used to update pheromone information), future work could examine the relative performance of the two approaches when local search is incorporated.

References

1. Bauer, A., Bullnheimer, B., Hartl, R.F., Strauss, C.: Minimizing total tardiness on a single machine using ant colony optimization. Cent. Eur. J. Oper. Res. 8, 125–141 (2000)
2. Blum, C., Sampels, M.: An ant colony optimization algorithm for shop scheduling problems. J. Math. Model. Algorithms 3, 285–308 (2004)
3. Colorni, A., Dorigo, M., Maniezzo, V., Trubian, M.: Ant system for job-shop scheduling. JORBEL 34, 39–53 (1994)

4. Stützle, T.: An ant approach to the flow shop problem. In: EUFIT 1998. 6th European Congress on Intelligent Techniques & Soft Computing, Aachen, Germany, pp. 1560–1564. Verlag Mainz (1998)
5. Montgomery, J., Fayad, C., Petrovic, S.: Solution representation for job shop scheduling problems in ant colony optimisation. In: Dorigo, M., Gambardella, L.M., Birattari, M., Martinoli, A., Poli, R., Stützle, T. (eds.) ANTS 2006. LNCS, vol. 4150, pp. 484–491. Springer, Heidelberg (2006)
6. Beasley, J.E.: OR-Library (2005),
   http://people.brunel.ac.uk/~mastjjb/jeb/info.html
7. van der Zwaan, S., Marques, C.: Ant colony optimisation for job shop scheduling. In: GAAL 1999. 3rd Workshop on Genetic Algorithms and Artificial Life (1999)
8. Montgomery, J., Randall, M., Hendtlass, T.: Structural advantages for ant colony optimisation inherent in permutation scheduling problems. In: Ali, M., Esposito, F. (eds.) IEA/AIE 2005. LNCS (LNAI), vol. 3533, pp. 218–228. Springer, Heidelberg (2005)
9. Montgomery, J., Randall, M., Hendtlass, T.: Solution bias in ant colony optimisation: Lessons for selecting pheromone models. Computers & Operations Research (in press), available online: doi:10.1016/j.cor.2006.12.014
10. Stützle, T., Hoos, H.: $\mathcal{MAX} - \mathcal{MIN}$ ant system. Future Gen. Comp. Sys. 16, 889–914 (2000)
11. Dorigo, M., Stützle, T.: The ant colony optimisation metaheuristic: Algorithms, applications and advances. In: Glover, F., Kochenberger, G. (eds.) Handbook of Metaheuristics. International Series in Operations Research and Management Science, vol. 57, pp. 251–285. Kluwer Academic Publishers, Boston, MA (2002)
12. Dorigo, M., Stützle, T.: Ant Colony Optimization. MIT Press, Cambridge (2004)
13. Blum, C., Sampels, M.: Ant colony optimization for FOP shop scheduling: A case study on different pheromone representations. In: 2002 Congress on Evolutionary Computation, pp. 1558–1563 (2002)
14. Montgomery, J., Randall, M., Hendtlass, T.: Search bias in constructive metaheuristics and implications for ant colony optimisation. In: Dorigo, M., Birattari, M., Blum, C., Gambardella, L.M., Mondada, F., Stützle, T. (eds.) ANTS 2004. LNCS, vol. 3172, pp. 390–397. Springer, Heidelberg (2004)

Analyzing the Role of "Smart" Start Points in Coarse Search-Greedy Search

Stephen Chen[1], Ken Miura[2], and Sarah Razzaqi[3]

[1] School of Information Technology, York University
4700 Keele Street, Toronto, Ontario M3J 1P3
sychen@yorku.ca
[2] Institute for Aerospace Studies, University of Toronto,
4925 Dufferin Street, Toronto, Ontario M3H 5T6
ken.miura@utoronto.ca
[3] Centre for Hypersonics, Division of Mechanical Engineering
University of Queensland, Brisbane, Australia 4072
s.razzaqi@uq.edu.au

Abstract. An inherent assumption in many search techniques is that information from existing solution(s) can help guide the search process to find better solutions. For example, memetic algorithms can use information from existing local optima to effectively explore a globally convex search space, and genetic algorithms assemble new solution candidates from existing solution components. At the extreme, the quality of a random solution may even be used to identify promising areas of the search space to explore. The best of several random solutions can be viewed as a "smart" start point for a greedy search technique, and the benefits of "smart" start points are demonstrated on several benchmark and real-world optimization problems. Although limitations exist, "smart" start points are most likely to be useful on continuous domain problems that have expensive solution evaluations.

Keywords: Heuristic Search, Fitness Landscapes, Coarse Search-Greedy Search.

1 Introduction

The easiest way to improve the performance of a greedy search technique is to run it several times and return the best solution. Effectively, this procedure leads to a random search in the (sub)space of local optima. Memetic algorithms can perform "a special kind of ... search over the subspace of local optima" [14] that can be particularly effective in globally convex search spaces [2, 6, 13]. However, if the number of local optima that can be generated is extremely small, then the population required by a memetic algorithm may not be possible.

The cost of finding local optima can be extremely high in certain real-world problems where solutions are evaluated by using a complex simulation (e.g. designing phased array ultrasonic transducers [4]). In these search spaces, it may be beneficial to use a coarse search technique to find and select the start points that will be used to seed a greedy search technique. Since the greedy search technique will optimize the

M. Randall, H.A. Abbass, and J. Wiles (Eds.): ACAL 2007, LNAI 4828, pp. 13–24, 2007.
© Springer-Verlag Berlin Heidelberg 2007

best preliminary solutions found by the coarse search process, there is an implicit assumption in coarse search-greedy search that the quality of the local optima are directly related to the quality of the initial (partially optimized) solutions.

The coarsest coarse search technique is random search. The use of random search as the coarse search technique also takes the relationship between start points and end points to an extreme – is there a relationship between the fitness of a random start point and the quality of its (nearby) local optima? On certain problems like the Travelling Salesman Problem (TSP), there is no such relationship. However, this relationship has been found (and exploited) on several benchmark and real-world optimization problems with continuous domains.

Start points found by random search are called "smart" start points (because they are better than random). When a relationship exists between the quality of a random start point and the quality of its local optimum, "smart" start points can be used to improve the performance of a greedy search technique. Conversely, "smart" start points provide no benefit to the performance of a greedy search technique on a problem like the TSP where there is no relationship between the quality of a random start tour and a (random) two-opt solution.

The benefits of "smart" start points clearly depend on many characteristics of the search space and the greedy search technique. For example, a problem on which the search technique can always find the globally optimal solution will not need "smart" start points. Conversely, a search technique that finds the nearest local optimum in a highly multi-modal search space will definitely benefit from having a better starting point. If many starting points can be explored, population-based search strategies like memetic algorithms are likely to be quite effective. However, if the number of starting points that can be explored is limited, then coarse search-greedy search may be more effective, and it is thus useful to understand the role of "smart" start points.

To prepare the context for "smart" start points, a brief review of related search techniques is presented in section 2. Section 3 focuses on the Travelling Salesman Problem where the use of "smart" start points provides no advantage over random tours. Sections 4 and 5 present results for benchmark optimization problems in the continuous domain, and section 6 demonstrates that these results can be meaningfully exploited on a real-world problem. A discussion of these results follows in section 7 before the conclusions in section 8.

2 Background

There have been many attempts to improve the performance of greedy search techniques by combining them with other (coarser) search techniques that "intelligently" select the start point(s) to be optimized. For example, simulated annealing can be viewed as performing a coarse search at higher temperatures for the start point that will eventually be optimized at lower temperatures [3]. Similarly, WoSP [8] can use a "higher energy" particle swarm (PSO) [10] (that does not converge fully) to find the start points for a separate greedy search technique.

The subspace of local optima can also be searched by using memetic algorithms. In a globally convex search space where the best locally optimal solutions share many similarities [2], a good start solution can be characterized as a solution that has many features in common with existing local optima. Start solutions with these characteristics can be created by using crossover operators (e.g. [6,14]).

The fundamental requirement for all of the above search techniques is time – time for an adequate cooling schedule, time for the particle swarm to (partially) converge, or time to generate a large population of local optima. Greedy heuristics frequently result from the need for time-compressed decisions. Fast computing has created an abundance of computational time (to employ more thorough search techniques) for many optimization problems. However, fast computing has also created the opportunity to optimize new problems that previously could not even be modelled.

In optimization problems where the computational cost of modelling and/or simulating a single solution is exceptionally high, there is still the need for highly efficient search techniques. If the number of local optima that can be generated is too small to create a viable population, then memetic algorithms will not be feasible. If the time constraints are satisfied by using rapid cooling schedules or coarser PSOs, then a relationship between the quality of a (somewhat) random solution and the quality of its corresponding local optimum will still be required. Therefore, there is value in knowing at the extreme if there is a relationship between the quality of a random solution and the quality of its (nearby) local optima.

3 TSP Data

In the Travelling Salesman Problem (TSP), the goal is to minimize the cost of visiting every city exactly once before returning home. Each solution of a TSP is a Hamiltonian cycle, so every (random) solution can be turned into the optimal solution through a finite series of two-opt swaps. It is proposed that it may be possible for any typical random solution (which has an average of one edge in common with the optimal solution) to be transformed into the optimal solution through a series of two-opt swaps that decreases the length of the tour after each swap. It could then be similarly possible to transform the same random solution into any other two-opt optimum through a (different) series of two-opt swaps that also decreases the length with each swap.

If each (typical) random solution can become any local optimum, then there should be no relationship between the quality (where higher quality means a shorter length) of a random TSP solution and a (random) two-opt solution that is generated from it. In the following experiment, 120 random solutions are generated for each of the 15 Euclidean TSP instances from TSPLIB that has between 1000 and 2000 cities. Random two-opt swaps are then applied, and any swap that increases the length is rejected and any swap that reduces the length is accepted. The quality of the final two-opt solutions is then compared for the 30 best and 30 worst random start tours.

For each set of 120 random solutions, the quality of the 30 best and 30 worst random solutions is shown in Table 1. Due to the clear separation between each set of best and worst solutions, the applied t-test easily shows that the best start solutions are indeed significantly better. However, there is no significant difference (to the one in twenty level) between the quality of two-opt solutions that are subsequently generated from the best and the worst random start solutions (see Table 2). In fact, the two-opt solutions generated from the worst initial tours are slightly better on 8 of the 15 TSP instances.

Table 1. Average (avg.) percent above known optimum and standard deviation (std. dev.) for the 30 best and 30 worst of 120 random TSP solutions. The two-tailed, homescedastic t-test (used for all experiments) confirms the clear separation of the two data sets.

Instance	Best		Worst		t-test
	avg.	std. dev.	avg.	std. dev.	
pr1002	2078%	22%	2178%	26%	0.0%
u1060	2434%	16%	2544%	24%	0.0%
vm1084	3402%	24%	3518%	21%	0.0%
pcb1173	2039%	12%	2127%	14%	0.0%
d1291	2898%	18%	3017%	23%	0.0%
rl1304	3500%	28%	3627%	21%	0.0%
rl1323	3444%	17%	3548%	25%	0.0%
nrw1379	2004%	12%	2091%	13%	0.0%
fl1400	7710%	94%	8052%	65%	0.0%
u1432	2055%	13%	2128%	10%	0.0%
fl1577	5211%	38%	5408%	41%	0.0%
d1655	2962%	23%	3057%	16%	0.0%
vm1748	4244%	27%	4380%	29%	0.0%
u1817	3122%	16%	3216%	13%	0.0%
rl1889	4463%	26%	4575%	20%	0.0%

Table 2. Data (similar to Table 1) for the two-opt solutions generated from the 30 best and 30 worst random start solutions. Values in bold represent an (insignificant) inverse relationship.

Instance	Best		Worst		t-test
	avg.	std. dev.	avg.	std. dev.	
pr1002	13.1%	1.1%	**12.7%**	1.0%	20.2%
u1060	12.7%	1.1%	12.8%	1.4%	71.7%
vm1084	12.4%	1.6%	13.0%	1.4%	12.7%
pcb1173	14.1%	1.0%	14.3%	1.2%	66.3%
d1291	16.9%	2.0%	**16.8%**	1.9%	76.0%
rl1304	15.0%	1.9%	15.2%	2.1%	82.2%
rl1323	14.5%	1.6%	**14.2%**	1.9%	59.2%
nrw1379	12.5%	0.9%	**12.3%**	0.8%	39.4%
fl1400	8.7%	1.7%	8.9%	2.1%	67.2%
u1432	14.1%	0.8%	14.1%	1.0%	94.2%
fl1577	14.4%	2.5%	**13.4%**	2.4%	13.0%
d1655	15.6%	1.3%	**15.3%**	1.4%	36.8%
vm1748	13.1%	1.0%	**12.6%**	1.3%	7.8%
u1817	17.9%	1.2%	18.1%	1.1%	45.1%
rl1889	14.9%	1.4%	**14.8%**	1.4%	78.3%

On the TSP, there is no relationship between the quality of a random start tour and the quality of a random two-opt solution that is generated from it. Therefore, there is no expectation that "smart' start points will help two-opt find better final solutions than random start tours. This expectation is confirmed by the following experiment in which 50 random tours are generated, the four best are optimized by two-opt, and the best solution found from these "smart" start points is compared against the best of four random two-opt solutions (e.g. the 1st, 31st, 61st, and 91st random two-opt solutions generated for the previous experiments). In Table 3, it can be seen that there is no significant difference between using random tours and "smart" start points, and that the "smart" start points even lead to slightly worse final solutions on 11 of the 15 TSP instances.

Table 3. Data (similar to Table 1) for the best of four random two-opt solutions and the solutions found with "smart" start points. Values in bold help confirm that "smart" start points provide no benefit on the TSP.

Instance	"Smart" Start Points		Four Random Tours		t-test
	avg.	std. dev.	avg.	std. dev.	
pr1002	11.6%	0.8%	**11.5%**	1.0%	48.5%
u1060	11.6%	0.8%	**11.5%**	0.9%	42.6%
vm1084	11.4%	0.8%	**10.9%**	1.1%	6.9%
pcb1173	13.0%	0.6%	**13.0%**	0.7%	82.0%
d1291	15.2%	1.4%	**15.0%**	1.4%	49.7%
rl1304	12.7%	1.6%	12.9%	1.6%	60.9%
rl1323	12.3%	1.3%	12.4%	1.1%	69.2%
nrw1379	11.7%	0.5%	**11.6%**	0.5%	77.6%
fl1400	6.2%	1.0%	6.6%	0.8%	6.0%
u1432	13.3%	0.5%	13.4%	0.5%	66.7%
fl1577	10.8%	1.9%	**10.5%**	1.6%	44.5%
d1655	14.5%	0.9%	**14.1%**	0.8%	5.7%
vm1748	11.7%	0.8%	**11.6%**	0.9%	61.7%
u1817	16.9%	0.9%	**16.6%**	1.2%	26.3%
rl1889	13.1%	1.0%	**13.0%**	0.9%	60.9%

4 Introductory Data

The relationship between the quality of starting points and their associated local optima is expected to be higher for optimization problems with continuous domains. In particular, it is possible to map any (non-maximal) start point to a single local optimum in a one-dimensional problem (see Figure 1). In higher dimensions, it may be possible to reach multiple optima with a contour-following greedy algorithm, but it is unlikely that the entire search space will be reachable from any (typical) random start point. Thus, a random start point "maps" to a set of possible local optima, and the quality of the random start point can be meaningfully related to the quality of the local optima that a greedy search technique can reach from it.

$$f(x) = |x| - \cos(5x)$$
$$x \in [-10, 10] \tag{1}$$

The sample function (1) is shown in Figure 1. From a given start point, contour following (or calculation) will lead to the nearest local optimum. Similar to the TSP data, the results shown in Table 4 compare the 30 best and 30 worst of 120 random points. The strong correlation between the quality of the start and end points is easily exploited by the "smart" start points to improve upon the performance of random start points.

5 Benchmark Data

The sample function in Figure 1 represents a trivial and highly idealized (globally convex) search space with smooth local optima wells of similar size. Since the outer local minima are worse solutions than the local maxima solutions that can lead to the

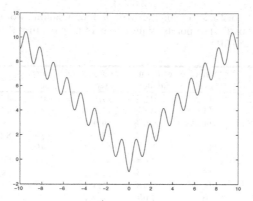

Fig. 1. Profile of sample function

Table 4. Average (avg.) and standard deviation (std. dev.) for the 30 best and 30 worst of 120 random points and their subsequent local optima on the sample function. A better starting point is a clear advantage for a highly local optimization technique, and this leads to a strong advantage for "smart" start points.

Start Points	30 Best		30 Worst		t-test
	avg.	std. dev.	avg.	std. dev.	
	1.01	0.75	8.39	1.00	0.0%
Local Optima	From 30 Best		From 30 Worst		t-test
	avg.	std. dev.	avg.	std. dev.	
	0.17	0.87	7.42	1.20	0.0%
	"Smart" Start Points		Four Random Points		t-test
	avg.	std. dev.	avg.	std. dev.	
	-0.79	0.48	0.55	1.46	0.0%

global minimum, there is an obvious relationship between the quality of a random start point and the quality of its corresponding local optimum. The following experiments on less-trivial benchmark problems attempt to determine if and when the above relationship exists for more realistic situations.

The three benchmark problems used in this study are Ackley, Rastrigin, and Schwefel. The Ackley and Rastrigin problems have similar search spaces to the sample function in one-dimension. However, these problems become much more interesting in p = 30 dimensions and when a standard optimization method like fmincon is used. The fmincon function is a greedy, gradient-based search technique that is available in the MATLAB® Optimization Toolbox.

$$f(x) = 20 + e - 20\exp(-0.2\sqrt{\frac{1}{p}\sum_{i=1}^{p} x_i^2}) - \exp(\frac{1}{p}\sum_{i=1}^{p}\cos(2\pi x_i)) \tag{2}$$

$$x \in [-30,30]$$

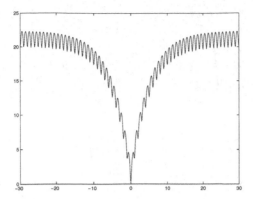

Fig. 2. Profile of Ackley function in one dimension

A profile of the Ackley function (2) in one dimension is shown in Figure 2. This function easily traps gradient-based search techniques, so there is little change between the function values for the start and end points. Since the end points are close to the start points, and since the search space has consistent undulations on top of a convex base function, there is an inherent relationship between the quality of the start points and the quality of the end points.

Using 120 random start points, the function values for the 30 best and 30 worst of these points is shown in Table 5. Due to the steepness of the search space, all of these start points appear to be in the broad plateau. There is a small range in magnitude among the quality of the start points, and the difference between the quality of the start points and the end points is also small. However, this difference is quite significant (as shown by the calculated t-tests), so there is still a clear benefit to using "smart" start points with fmincon on the Ackley function.

$$f(x) = 10p + \sum_{i=1}^{p}(x_i^2 - 10\cos(2\pi x_i))$$
$$x \in [-5.12, 5.12]$$

(3)

Table 5. Data (similar to Table 4) for the Ackley function. Consistency in the performance of the local optimization technique leads to a significant benefit to using "smart" start points.

Start Points	30 Best		30 Worst		t-test
	avg.	std. dev.	avg.	std. dev.	
	20.8	0.17	21.3	0.08	0.0%
fmincon Solutions	From 30 Best		From 30 Worst		t-test
	avg.	std. dev.	avg.	std. dev.	
	19.16	0.24	19.52	0.09	0.0%
	"Smart" Start Points		Four Random Points		t-test
	avg.	std. dev.	avg.	std. dev.	
	18.83	0.21	19.15	0.23	0.0%

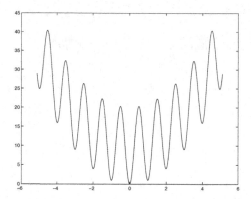

Fig. 3. Profile of the Rastrigin function in one dimension

A profile of the Rastrigin function (3) in one dimension is shown in Figure 3, and it should be noted that this profile is quite similar to that of the sample function shown in Figure 1. However, compared to a contour-following algorithm that moves to the nearest local minimum, the next test point used by `fmincon` may actually be in a different "valley" of the search space. Thus, `fmincon` should perform better than contour following on the Rastrigin function (i.e. be less dependent on the start point), and there should be less of a relationship between the quality of start and end points.

The results of the experiments for `fmincon` on the Rastrigin function are given in Table 6. Although there is a significant difference between the quality of end points for the best and worst start points, there is also a high variation in the consistency of `fmincon`. On problems with a high variation, random restart of the greedy search technique should be very effective – a set of random solutions will likely contain solutions that are both much better and much worse than the average. This effectiveness in random restart (and/or the high variation in the performance of the greedy search technique) causes the benefits of "smart" start points to become insignificant on the Rastrigin function.

$$f(x) = 418.9829 + \sum_{i=1}^{p} x_i \sin(\sqrt{|x_i|})$$

$$x \in \left[-512.03, 511.97\right]$$

(4)

Table 6. Data (similar to Table 4) for the Rastrigin function. The high variation in the quality of end points allows random start points to be nearly as effective as the "smart" start points.

Start Points	30 Best		30 Worst		t-test
	avg.	std. dev.	avg.	std. dev.	
	475.3	26.4	629.1	27.9	0.0%
`fmincon` Solutions	From 30 Best		From 30 Worst		t-test
	avg.	std. dev.	avg.	std. dev.	
	161.7	47.7	228.2	48.9	0.0%
	"Smart" Start Points		Four Random Points		t-test
	avg.	std. dev.	avg.	std. dev.	
	149.5	32.9	152.1	38.1	78.0%

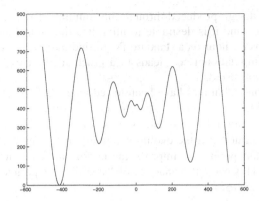

Fig. 4. Profile of the Schwefel function in one dimension

A profile of the Schwefel function (4) in one dimension is shown in Figure 4. This function is particularly challenging because the second best minimum (which traps many search techniques) is very far from the global minimum. It appears that fmincon is such a search technique that is easily trapped – it appears to produce many final solutions that have values around the second best minimum (see Table 7). Subsequently, there is no (significant) relationship between the quality of start and end points, and there is no benefit to using "smart" start points on the Schwefel function.

Table 7. Data (similar to Table 4) for the Schwefel function. A weak (and inverse) correlation between the quality of the start and end points leads to no possibility of a benefit to using "smart" start points.

Start Points	30 Best		30 Worst		t-test
	avg.	std. dev.	avg.	std. dev.	
	11363	446	13995	583	0.0%
fmincon Solutions	From 30 Best		From 30 Worst		t-test
	avg.	std. dev.	avg.	std. dev.	
	5502	627	**5351**	747	40.0%
	"Smart" Start Points		Four Random Points		t-test
	avg.	std. dev.	avg.	std. dev.	
	5138	694	**4680**	518	0.5%

6 Real-World Data

There are many ways to find start points to seed a greedy search technique (e.g. [8,14]). The reason to use random search to find "smart" start points is because these other search techniques can be prohibitively expensive on certain problems. For example, there are problems on which the evaluation of a solution involves a complex and time-consuming simulation.

The following data are from optimizing the design of phased array ultrasonic transducers. In this problem, each real-valued search point is converted into a design for an ultrasonic transducer through the simulation and evaluation of a physics-based model [7,11]. The design produced through this simulation will consist of an integer number of elements, and it is desirable to minimize this number. From a given start point, it can take over 1 hour on a standard PC to find a locally optimal transducer design. This high computational cost leads to a practical limit on the number of start points that can be optimized.

Two optimization techniques have been tested extensively on the ultrasonic transducer design problem – gradient descent in the form of fmincon from the MATLAB® Optimization Toolbox and a $(1+\lambda)$-evolution strategy. Previous experiments have demonstrated that the evolution strategy performs better than fmincon, and that "smart" start points can improve the performance of this evolution strategy (see Table 8) [4]. However, the analysis of "smart" start points has not previously been extended to fmincon.

Table 8. Data (similar to Table 4) for the ultrasonic transducer design problem. Although quite robust, the performance of the $(1+\lambda)$-evolution strategy still receives significant benefits from using "smart" start points.

Start Points	30 Best		30 Worst		t-test
	avg.	std. dev.	avg.	std. dev.	
	116.5	63.3	736.3	83.2	0.0%
$(1+\lambda)$-ES Solutions	From 30 Best		From 30 Worst		t-test
	avg.	std. dev.	avg.	std. dev.	
	31.7	3.1	34.1	5.7	4.2%
	"Smart" Start Points		Four Random Points		t-test
	avg.	std. dev.	avg.	std. dev.	
	30.1	3.2	31.3	2.9	2.2%

An evolution strategy (ES) tends to perform better than gradient-based search techniques in highly multi-modal search spaces because it is less prone to getting trapped in (poor) local optima [1]. Subsequently, "smart" start points should provide much greater benefits to fmincon. Using the same 120 random start points, the results for fmincon on the 30 best and 30 worst of these points is shown in Table 9.

Table 9. Data (similar to Table 4) for the ultrasonic transducer design problem. The inconsistent performance of fmincon leads to greater benefits for using "smart" start points. These benefits essentially compensate for the greater robustness of the evolution strategy.

Start Points	30 Best		30 Worst		t-test
	avg.	std. dev.	avg.	std. dev.	
	116.5	63.3	736.3	83.2	0.0%
fmincon Solutions	From 30 Best		From 30 Worst		t-test
	avg.	std. dev.	avg.	std. dev.	
	42.2	16.2	90.0	78.1	0.2%
	"Smart" Start Points		Four Random Points		t-test
	avg.	std. dev.	avg.	std. dev.	
	30.2	2.5	33.0	4.4	0.4%

The results from the random start points demonstrate that the $(1+\lambda)$-ES is a much more robust search technique and that fmincon frequently gets trapped in poor local optima. Subsequently, the performance of fmincon is much more dependent upon the quality of its initial starting point than the performance of the $(1+\lambda)$-ES. This strong relationship between the quality of the start and end points leads to a similarly strong benefit to using "smart" start points, and the subsequent performance of fmincon is essentially the same as the $(1+\lambda)$-ES when "smart" start points are used. For this specific optimization problem, there is a similar advantage to finding good start points as there is to developing a more effective and robust search technique.

7 Discussion

The improvement of greedy search techniques can follow one of two primary strategies – escape from local optima to find better ones (e.g. simulated annealing) or explore multiple optima independently (e.g. memetic algorithms). Results in the literature suggest that exploring multiple optima (e.g. [12]) is more popular and successful than attempting to escape from local optima (e.g. [9]).

The use of multiple runs introduces a new design consideration – a selection strategy for the start points is required. This strategy may be trivial (e.g. random search) or more complex (e.g. WoSP [8]). "Smart" start points are on the more simplistic end of the spectrum, so their practical benefits are likely limited to optimization problems with extremely expensive evaluations.

As more function evaluations become available, the justification to use random search as the coarse search strategy will lessen. However, the search space feature required for "smart" start points to be effective (i.e. a strong correlation between the quality of the start and end points) is likely to be an important indicator in the effectiveness of other coarse search-greedy search implementations. This search space feature has been successfully exploited in a PSO-ES coarse search-greedy search technique for the design of phased array ultrasonic transducers [5].

8 Conclusions

A relationship between the quality of a random start point and its (nearby) local optima can exist for some optimization problems. When this relationship exists, the use of "smart" start points (found by random search) can perform better than the random restart of a greedy search technique. This performance improvement can be useful on optimization problems with expensive solution evaluations. More importantly, this relationship can be a useful indicator of success for other coarse search-greedy search implementations.

Acknowledgements

This work has received funding support from the Natural Sciences and Engineering Research Council of Canada and the Atkinson Faculty of Liberal and Professional Studies, York University.

References

1. Beyer, H.-G., Schwefel, H.-P.: Evolution Strategies: A comprehensive introduction. Natural Computing 1, 3–52 (2002)
2. Boese, K.D.: Models for Iterative Global Optimization. Ph.D. diss., Computer Science Department, University of California at Los Angeles (1996)
3. Brünger, A.T., Krukowski, A., Erickson, J.W.: Slow-cooling protocols for crystallographic refinement by simulated annealing. Acta Crystallographica A46, 585–593 (1990)
4. Chen, S., Razzaqi, S., Lupien, V.: An Evolution Strategy for Improving the Design of Phased Array Transducers. In: Proceedings of the 2006 IEEE Congress on Evolutionary Computation, pp. 2859–2863. IEEE Press, Los Alamitos (2006)
5. Chen, S., Razzaqi, S., Lupien, V.: Towards the Automated Design of Phased Array Ultrasonic Transducers – Using Particle Swarms to find "Smart" Start Points. In: Okuno, H.G., Moonis, A. (eds.) Proceedings of the 20th International Conference on Industrial, Engineering and Other Applications of Applied Intelligent Systems. LNCS, vol. 4570, pp. 313–323. Springer, Heidelberg (2007)
6. Chen, S., Smith, S.F.: Putting the "Genetics" back into Genetic Algorithms (Reconsidering the Role of Crossover in Hybrid Operators). In: Banzhaf, W., Reeves, C. (eds.) Foundations of Genetic Algorithms 5, Morgan Kaufmann, San Francisco (1999)
7. Hassan, W., Vensel, F., Knowles, B., Lupien, V.: Improved Titanium Billet Inspection Sensitivity through Optimized Phased Array Design, Part II: Experimental Validation and Comparative Study with Multizone. In: AIP Conference Proceedings. Quantitative Nondestructive Evaluation, vol. 820, pp. 861–868. AIP (2006)
8. Hendtlass, T.: WoSP: A Multi-Optima Particle Swarm Algorithm. In: Proceedings of the 2005 IEEE Congress on Evolutionary Computation, vol. 1, pp. 727–734. IEEE Press, Los Alamitos (2005)
9. Johnson, D.S., McGeoch, L.A.: The Traveling Salesman Problem: A Case Study in Local Optimization. In: Aarts, E.H.L., Lenstra, J.K. (eds.) Local Search in Combinatorial Optimization, pp. 215–310. John Wiley and Sons, Chichester (1997)
10. Kennedy, J., Eberhart, R.C.: Particle Swarm Optimization. In: Proceedings of the IEEE International Conference on Neural Networks, vol. 4, pp. 1942–1948. IEEE Press, Los Alamitos (1995)
11. Lupien, V., Hassan, W., Dumas, P.: Improved Titanium Billet Inspection Sensitivity through Optimized Phased Array Design, Part I: Design Technique, Modelling and Simulation. In: AIP Conference Proceedings. Quantitative Nondestructive Evaluation, vol. 820, pp. 853–860. AIP (2006)
12. Merz, P., Freisleben, B.: Memetic Algorithms for the Traveling Salesman Problem. Complex Systems 13(4), 297–345 (2001)
13. Mühlenbein, H.: Evolution in Time and Space–The Parallel Genetic Algorithm. In: Rawlins, G. (ed.) Foundations of Genetic Algorithms, Morgan Kaufmann, San Francisco (1991)
14. Radcliffe, N.J., Surry, P.D.: Formal memetic algorithms. In: Fogarty, T.C. (ed.) Evolutionary Computing. LNCS, vol. 865, pp. 1–16. Springer, Heidelberg (1994)

Concealed Contributors to Result Quality — The Search Process of Ant Colony System

Irene Moser

Complex Intelligent Systems Laboratory
Centre for Information Technology Research
Faculty of Information & Communication Technologies
Swinburne University of Technology, Melbourne, Australia
imoser@swin.edu.au

Abstract. Stochastic solvers are researched primarily with the goal of providing 'black box' optimisation approaches for situations where the optimisation problem is too complex to model and therefore impossible to solve using a deterministic approach. Sometimes, however, problems or their instances have characteristics which interact with the solver in undocumented and unpredictable ways. This paper reviews some pertinent examples in the literature and provides an experiment which demonstrates that ant colony optimisation has arcane mechanisms which are partly responsible for results which are currently attributed to the pheromone-based learning.

1 Introduction

Employing heuristic (i.e. stochastic) methods to find near-optimal solutions is motivated when time is more important than the knowledge of the quality the solver can provide [1], or when the problem space is too complex to analyse. However, considerable research effort is being dedicated to the discovery of good matches between solvers and problems, indicating that although stochastic solvers cannot guarantee the quality of a result, there are still expectations for them to produce results of acceptable quality reliably. Reliability presumes that the information about the solver's performance on a problem covers all problem instances, which can only be achieved when there is sufficient information about both the solver and the problem.

Different algorithmic approaches make use of different problem features. This is the principal reason why researchers explore different algorithms to be able to issue recommendations on which algorithm is best applied to which kind of problem. It is easy to see that the search space of an optimisation problem is shaped by search mechanisms of the solver [12]. Fisher [5] demonstrates the influence of the design on the form of the search landscape, which can also change drastically through relatively trivial changes in the constraints or objective function.

It is often not immediately apparent how the solver interacts with the problem. Stochastic solvers are deliberately designed to employ intrinsic mechanisms for handling the balance between exploration and exploitation, the two basic

M. Randall, H.A. Abbass, and J. Wiles (Eds.): ACAL 2007, LNAI 4828, pp. 25–35, 2007.

elements that enable search space sampling in an 'informed' way. Often these elements are employed at the same time in different proportions, as is usually the case in the constructive steps of Ant Colony Optimisation (ACO). Less frequently the proportions between these elements in a move are chosen deliberately, as in some forms of GA elitism. These basic principles are well understood, but the exact mechanisms responsible for good results are not always obvious. If good results are obtained, researchers are often less interested in investigating why this is the case and how the results have come about [10].

2   Algorithm Unpredictabilities Uncovered

Many, possibly most, publications exploring the abilities and properties of stochastic algorithms contain experiments which result in some unexpected elements of algorithm behaviour. Solving example problems with a promising new procedure, authors often find that these perform very well on some problem instances while not providing the same quality on others. While authors usually seem to be more eager to report good results, unexpected behaviour is at least as informative, if the unexpected results are explored thoroughly. Some explicit endeavours to uncover unexpected behaviour are reported below.

Undocumented traits are typically only observed in a limited number of instances. For example, Particle Swarm Optimisation (PSO) was shown to have a bias toward searching along the dimension axes. To the best of our knowledge, Janson and Middendorf [7] were the first to make this observation. They did so more than a decade after PSO was first conceived, showing that most continous functions PSO was usually used to solve were aligned along the dimensional axes. Rotating the same functions to move the optima away from the axes lead to experiments where PSO could not produce the same level of quality.

As an approach to solving dynamic problems with a Genetic Algorithm (GA), Cobb [2] proposed triggered hypermutation. The increased mutation rate is applied when a change of fitness is detected in the best individuals of the current population. As has subsequently been pointed out, on some occasions, the change of problem state does not affect any location represented in the current population [6].

Another illustrative example from the area of ACO is given by Merkle [9]. He considers a permutation problem (which is not a cycle) with x components. To incur the minimal cost, the first two components can be assigned to any position, while all other components can only be assigned to a single position in the sequence without incurring a higher cost. The simplified cost matrix (the original matrix [9] has 50 elements) is shown in Fig. 1.

Solving the problem with ACO, Merkle observes that the optimal solution is not found when the components are assigned to positions first-to-last. However, the optimal solution is found easily when the solutions are built last-to-first. It is easy to see why this is the case. As all positions incur the same cost for the first two elements, no quality guidance is available if the uppermost two components are the first to be assigned. Thus they are likely to occupy positions which are

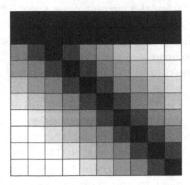

Fig. 1. Pheromone map for permutation problem described by Merkle [9], simplified

least-cost for subsequent components. As ant-based methods are constructive algorithms, components are assigned one by one. If the majority of the problem instances has only one least-cost position for each of the components, there may be no awareness of this predicament.

3   Algorithm

For the experiments in this work, a very successful variation of ACO, Ant Colony System (ACS) is used. It was first introduced by Dorigo and Gambardella [3], a more detailed discussion has since been published by Dorigo and Stützle [4]. Because of its greedy solution construction process, it is particularly well suited to Travelling Salesperson Problems (TSP) in Euclidean space. Like all ACO algorithms, it maintains pheromone values associated with solution components. In the case of the TSP, the solution components are the links or edges between two nodes or cities.

The pheromone variables are initialised to a very small non-zero value which traditionally has been defined as the inverse of the length of an upper bound solution, such as a tour found by a nearest-neighbour heuristic. It has been observed in many preliminary experiments, however, that the algorithm is not sensitive to this value.

An ACS iteration is defined as a cycle in which a definable number of solutions are built. At the end of the construction process, the links belonging to the best-known tour (not necessarily found during the same iteration) have their pheromone variables τ updated according to Eq. 1. The evaporation factor $\rho = 0.1$ balances the effects of recent experience with the additions brought about by earlier good results.

$$\tau(t+1) = (1-\rho) * \tau(t) + \rho * \Delta; \Delta = \frac{1}{\text{tour length}} \tag{1}$$

This positive feedback from the current most successful solution is employed in the search for better solutions. For the choice of next step, the pheromone variable is balanced against a heuristic, which in the case of the Euclidean TSP has traditionally been the inverse of the edge's length η. The factor β (set to 2 in our experiments as proposed in [3]) has been introduced as a weighting between pheromone and heuristic value.

The maximum weighted product of pheromone and heuristic value is employed *most of the time*, i. e. the variable q_0 which expresses the balance between choosing the best available next link and choosing the next link probabilistically has often been found to have its best setting at 0.9. It is compared to a uniform random number q at each step to determine the whether the greedy or the probabilistic rule is to be applied to the choice of next link between the current node i and a node j which is not yet included in the solution. Eq. 2 formalises this transition rule. A 90% greedy choice has initially [3] been found to produce the best performance on Euclidean TSPs, an observation confirmed by our studies.

$$p_{ij} = \begin{array}{l} \max \tau_{ij} * \eta_{ij}^{\beta} \quad \text{if } q < q_0 \\[2mm] \dfrac{\tau_{ij} * \eta_{ij}^{\beta}}{\sum \tau_{ij} * \eta_{ij}^{\beta}} \quad \text{otherwise} \end{array} \tag{2}$$

Whenever a link is chosen according to these rules and added to the current solution, its pheromone value is reduced using the ρ factor. Consequently, toward the end of a cycle, the best-known solution is followed to a decreasing extent. The number of useful tours to be built during a cycle is therefore naturally limited and can be calculated as demonstrated in [3], where as few as 10 tours per cycle were adopted. In the current work, we allowed a wasteful 50 to be able to record how early solution discoveries where likely to take place.

The flexible self-adjustment of the pheromone values is demonstrated in Fig. 2. The two plots follow the pheromone development on two edges. One of the two links has a lower heuristic value η. Over the first eight cycles, this link has its pheromone value augmented to compensate for the lower heuristic value. After the first pheromone update, the link's product of pheromone and heuristic is not high enough for the link to be chosen as often as to have its pheromone level evaporated to the initial level. Therefore, the pheromone is adjusted to a higher level over several cycles and retains that level during cycles 4 – 16 (measurements 8 – 32, as there are two measurements per cycle). At this stage (cycle 17 or measurement 34), the link is excluded from the best-known solution, to be included a second time in cycle 29 for a short period. While it is excluded, no updates are carried out and the pheromone level quickly erodes to the initial level.

In essence, the principle of the ACS paradigm, found to perform very successfully on Euclidean TSPs of less than 100 nodes, is understood to consist of two mechanisms balancing exploration and exploitation. The first mechanism is represented by the 90% greedy choice of including the best next move according

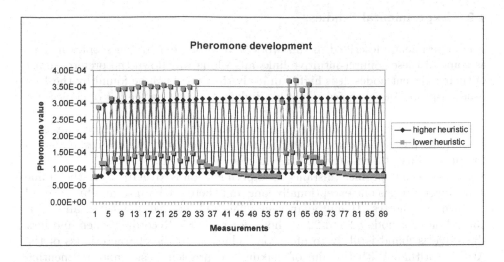

Fig. 2. Measuring the pheromone values of two links before and after the pheromone updates during an ACS trial. It is clear that one of the two links was part of the best-known solution throughout the trial.

to the maximum weighted product of heuristic and pheromone. The second is a 10% probabilistic exploration phase which makes small changes to the learned optimum to search its neighbourhood for new optima. Our experiments are designed to test whether this principle is actually solely responsible for the ACS results.

4   Solving a TSP Using ACS

4.1   Some Characteristics of Euclidean TSPs

The fact that the best algorithms to solve the TSP are all iterative like the very successful Lin-Kernighan approach [8] is a good indicator for the fact that the optima in this problem are all located in close proximity. It is easy to see why the greediest of the ACO implementations has been performing best on this problem.

Analysing some geographical instances from the online TSP library [11] such as berlin52 or kroA100, it becomes clear that the optimal solution to these follows the nearest-neighbour paradigm for around 50% of the links. If for each node, the links to all other nodes are listed by proximity, the first and second choices (first- and second-closest nodes) would be the nearest-neighbour choices for each node. Only occasionally does the ideal tour require the use of a low-ranking choice, which is counter-intuitive according to the heuristic value. Finding these counter-intuitive links is the crux of the optimisation task the algorithm has to perform.

4.2   Experimental Studies

The experiments described in this work were prompted by the suspicion that, as some of these counter-intuitive links are a long way down the preference list of their adjacent nodes, it is highly unlikely that these can be found by the 10% random-proportional choice alone. If a link is e.g. 13th on the list of choices for next step, the probability for it to get chosen *in connection with other links of the best tour* is very low. However, if most of the other options lead to nodes that are already part of the current solution, the reduced number of choices augments the probability of a counter-intuitive link being included.

An interesting problem instance to experiment on is kroA100 from the online TSP library. It has one exceptionally long and therefore hard to find link needed to form the shortest Hamiltonian cycle. The link between nodes 82 and 85 is 13th choice for node 82 and 22nd choice for node 85. Recording when and how this edge is found is likely to offer some clues as to the characteristics of the ACS algorithm. With the aim of making the problem easier and augmenting the probability of the optimum being found, 60% of the nodes were removed to leave the instance whose optimal solution is shown in Fig. 3. The numbers of the nodes retained are visible in Fig. 5.

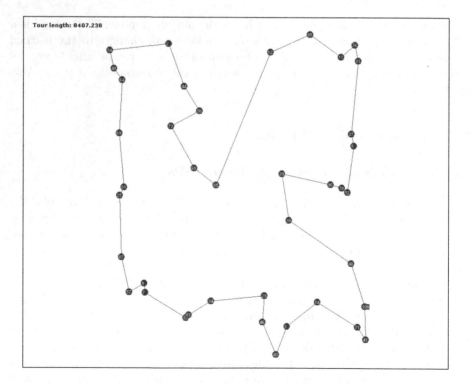

Fig. 3. Example problem used - 40 nodes from the kroA100 problem and the optimal solution

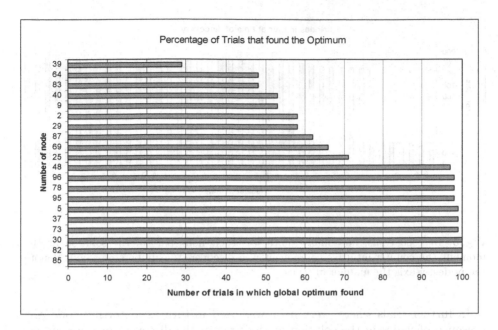

Fig. 4. Number of times (in 100 trials) the global optimum was found starting from the same node every time. Only the ten worst-performing starting nodes and the ten best-performing are shown. The intermediary 20 nodes have been omitted.

The reduction of choices as the solution building proceeds and fewer nodes are available is suspected to contribute significantly to the discovery of long links needed for the optimal tour. To verify this hypothesis, 40 different experiments are set up with 100 trials each. In each of the experiment, all tours are built starting from the same node. Observing whether there is a difference between where the algorithm starts the construction and the discovery of the optimum can be expected to offer clues as to whether a reduced choice has an influence. Fig. 4 shows how many times in 100 trials the optimum was found when starting from a given node. Only the ten best-performing and the ten worst-performing nodes are shown, as the crucial information lies in the disparity between these extremes.

The graph shows a clear correlation, with only three of the nodes leading to the discovery each time, while the worst-performing starting node has a success rate of only 29%. From the nodes which have a 100% success rate, two are linked by the counter-intuitive link which has been seen as the major obstacle to the discovery of the best tour. This seems illogical, but it is easily explained by the fact that the counter-intuitive link is found last, as the algorithm initially starts building the tour *away from it*. By the time the construction reaches the problematic link from the other end, there are no other choices left.

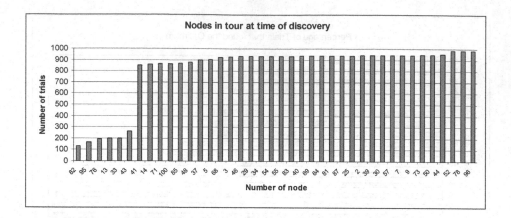

Fig. 5. In 1000 trials, how many times was each node part of the solution at the moment the counter-intuitive component was discovered? Numbers on x-axis denote the nodes, y-axis the number of trials.

In further trials where every node was used in turn as a starting node, the presence of nodes in the solution at the time of the discovery of the counter-intuitive link was recorded. When the trials use different starting nodes, the optimum is found without fail in all of the 1000 trials. The number of times each node was present in the half-finished solution at the time of the first inclusion of the sought-after connection between 82 and 85 is shown in Fig. 5. The graph reveals a striking pattern: Six of the forty nodes were present approximately one fifth of the time while the other thirty-six were present four fifths of the time. This suggests that there are few and very distinct part solutions which enable the discovery of the counter-intuitive link. This supports our hypothesis, as it suggests that the exclusion of a set of distinct nodes (the nodes which are likely to compete for the next step) tends to lead to the discovery of a crucial element.

Fig. 5 also reveals that among the adjacent nodes of the sought-after link, 85 is present 868 times and 82 is included 132 times. As we are examining the event of including the link connecting them, it is not surprising that the two numbers sum to 1000. However, the fact that it is discovered vastly more often from node 85 seems counter-intuitive, as the sought-after link ranks 22nd among the choices of 85.

To find an answer to this question, the 'approach' to the critical situation was examined more closely. All part solutions at the stage of discovery were examined for the exact sequence prior to arriving at node 85. Following the nodes backwards from 85, the same sequence of 11 nodes appears in the record in over 60% of the cases. Fig. 6 shows the sequence immediately prior to the approach to node 85, from which the counter-intuitive link is subsequently discovered. The path between the starting node (which is often, but not always, node 37 or 5) up to node 25 can vary, but as Fig. 5 shows, the solution is likely to include all but six nodes at this stage.

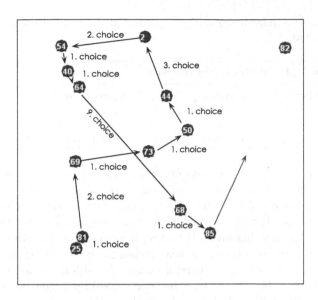

Fig. 6. Path to node 85 immediately before the discovery of the counter-intuitive edge in over 60% of the trials

Examining the final steps before the discovery more closely, it becomes clear that no counter-intuitive move has to be made during these last 11 steps to node 85, i.e. none of these steps requires any pheromone. Fig. 6 reveals the ranking for each of the links. Note that the ranking would be different if the choice was made from the opposite node, i.e. the path was created starting from node 85. If a move was first choice for the current node, there is no question as to why it was chosen. Therefore, Table 1 lists the reasons for the inclusion of each link which ranked lower than one for the node from which it was chosen. All of the steps can be explained with the help of the heuristic value alone.

At the time of the inclusion, every one of them was the shortest available link. No pheromone was needed; therefore the ACS learning mechanism was not employed here. Whether any steps involving inferior ranks where included earlier in the tour is a question that cannot be answered, as the solutions diverge somewhat in the sequence prior to node 25.

However, using node 25 as a starting node, the optimal solution is discovered in 70% of the trials (as shown in Fig. 4). As starting from node 25 is likely to lead to the sequence in Fig. 6 without other nodes present in the tour, the situation is not ideal for finding the counter-intuitive edge. The fact that even with node 25 as a starting node, the optimum is found in 70% of the trials indicates that the well-understood mechanism of 90% exploitation and 10% exploration is indeed partly responsible for the algorithm's performance.

Looking at how early in the cycle the counter-intuitive component is discovered, there is a significant difference in the average number of tours built until the discovery. Depending on the starting node, the link between nodes 82 and 85 may

Table 1. Reasons for links shown in Fig. 6 were included. Only links with a rank bigger than one are included.

Link	Rank of Link	Reason for Inclusion
81 → 69	2nd choice	1st choice (link to 25) already included
44 → 2	3rd choice	1st and 2nd choice (link to 50 and 73) already included
2 → 54	2nd choice	1st choice (link to 44) already included
64 → 68	9th choice	All higher ranks included (link to 81 is 8th choice)

be discovered on average as early as during the third or as late as during the seventeenth tour construction. However, without examining the exact part solutions that led to the discovery, it is not possible to deduce the reasons for the differences.

One might argue that the feature of a very long and counter-intuitive link is typical only for this instance. However, if no counter-intuitive edges were needed for the optimal solution, a nearest-neighbour heuristic would make a better choice as a solver. It is therefore likely for similar situations to occur in other problems, where there may be more, possibly higher-ranking inferior choices which have to be included to create the optimal solution.

5 Conclusion

Stochastic algorithms are often explored experimentally in the literature for years until some of their crucial intrinsic features are uncovered. These features may manifest themselves when solving problems or instances whose characteristics are not uncommon and cannot be dismissed as outliers.

In the current work, some experimental studies have revealed a simple pattern in the behaviour of ACS, ten years after the algorithm was first introduced. A subsequently published handbook of ACO [4] by authors whose experience of the algorithm cannot be doubted yields only a few passing hints with regards to the feature discussed in this work. It contains the repeated recommendation of starting the solution construction from a different random node each time. While we have found that the recommendation is justified, the effects it has been found to have might deserve a more thorough discussion.

In the field of stochastic algorithms, authors often research the application of a particular algorithms to a number of problems in the hope of offering a general-purpose black box approach for intractable or combinatorially complex problems. However, it seems that without a deeper understanding of the problem, its instances and the mechanisms of the algorithms, there can be no guarantee as to the quality of the result.

References

1. Barr, R.S., Golden, B.L., Kelly, J.P., Resende, M.G.C., Stewart, W.R.: Designing and Reporting on Computational Experiments with Heuristic Methods. Journal of Heuristics 1, 9–32 (1995)

2. Cobb, H.G.: An Investigation into the Use of Hypermutation as an Adaptive Operator in Genetic Algorithms Having Continuous Time-Dependent Nonstationary Environments. Technical Report, Naval Research Laboratory, Washington (1990)
3. Dorigo, M., Gambardella, L.: Ant Colony System: A cooperative learning approach to the traveling salesman problem. IEEE Transactions on Evolutionary Computation 1, 53–66 (1997)
4. Dorigo, M., Stützle, T.: Ant Colony Optimization. The MIT Press, Cambridge, MA (2004)
5. Fisher, D.S.: Dynamics and domain walls: Is the 'landscape paradigm' instructive? Physica D 107, 204–217 (1997)
6. Grefenstette, J.J.: Genetic algorithms for changing environments. Parallel Problem Solving from Nature 2, 137–144 (1992)
7. Janson, S., Middendorf, M.: On Trajectories of Particles in PSO. In: Proceedings of the 2007 IEEE Swarm Intelligence Symposium (2007)
8. Lin, S., Kernighan, B.W.: An effective heuristic algorithm for the traveling salesman problem. Operations Research 21, 498–516 (1973)
9. Merkle, D.: Ameisenalgorithmen – Optimierung und Modellierung. PhD thesis, Institut für Angewandte Informatik und Formale Beschreibungsverfahren, Universität Karlsruhe (TH) (in German) (2002)
10. Mitchell, M., Holland, J.H., Forrest, S.: When Will a Genetic Algorithm Outperform Hill Climbing? Advances in Neural Information Processing Systems 6 (1994)
11. http://www.iwr.uni-heidelberg.de/groups/comopt/software/TSPLIB95/
12. Younes, A., Calamai, P., Basir, O.: Generalized Benchmark Generation for Dynamic Combinatorial Problems. In: Proceedings of the 2005 workshops on Genetic and evolutionary computation, pp. 25–31 (2005)

Ants Guide Future Pilots

Sameer Alam, Minh-Ha Nguyen, Hussein A. Abbass, and Michael Barlow

The Artificial Life and Adaptive Robotics Laboratory, School of ITEE,
University of New South Wales @ Australian Defence Force Academy
Canberra, Australia
{s.alam,m.nguyen,h.abbass,spike}@adfa.edu.au

Abstract. In this paper an Ant Colony Optimization (ACO) approach
is extended to the safety and time critical domain of air traffic management. This approach is used to generate a set of safe weather avoidance
trajectories in a high fidelity air traffic simulation environment. Safety
constraints are managed through an enumeration-and-elimination procedure. In this procedure the search space is discretized with each cell
forming a state in graph. The arcs of the graph represent possible transition from one state to another. This state space is then manipulated
to eliminate those states which violate aircraft performance parameters. To evolve different search behaviour, we used two different approaches (dominance and scalarization) for updating the learned knowledge (pheromone) in the environment. Results shows that our approach
generates set of weather avoidance trajectories which are inherently safe.

1 Introduction

Ant Colony Optimization (ACO) is a population based optimization technique
based on social behaviour of real ant colonies. It is an iterative, probabilistic,
meta-heuristic for finding solutions to hard combinatorial optimization problems
and shows several desirable properties for application in the transportation domain [15]. ACO approaches are applied extensively to benchmark problems like
travelling salesman problem (TSP), the quadratic assignment problem (QAP),
and job shop scheduling problem [4]. ACO algorithms use simulated pheromone
as a collective form of self adaptation to produce increasingly better results.
ACO techniques are also extended for multi-objective optimization and examined notably in dynamic TSP and vehicle routing problems. Some of its variants
are applied on highly constrained problems in transport and telecommunication
domain [6]. One desirable area of extending ACO based approaches is safety critical domains such as Air Traffic Management [15]. However, it is not yet clear
how to design an effective ACO algorithm for such problems.

In recent years, there has been a quantum increase in weather related air traffic
delays. Weather disturbances are the leading cause of delays in air traffic and
account for approximately 70% of all delays in US National Airspace alone [11].
Previous approaches for weather avoidance systems are based on heuristic search
techniques such as A-Star [3], Depth First Search [9], etc. These approaches

M. Randall, H.A. Abbass, and J. Wiles (Eds.): ACAL 2007, LNAI 4828, pp. 36–48, 2007.

suffers from several drawbacks such as: the algorithm reports the first-found trajectory, which is often sub-optimal, secondly the algorithm does not take into consideration the aircraft performance envelop and other airspace constraints and most importantly these approaches were not examined in a high fidelity air traffic environment.

With the proposed flexibility given to the future pilots in trajectory planning [8], the previous approaches are deemed unsuitable. To incorporate safety constraints we pre-process the search space based on aircraft performance envelop. This has two advantages, first it reduces the search space so that algorithms can explore more and secondly this leads to a safety inherent design for trajectory planning. We use goal directed search in ACO with multiple objectives to generate a set of non-dominated solution trajectories, instead of a single solution approach. We have used high fidelity air traffic simulation environment for our experiments with realistic air traffic data and weather patterns obtained from meteorological data.

For a variety of optimization problems, the quality of the solutions constructed by ACO algorithms can be substantially improved through local search [6]. This performance improvement depends upon how the local search interacts with the design feature and parameter setting of the ACO [10]. We have blended the exploration feature of population based search (ACO) with the quick convergence feature of the informed heuristic search (A-Star) and use it as our goal directed search approach.

To incorporate multi-objective optimization in ACO we have used an intuitive means of incorporating iterative weight update mechanisms (for different objectives) in the A-Star algorithm and used the A-Star objective function as the visibility parameter of the ACO algorithm. We used different weather patterns and carried out performance assessment of solutions generated by the ACO algorithm using *median attainment surface* [17]. We also evaluated several implementation options for the ACO algorithm in terms of different pheromone update mechanism with different combinations of exploration-exploitation and heuristic desirability parameters.

The paper is organized as follows, in the next section we explain the weather avoidance problem and state space pre-processing mechanism followed by the design of the algorithm and a discussion of the two pheromone update strategies employed. We then explain the experimental design and performance measures used and we conclude with results and discussions.

2   Weather Avoidance in Air Traffic Management

2.1   Search Space

Finding weather avoidance trajectories can be seen as path planning in three dimensions, which is a well known NP-hard problem [5]. In air traffic management this problem attains unique dimensions due to safety and airspace constraints posed on it. Apart from the hard safety constraints in terms of aircraft performance envelop, the objectives are to minimize severe weather cells impact, minimize changes in heading, minimize changes in altitude (climb & descent), and minimize the distance travelled in a given route. The weather cells can spread in an area extending up to 1000 square nautical mile. Finding feasible solution in a such a large search space can be damning form safety and time perspective. If number of states in search space can be reduced without compromising the solution quality, we hypothesized this reduction of state space size will result in faster convergence. In worst case the algorithm will emphasize on learning transitions which are valid.

Classical approach of handling safety constraints uses penalty value. However, if there are large number of parameter, there is no guarantee of feasibility of the solutions. By pre-processing unfeasible transition in the search space, search is guaranteed to produce feasible solutions We eliminate the state space recursively starting from entry point in the 3-D grid (explained later) doing forward recursion on different layers. At each layer we eliminate states which violate the aircraft performance parameter, then we move to the following layer. Following this approach we can guarantee that the resultant state space contains feasible transitions only.

2.2 Weather Avoidance

The problem can then be stated as follows: Given an 3-D matrix (Airspace) of dimensions i (latitude) × j (longitude) × k (altitude), an entry point x (start manoeuvre point) and an exit point y (end manoeuvre point) find the set of routes between x and y on given objectives. We have used ATOMS [2], a high

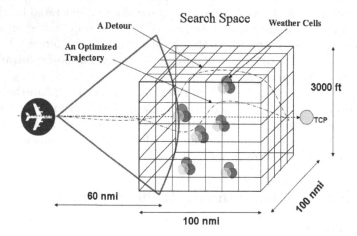

Fig. 1. Conceptual representation of weather avoidance algorithm design. Airspace is discretized into 3-D grid, at a distance of 60nm (weather radar range) from the aircraft. Optimized trajectories lead to reduction in fuel and time. TCP in the figure indicates next trajectory change point, which is the reference point for computing the exit point in the grid.

fidelity purpose build Air Traffic Management Simulator to simulate bad weather cells and air traffic. Weather cells were simulated by assigning a radar reflectivity, (a measure of thunderstorm intensity) between 5 and 50 generated by a uniform distribution random number generator, to the discretized cells of the airspace. Each cluster of bad weather comprises of 6 to 12 thunderstorm cells of dimension 10nmi × 10nmi × 3000ft [12]. We generated several weather scenarios ranging from 6 to 12 cells in different patterns on the intended trajectory of a flight in a high fidelity air traffic simulator [2]. As shown in the figure 1, the algorithm upon detection of weather cells within 60nm, (weather radar range) generates a three dimension volume around them. This airspace volume is of dimension 100nmi X 100nmi X 3000 ft.

This search space is then discretized in a grid of $10 \times 10 \times 3$ (300 nodes), it gives enough volume (1000 sq nautical mile) to cover the entire weather pattern. Each cell in the grid forms a state in the graph. The arcs of the graph represent possible transition from one state to another. The state space is then processed by removing those states which violate aircraft performance parameters such as bank angle, turn angle, operating altitude, maximum rate of climb or descent etc. This performance data is derived from the Eurocontrol's [2] Base of Aircraft Data (BADA) [1] based on the respective aircraft's state information (speed, altitude, heading etc.). The pseudo code for the same is presented in algorithm 1. In a grid of 3 X 10 X 10, we have 300 nodes with 4788 arcs which after processing were reduced 163 nodes with 2591 arcs.

The resulting three dimensional grid is stored in a 3-D array data structure as an enumerated state space where each element of the array represent a point in the 3-D grid. Every array element stores the information about its position (latitude, longitude and altitude) and all the immediate next links (which do not violate aircraft performance envelop) from that point in the grid. Further for each link the array element stores the heading change required, altitude change required, distance between the two and the distance to exit point from that link. The ACO perform search on this pre-processed state space.

3   Goal Directed Search : ACO with A-Star

To incorporated goal directed search in ACO we have used A-Star algorithm. The details are as follows:

3.1   A Star

The A Star (A*) [13] algorithm is an informed heuristic search technique which minimizes the estimated path cost to a goal state (destination). At node n, the A* algorithm will choose the next state which minimizes the function $f(n) = g(n) + h(n)$, where $g(n)$ gives the path cost from the current node to the next

---

[2] Eurocontrol is the European organization for the safety of air navigation. It currently numbers 31 member states. Eurocontrol has as its primary objective the development of a seamless, pan-European air traffic management (ATM) system.

Algorithm 1. The pseudo code for state space processing

Require: Aircraft Performance Database D :(Rate of Heading Change, Rate of Climb
  & Descent, Altitude Ceiling, Max. Bank Angle, Max. Turn Angle, Max. Vertical
  Speed)
  Grid G with i × j × k dimensions
  Exit node position in the grid(sink)
1: **for** Node in G **do**
2:     Compute the turn angle, altitude change, rate of heading change require, rate
       of altitude change required, vertical speed required from each node (xyz) to
       the grid point (ijk) in G
3:     **if** Transition from node xyz to grid point ijk violates D **then**
4:         Eliminate the link
5:     **else**
6:         Retain the link
7:     **end if**
8: **end for**
9: **return** Grid G with links within safety envelop

---

node j, and $h(n)$ is the estimated cost from the next node to the destination node
(exit point). We define cost function $g(n)$ as the sum of the normalized weather
intensity (WF), normalized heading change (HF), normalized altitude change
(AF) and normalized distance (DF) to the next node j. The heuristic function
$h(n)$ is defined by the normalized estimated cost on the above objectives from
the next node j to the exit point in the search grid. We then form $\Upsilon(n)$ as the
weighted sum of the two objectives.

$$\Upsilon(n) = (WF \times \omega_u + HF \times \omega_v + AF \times \omega_w + DF \times \omega_l) + (WF_{est} \times \omega_p + HF_{est} \times \omega_q + AF_{est} \times \omega_r + DF_{est} \times \omega_s)$$

where $\omega_u, \omega_v, \omega_w, \omega_l, \omega_p, \omega_q, \omega_r, \omega_s$ are dynamically initialized polar weights on
the surface of a unit sphere [7] and $WF_{est}, HF_{est}, AF_{est} and DF_{est}$ are the esti-
mated cost of weather cell impact, heading change, altitude change and distance
traveled respectively of reaching from the next node to exit node.

3.2   ACO

In ACO, the transition rule which is the probability for an ant k on node i to
choose node j while building its tour is given according to the following rule [6]

$$j = \begin{cases} arg\,max_{u \in J_i^k}[\tau_{iu}(t)] \times [\eta_{iu}]^\beta & \text{if } q \leq q_0 \\ J & \text{if } q > q_0 \end{cases} \tag{1}$$

where $J \in J_i^k$ is a node that is randomly selected according to the following
probability

$$p_{ij}^k(t) = \frac{[\tau_{ij}(t)] \times [\eta_{ij}]^\beta}{\Sigma_{l \in J_i^k}[\tau_{ik}(t)] \times [\eta_{ik}]^\beta} \tag{2}$$

where τ_{ij} is the pheromone value between the two nodes i and j, β controls
the relative weight of η_{ij}, and q is a random variable uniformly distributed over
$[0, 1]$.

Parameter q_0 in equation 2 is a tunable parameter $(0 \leqslant q_0 \leqslant 1)$, where $q \leqslant q_0$ corresponds to an exploitation of the heuristic information of given objectives and the learned knowledge memorized in terms of pheromone trails, whereas $q > q_0$ favors more exploration of the search space. We tune this parameter in the interval $[0\ 1]$, to evolve different search behavior.

The visibility parameter η_{ij} in equation 2 represents the heuristics desirability of choosing node j when at node i, it can be used to direct the search behavior of ants by tuning β in the interval $[0\ 1]$.

We have incorporated the inverse (since it is a minimization problem) of the A-Star evaluation function $\Upsilon(n)$ for the visibility parameter $\eta_{i,j}$.

$$\eta_{i,j} = \frac{1.0}{\Upsilon(n)} \tag{3}$$

To evolve different search behaviour we assigned weights iteratively for the A* objective function where for each objective a dynamic weight is assigned. In every iteration the weights are re-initialized using the hypercube rejection method [7] where weights are spread on the surface of a unit sphere. This results in different weights (relative importance) assigned to the objectives in each iteration resulting in diverse solution paths.

The heuristics information represented by the visibility parameter η_{ij} is not static as it is in the case of ACO algorithm, it changes due to the dynamic weight initialization of A-Star objectives in every iteration. This parameter is also tuned in the interval of 0 to 1. A value of 1 will result in a typical A-Star behaviour, and a value of 0 is indicative of the use of pheromone information only.

3.3   Different Search Behaviour Resulting from Different Pheromone Update Mechanism

We have investigated two different pheromone update strategies:

Pheromone update based on dominance : All valid solutions N from the current set P and archive set A (which stores the non-dominated solution obtained so far) are allowed to update the pheromone matrix. The pheromone update strategy is based on the SPEA2 [16] mechanism, where each individual solution i in the archive A and the current set P is assigned a strength value $S(i)$, representing the number of solutions it dominates:

$$S(i) = \mid j \mid j \epsilon P_t + A_t \wedge i \succ j \mid \tag{4}$$

where $\mid \cdot \mid$ denotes the cardinality of a set, + stands for multi set union and the symbol $\succ$ corresponds to the Pareto dominance relation extended to individuals $(i \succ j$ if the vector encoded by i dominates the vector encoded by j). Based on the S value, the raw fitness $R(i)$ of an individual i is calculated by summing the strengths of its dominators in both archive and current population.

$$R(i) = \sum_{j \epsilon P_t + A_t, j \succ i} S(j) \tag{5}$$

$R(i) = 0$ corresponds to a non-dominated individual, while a high $R(i)$ value means that i is dominated by many individuals. We then update the pheromone quantity as:

$$\forall \quad N \in [P, A] : \tau_{xy}(t) \leftarrow (1 - \rho) \cdot \tau_{xy}(t) + \rho \cdot (1 - R(i)) \tag{6}$$

where ρ is the pheromone decay parameter, and (x, y) represent the links in the solution path. We call this approach ACO-D.

Pheromone update based on scalarization: All the valid solutions N in the current population P are allowed to update the pheromone matrix based on the quality of the solution as:

$$\forall \quad N \in [P] : \tau_{xy}(t) \leftarrow (1 - \rho) \cdot \tau_{xy}(t) + \rho \cdot (1 - \psi(n)) \tag{7}$$

The quality of the solution is determined in terms of the scalarization $\psi(n)$ of n objectives. This mechanism is analogous to real world ants which deposits a higher quantity of pheromone returning from a rich food source. We call this approach ACO-S.

For the management of pheromone evaporation, pheromone trails and pheromone limits we have followed Max-Min Ant System (MMAS) [14] methodology, which sets the initial pheromone to a maximum value and ensures that pheromone information remains in a defined bound, preventing local optima and early convergence of solutions.

4   Experiment Design and Performance Measures

ATOMS was employed to generate weather patterns based on meteorological data. Flights based on recorded air traffic data and flight plan was simulated between two airports in the middle-east region. The aircraft was B747-400 with cruise altitude 27kft. The algorithm was examined on three different weather patterns: distributed dense weather cells, distributed sparse weather cells and clustered weather cells. The solution trajectories generated in terms of grid index were converted into artificial waypoint with associated latitude, longitude and altitude and fed into the flight management system of the aircraft, which is then flown in auto-pilot mode to fly the desired trajectory.

To evolve different search behaviours and understand performance of ACO algorithm with heuristic search under various parameter configurations, we have examined the following combinations:

1. Use of different combinations of values of the exploration-exploitation parameter (q_0) and heuristic desirability parameter(β).
2. Pheromone update mechanism based on *Dominance* v.s. pheromone update mechanism based on *Scalarization*.

We performed preliminary experiments to determine both the size of the ant colony, and the number of iteration required for convergence to a solution. We found that 30 ants with 300 iteration provides a good solution convergence. Based on MMAS rules we set $\rho = 0.9$ for the pheromone evaporation. τ_{max} was set to a theoretically largest value [14]. Combinations of the following values of parameter $\beta = \{0.1, 0.3, 0.5, 0.7, 0.9\}$ and parameter $q_0 = \{0.1, 0.3, 0.5, 0.7, 0.9\}$ were examined to understand their effect on ACO with heuristic search performance.

To measure the performance of these aspects we have used the Median Attainment Surface[17], which quantifies how much an algorithm A is better (strictly dominates) than algorithm B. We also used ANOVAN analysis which performs multi-way (n-way) analysis of variance (ANOVA) for testing the effects of multiple factors (grouping variables) on the mean of the given vector.

5   Experiments and Results

The ACO-S and ACO-D were tested on three different weather patterns (distributed dense weather cells, distributed sparse weather cells and clustered weather cells), as shown in the figure 2. Each experiment was run 20 times for each instance for 300 iterations. Figure 3 shows a 3D view of the set of solution trajectories generated by the ACO algorithm for a particular weather pattern. The solution set contains trajectories with different trade offs. For all the given combinations of q_0 (exploration-exploitation) and β (heuristic desirability) parameter, performance measures were computed, based on the outcome we plotted color map which displays the solution quality over other configurations, where darker shade indicates a good performance of the strategy parameter combination for the implied configuration in relation to all other combinations, and a lighter shade shows bad performance.

Fig. 2. Air traffic simulator snap shot of the ACO generated avoidance trajectory in the different weather scenarios. The three weather patterns with thunderstorm cells with varying radar reflectivity can be seen. Flight is displayed along with its call sign, altitude, speed and avoidance trajectory. OERK and OKBK shown in the figure are two airports in the middle-east region. The area displayed is approx. 2500 nm × 1000nm.

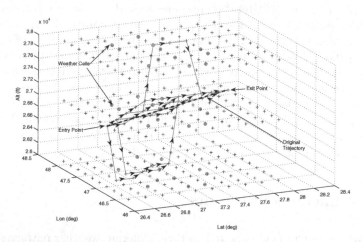

Fig. 3. A set of solution trajectories generated by ACO from grid entry point to grid exit point. Some of the trajectories that passes through high intensity thunderstorm cells (displayed as circles) are optimal on heading changes and distance traveled.

5.1   Median Attainment Surface

We computed the *median attainment surface*(MAS) for the non-dominated set of solutions generated by the ACO with informed heuristics search for the two approaches and for different parameter combinations. As shown in figure 4 and judging from the average darkness, some interesting observations can be made out of them. In dominance based pheromone update mechanism, we can see that when q_0 is low (q_0=0.1 and q_0=0.3) then high exploration (β=0.7) of search space gives good results. When q_0 is medium (q_0=0.5) then high to very high exploration($\beta \geq 0.7$) is results in good solutions. When q_0 is high to very high ($q_0 \geq 0.7$) then very high exploration (β=0.9) is undesirable. Best Solution were obtained for parameter combination of q_0=0.7 and β=0.5. In scalarization based pheromone update mechanism, we can see that with increasing value of q_0 we get very good solutions and β has very marginal effect on the solution quality. However, for all the good solutions we can see that parameter β has a value of 0.5. In ACO-D over ACO-S, the color map indicates that high to very high value of exploitation ($q_0 \geq 0.7$) in ACO-D gives good solution quality over ACO-S and similarly in ACO-S over ACO-D, with high exploitation ACO-S gives good solution quality over ACO-D. This indicates that very high exploitation of learned knowledge in the search space coupled with medium to high use of heuristic information can lead to good quality solutions for this kind of problems.

To get a more conclusive picture we then used ANOVAN analysis on MAS data. Figure 5 shows the ANOVAN analysis of ACO-S and ACO-D with different factors q_0 and β; each sub–figure presents the means and 95% confidence interval of the means of each data groups. Figure 5 top left and right shows ANOVAN for

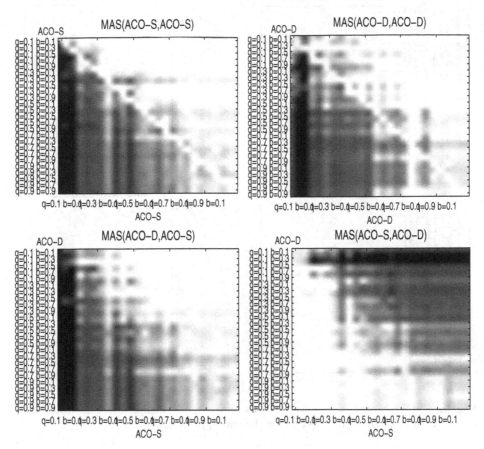

Fig. 4. MAS performance of ACO-S with in its own set (top-left), ACO-D with in its own set (top-right), ACO-D (y-axis) compared with ACO-S (x-axis) (bottom-left) and ACO-S (x-axis) compared with ACO-D (y-axis) (bottom-right). Darker shade indicates a good performance of the strategy parameter combination for the implied configuration in relation to all other combinations, and a lighter shade shows bad performance.

ACO-S, ACO-D for various parameter combination. It can be seen that the best strategy in both the approaches is high exploitation of embedded information in the environment. with increased exploitation we get good results, but if too much exploitation of heuristic is done the solution quality decreases. In ACO-S this however depends on medium favour for heuristic information available about the system. Whereas in ACO-D this depends on high favour for heuristic information available about the system. Thus in dominance based approach heuristic information plays a greater role than in scalarization based approach. In general we can say that very high exploitation of learned knowledge coupled with medium to high exploitation of heuristic information available about the system leads to good quality solution.

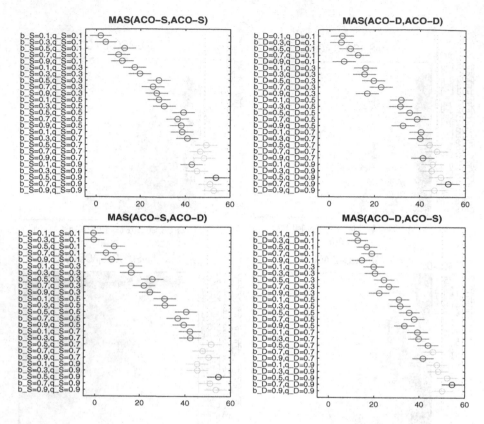

Fig. 5. ANOVAN analysis (showing means and 95% confidence interval) grouped on different set of q_0 and β parameters for the two approaches.

6   Conclusion and Future Work

A safety inherent design coupled with goal directed search can provide a good framework towards safety critical problems such as weather avoidance in air traffic environment. Pre-processing the state space before search helps in reducing the search space and making the search manageable for a safety and time critical domain. ACO with A-Star provided a good mechanism of incorporating multiple objectives and combining the virtues of population based approach with informed heuristics search. The algorithm generates a set of solution trajectories in different weather patterns successfully. The trajectories generated were all flyable, such that they do not compromise the performance envelop of the aircraft. The solution quality is strongly affected by exploitation of the embedded information in the environment rather than the available heuristics. High heuristic desirability leads to poor quality solution indicating early convergence to sub-optimal solution due to more weight on the local information. Overall the best strategy was to exploit more what is learned about the search space and use average local information. We will be further investigating this approach in a dynamic weather environment with neighbouring air traffic.

Acknowledgements

This work is supported by the Australian Research Council (ARC) Centre for Complex Systems grant number CEO0348249.

References

1. User manual for base of aircrfat data (BADA): Technical Report Rev No:3.6, EU-ROCONTROL Experiment Center, Bretigny, France (2004)
2. Alam, S., Abbass, H.A., Barlow, M.: Air traffic operations and management simulator ATOMS. IEEE Trans. Intelligent Transportation System (in press, 2008)
3. Bokadia, S., Valasek, J.: Severe weather avoidance using informed heuristic search. In: AIAA Guidance, Navigation, and Control Conference and Exhibit, Montreal, Canada, August 6-9, 2001, vol. 4232 (2001)
4. Bonabeau, E., Dorigo, M., Theraulaz, G.: Swarm Intelligence: From Natural to Artificial Systems. Oxford University Press, New York (1999)
5. Canny, J., Reif, J.: New lower bound techniques for robot motion planning problems. In: Proceedings of the 28th IEEE Symposium on Foundations of Computer Science, New York NY, pp. 49–60 (1987)
6. Dorigo, M., Stutzle, T.: Ant Colony Optimization. MIT Press, Cambridge, MA (2004)
7. Gentle, J.E.: Random Number Generation and Monte Carlo Methods, 2nd edn. Springer, New York (2003)
8. Hoekstra, J.M., WvanGent, R.N.H., Ruigrok, R.C.J.: Designing for safety: the 'free flight' air traffic management concept. Reliability Engineering & System Safety 75(2), 215–232 (2002)
9. Krozel, J., Penny, S., Prete, J., Mitchell, J.S.B.: Comparison of algorithms for synthesizing weather avoidance routes in transition airspace. In: Proceedings of the AIAA Guidance, Navigation, and Control Conference, Providence, RI (August 2004)
10. Lopez-Ibanez, M., Paquete, L., Stutzle, T.: On the design of aco for the biobjective quadratic assignment problem. In: Dorigo, M., Birattari, M., Blum, C., Gambardella, L.M., Mondada, F., Stützle, T. (eds.) ANTS 2004. LNCS, vol. 3172, pp. 214–225. Springer, Heidelberg (2004)
11. National Research Council: Weather forecasting accuracy for FAA traffic flow management. Technical report. National Academic Press, Washington, DC (2003)
12. Peter, F.L.: Aviation Weather. Jepesson Sanderson, Inc. (1995)
13. Russell, S.J., Norvig, N.: Artificial Intelligence A Modern Approach. Prentice-Hall, Englewood Cliffs (1995)
14. Stutzle, T., Holger, H.H.: MAX - MIN Ant System. Future Generation Computer Systems 16, 889–914 (2000)
15. Teodorovic, D.: Transport modeling by multi-agent systems: a swarm intelligence approach. Transportation Planning and Technology 26(4), 289–312 (2003)

16. Zitzler, E., Laumanns, M., Thiele, L.: SPEA2: Improving the strength pareto evolutionary algorithm. Technical Report TR-103, Computer Engineering and Communication Networks Lab TIK, SFIT ETH, Zurich, Switzerland (May 2001)
17. Zitzler, E., Thiele, L., Laumanns, M., Fonseca, C.M., da Fonseca, V.G.: Performance assessment of multiobjective optimizers: An analysis and review. IEEE Trans. Evol. Comput. 7(2), 174–188 (2003)

Information Transfer by Particles in Cellular Automata

Joseph T. Lizier[1,2], Mikhail Prokopenko[1], and Albert Y. Zomaya[2]

[1] CSIRO Information and Communications Technology Centre, Locked Bag 17,
North Ryde, NSW 1670, Australia
[2] School of Information Technologies, The University of Sydney, NSW 2006, Australia
`jlizier@it.usyd.edu.au`

Abstract. Particles, gliders and domain walls have long been thought to
be the information transfer entities in cellular automata. In this paper we
present local transfer entropy, which quantifies the information transfer
on a local scale at each space-time point in cellular automata. Local
transfer entropy demonstrates quantitatively that particles, gliders and
domain walls are the dominant information transfer entities, thereby
supporting this important conjecture about the nature of information
transfer in cellular automata.

1 Introduction

Design and analysis of complex nonlinear behavior in artificial life systems has
recently begun to consider the concept of *information transfer* (e.g. via influence of agents over their environments [1], in co-ordination between individual
modules of modular robots [2], and inducing neural structure in robots [3]).
Nowhere is the consideration of information transfer more clear than in studies
of cellular automata (CAs), where the emergent structures known as particles,
gliders and domain walls have long been suggested to be the information transfer entities therein [4,5]. Importantly, information transfer is also viewed as an
important component of complex behavior beyond the field of artificial life (e.g.
self-organization caused by dipole-dipole interactions in microtubules [6]).

Despite the abundance of complexity measures though (e.g. [7,8]), quantitative studies of information transfer in complex systems are noticeably absent. We
derive a measure of local information transfer from the *transfer entropy* [9], an existing averaged information-theoretical measure. Local transfer entropy charac-
terizes the information transfer into each spatiotemporal point in a given system
rather than providing a global average over all points in an information channel.
Local transfer entropy facilitates close study of parameters of the average transfer entropy, and is independently useful in highlighting or filtering "hot-spots"
in information channels. We apply local transfer entropy to Elementary Cellular
Automata (ECAs), a class of simple yet powerful discrete dynamical models. Local transfer entropy profiles for ECAs highlight the particles, gliders and domain
walls as the dominant information transfer entities, importantly providing the
first quantitative evidence for this widely-accepted conjecture about the nature
of information transfer in CAs.

M. Randall, H.A. Abbass, and J. Wiles (Eds.): ACAL 2007, LNAI 4828, pp. 49–60, 2007.

2   Information Transfer in Cellular Automata

We begin by introducing cellular automata (CAs), a renown example of complex systems, and discuss the importance of information transfer therein so as to contextualize our motivation. CAs are discrete dynamical systems consisting of an array of cells which each synchronously update their state as a function of the states of a fixed number of spatially neighboring cells using a uniform rule. While the behavior of each cell is simple, their (non-linear) interactions can lead to quite intricate global behavior. As such, CAs have become a well-studied example of complex behavior, and been used to model a wide variety of real world phenomena [4]. *Elementary CAs*, or *ECAs*, are a simple variety of 1D CAs using binary states, deterministic rules and one neighbor on either side (i.e. cell range $r = 1$). An example evolution of an ECA may be seen in Fig. 1a. Wolfram [10] provides a more detailed introduction to CAs, and defines the Wolfram rule number convention used here for describing update rules.

An important outcome of Wolfram's well-known attempt to classify the asymptotic behavior of CA rules into four classes [10] was a focus on emergent structure in CAs: *particles*, *gliders* and *domains*. A domain (formally defined within computational mechanics [11]) is a set of background configurations in a CA, any of which will update to another such configuration in the absence of a disturbance. A domain may be a *regular*, with periodic repetition, or is otherwise *irregular*. Particles are moving elements of coherent spatiotemporal structure; gliders are periodically-repeating particles. Formally, particles are defined as a boundary between two domains [11]; they can also be termed as *domain walls*, though this is typically used with reference non-periodic particles.

The continuing focus on the dynamics of propagating and static structures and their interactions (e.g. [8]) underlines the importance of information transfer in CAs. Particles are often said to form the basis of information transmission, and their interactions or collisions the basis of information modification (e.g. [5]). In particular, we find these analogies in analyses of CAs performing intrinsic, universal or other specific computation [11,12], and discussions on the nature of particles and their interactions [12,13]. However, no study has quantified the information transfer on average within specific channels or at specific spatiotemporal points in a CA, nor quantitatively demonstrated that particles (and gliders and domain walls) are in fact information transfer agents, therefore leaving these suggestions as conjecture only. We expect that a measure of local information transfer into each spatiotemporal point in CAs would reveal particles as the dominant information transfer agents.

Such spatiotemporal profiling can be viewed as a filtering for regions of interest in CAs, several methods of which exist: finite state transducers [11], frequency of rule execution [8], local statistical complexity and local sensitivity [14], and local information [15]. All of these successfully highlight particles, and so filtering is not a new concept. However the use of information transfer profiling could provide the first thoroughly quantitative evidence that particles are the information transfer elements in CAs. Additionally, it would provide several filtered views by examining each spatiotemporal direction of information transfer in the CA, and should reveal interesting differences in the parts of the structures highlighted. In the following sections, we present the information-theoretical foundations for and subsequently derive our local measure of information transfer.

3   Information-Theoretical Foundations

The measure of information transfer used here, transfer entropy, is defined using information-theoretical quantities. Importantly, information theory (e.g. see [16]) is known to be a useful framework for the design and analysis of complex self-organized systems (see [17] and specific examples in [2,1,3]).

The fundamental quantity is the *Shannon entropy*, the uncertainty associated with any measurement x of a random variable X (logarithms are in base 2, giving units in bits):

$$H(X) = -\sum_x p(x) \log p(x). \tag{1}$$

The *joint entropy* of two random variables X and Y is a generalization to quantify the uncertainty of their joint distribution:

$$H(X, Y) = -\sum_{x,y} p(x, y) \log p(x, y). \tag{2}$$

The *conditional entropy* of X given Y is the average uncertainty that remains about x when y is known

$$H(X|Y) = -\sum_{x,y} p(x, y) \log p(x|y). \tag{3}$$

The *mutual information* between X and Y measures the average reduction in uncertainty about x that results from learning the value of y, or vice versa:

$$I(X; Y) = \sum_{x,y} p(x, y) \log \frac{p(x, y)}{p(x)p(y)}. \tag{4}$$

$$I(X; Y) = H(X) - H(X|Y) = H(Y) - H(Y|X). \tag{5}$$

The *conditional mutual information* between X and Y given Z is the mutual information between X and Y when Z is known:

$$I(X; Y|Z) = H(X|Z) - H(X|Y, Z). \tag{6}$$

In the following section, we describe the use of information theory to define transfer entropy as a directional, dynamic measure of information transfer, and present local transfer entropy to quantify information transfer at each point in space and time in a given system.

4   Local Transfer Entropy

Mutual information has been a de facto candidate for measuring information transfer in complex systems, e.g. [5]. Yet mutual information as an information transfer contains no directionality; attempts to address this include using a time-lag between the "source" and "destination" values (known as *time-lagged mutual information*). However, this ignores the fundamental problem that it measures *statically* shared information [9].

To address these inadequacies Schreiber introduced transfer entropy in [9], the deviation from independence (in bits) of the state transition of an information destination X from the (previous) state of an information source Y[1]:

$$T_{Y \to X} = \sum_{z_n} p(z_n) \log \frac{p(x_{n+1}|x_n^{(k)}, y_n)}{p(x_{n+1}|x_n^{(k)})}. \tag{7}$$

Here n is a time index, z_n represents the state transition tuple $(x_{n+1}, x_n^{(k)}, y_n)$, and $x_n^{(k)}$ represents the k past values of x from and including time n (with $k = 1$ being the default choice). Reference [18] points out that the transfer entropy can be viewed as a *conditional* mutual information (6), that is the average information in the source about the next state of the destination that was not already contained in the destination's past.

To derive a local transfer entropy measure, we note that $p(z_n)$ may be expressed as the ratio of the count of observations $c(z_n)$ of z_n, to the total number of observations N: $p(z_n) = c(z_n)/N$. (Note that perfect estimation of the probability distribution function $p(z_n)$ would require an infinite number of observations). We replace the count by its definition to get:

$$p(z_n) = \left(\sum_{m=1}^{c(z_n)} 1 \right) /N. \tag{8}$$

Substituting (8) into (7), we then bring the log term inside this inner sum, leaving a double sum over each observation m for each possible tuple z_n. We combine these into a single sum over all N observations, and see that the transfer entropy metric is a global *average* (or expectation value) of a *local* transfer entropy at each observation:

$$T_{Y \to X} = \frac{1}{N} \sum_{n=1}^{N} \log \frac{p(x_{n+1}|x_n^{(k)}, y_n)}{p(x_{n+1}|x_n^{(k)})}, \tag{9}$$

$$i.e. \ T_{Y \to X} = \langle t_{Y \to X}(n+1) \rangle. \tag{10}$$

---

[1] Schreiber's presentation [9] considers the transfer from l previous states of the source variable. Here, we use $l = 1$: only the one previous source value has a direct causal influence on the destination in CAs, and we consider the information transfer in this causal relationship only at the given time step.

For systems such as CAs with homogeneous spatially-ordered agents, we instead represent the local information transfer to cell X_i from X_{i-j} at time $n+1$ as:

$$t(i, j, n+1, k) = \log \frac{p(x_{i,n+1}|x_{i,n}^{(k)}, x_{i-j,n})}{p(x_{i,n+1}|x_{i,n}^{(k)})}. \tag{11}$$

$t(i, j, n, k)$ is defined for every spatiotemporal destination (i, n), for every information channel or direction j where sensible values for CAs are within the cell range, $|j| \leq r$. For such systems, it is appropriate to estimate the probability distributions used in (11) from all spatiotemporal observations (i.e. from the whole CA).

It is important to note that the destination's own historical values can indirectly influence it via the source, which may be mistaken as an independent flow from the source. Such self-influence is a non-traveling form of information (like standing waves), eliminated from the measurement by conditioning on the destination's history $x_{i,n}^{(k)}$. However any self-influence transmitted prior to these k values will not be eliminated; we generalize comments on the entropy rate in [9] to suggest that the asymptote $k \to \infty$ is most correct for agents displaying non-Markovian dynamics. Local transfer entropy is then formally defined as:

$$t(i, j, n+1) = \lim_{k \to \infty} \log \frac{p(x_{i,n+1}|x_{i,n}^{(k)}, x_{i-j,n})}{p(x_{i,n+1}|x_{i,n}^{(k)})}, \tag{12}$$

though we acknowledge that its computation is not feasible in general, and retain the notation $t(i, j, n, k)$ for estimation with finite k.

While the averaged transfer entropy metric is constrained between 0 and $\log b$ (where b is the number of possible states of the destination element), it is important to note that the local transfer entropy is not constrained so long as it averages into this range. This means that is can be measured to be greater than $\log b$, indicating a very significant information transfer, and can also in fact be measured to be negative. Local transfer entropy is negative where (given the destination's history) the source element is actually *misleading* about the next state of the destination. It is possible for the source to be misleading in this context where other information sources influence the destination.

We label the special case $j = 0$ as *self-information transfer*, where the source element is the immediate past of the destination. By convention, we condition this calculation on the k values before the previous source value so as not to condition on the source. Self-information transfer is not a particularly meaningful quantity in and of itself, however it helps to form a useful visually filtered image with transfer entropy profiles for other values of j in the *summed information information transfer profile* $t_s(i, n, k) = \sum_{j=-r}^{r} t(i, j, n, k)$, where r is the range of information contributors (i.e. the cell range r for CAs).

5   Results and Discussion

The local transfer entropy $t(i, j, n, k)$ was measured for all space-time points in instances of several important ECA rules. We ran each instance from an initial randomized state of 8 000 cells, the first 30 time steps eliminated to allow the CA to settle, and a further 500 time steps captured for investigation. Periodic boundary conditions were used and results were confirmed by at least 10 runs from different initial states. All figures presented here were generated using modifications to [19].

The raw states for rule 110 (a classically complex rule) are displayed in Fig. 1a. As base cases we measured $t(i, j, n, k = 1)$ (i.e. for the default destination conditioning length $k = 1$), and time-lagged *local mutual information* (constructed in the same way as our local metric of transfer entropy) which is equivalent to $t(i, j, n, k = 0)$. Despite the known existence of particles and collisions [8,14] for this rule these measures were unable to quantitatively distinguish such structure with any more clarity than the raw CA plot itself (results not shown). The comparison case of local mutual information mirrors that performed with the globally averaged measures in [9], but the local profiles allow a more detailed comparison here than their averages do.

We continue to measure local transfer entropy $t(i, j, n, k)$ with larger values of k (the examples in [9] all used $k = 1$ in less coupled systems). For $k \geq 6$ the particle regions contain distinguishably more information transfer than the regular domains: this is shown for $k = 6$ in the information transfer profiles for $j = 1$ (i.e. one cell to the right per unit time step) and $j = -1$ (i.e. one cell to the left per unit time step) in Fig. 1c and e respectively. As expected, higher values of local transfer entropy are attributed by each measure to the gliders moving in the *same* macroscopic direction of motion as the direction of information transfer being measured. In contrast, notice that negative local transfer entropy is attributed to gliders with a macroscopic direction of motion *orthogonal* to the direction of information transfer being measured (see Fig. 1d and f); this is because sources of information from the orthogonal direction to the glider, which are still part of the domain, are misleading about the next state of the destination. Fig. 1b displays the summed profile $t_s(i, n, k = 6)$: this stark contrast between the gliders and the domain gives a filtered plot very similar to those produced for rule 110 by other filtering techniques (see [8,14]). Relying on the average transfer entropy values as solitary numbers does not provide us the same level of detail (e.g. for transfer one cell to the right per unit time $\langle t(i, j = 1, n, k = 1) \rangle = 0.21$ bits and $\langle t(i, j = 1, n, k = 6) \rangle = 0.12$ bits).

Importantly, note that since regular domains are temporally periodic [13], with period say p, local transfer entropy measurements with $k \geq p$ in an *infinite* such domain would not detect any additional information from the neighbors about the next state of the destination cell than is contained in its p previous states. That is to say, each cell's individual dynamics would appear to be deterministically Markovian. However the presence of gliders can render the probability distributions of (11) to measure small but non-zero information transfer at certain points in the regular background domain (small enough to appear to be zero

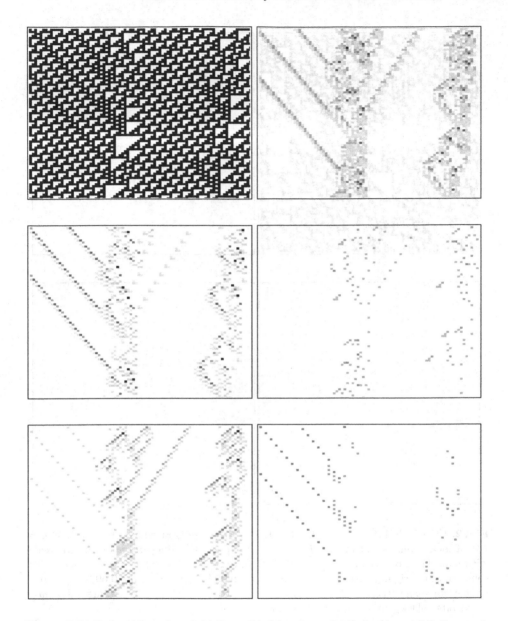

Fig. 1. ECA Rule 110: a. (*top left*) Raw CA (time is vertical); b. (*top right*) Summed profile $t_s(i, n, k = 6)$ of local transfer entropies $t(i, j, n, k = 6)$, positive values only shown, grayscale (all with 16 colors) with max. 7.80 bits (black), min. 0.00 bits (white); c. (*middle left*) $t(i, j = 1, n, k = 20)$ (one cell to the right), positive values only, max. 4.95 bits (black), min. 0.00 bits (white); d. (*middle right*) $t(i, j = 1, n, k = 20)$, negative values only, max. 0.00 bits (white), min. -1.94 bits (black); e. (*bottom left*) $t(i, j = -1, n, k = 20)$ (one cell to the left), positive values only, max. 6.62 bits (black), min. 0.00 bits (white); f. (*bottom right*) $t(i, j = -1, n, k = 20)$, negative values only, max. 0.00 bits (white), min. -2.05 bits (black).

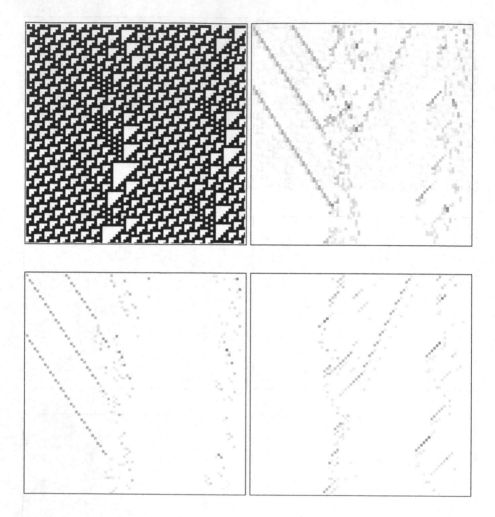

Fig. 2. ECA Rule 110: a. (*top left*) Raw CA; b. (*top right*) Summed profile $t_s(i, n, k = 16)$ of local transfer entropies $t(i, j, n, k = 16)$, positive values only shown, grayscale with max. 11.62 bits (black), min. 0.00 bits (white); c. (*bottom left*) $t(i, j = 1, n, k = 16)$ (one cell to the right), positive values only, max. 9.99 bits (black), min. 0.00 bits (white); d. (*bottom right*) $t(i, j = -1, n, k = 16)$ (one cell to the left), positive values only, max. 10.43 bits (black), min. 0.00 bits (white).

in Fig. 1). These small values in the domain are effectively an indication of the absence of a glider, that is that the domain shall continue. These non-zero values in the domain tend to be stronger in the wake of real gliders: because secondary gliders often follow other gliders, there is a stronger indication of their absence.

While the results in Fig. 1b visually match previous filtering work, each individual cell's dynamics are in fact non-Markovian, so using $k \to \infty$ would provide a more correct estimation of the information transfer in this system. Achieving

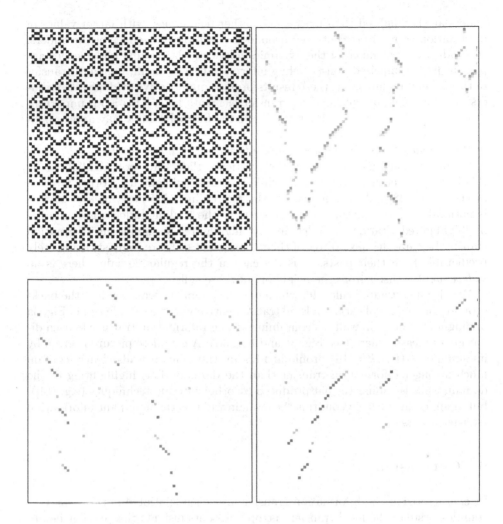

Fig. 3. ECA Rule 146: a. (*top left*) Raw CA; b. (*top right*) Summed profile $t_s(i, n, k = 16)$ of local transfer entropies $t(i, j, n, k = 16)$, positive values only shown, grayscale with max. 13.50 bits (black), min. 0.00 bits (white); c. (*bottom left*) $t(i, j = 1, n, k = 16)$ (one cell to the right), positive values only, max. 13.50 bits (black), min. 0.00 bits (white); d. (*bottom right*) $t(i, j = -1, n, k = 16)$ (one cell to the left), positive values only, max. 10.66 bits (black), min. 0.00 bits (white).

the limit $k \rightarrow \infty$ is computationally infeasible, but reasonable estimates of the probability distributions can be made for finite values of k in finite CA runs: we therefore increase the destination conditioning length under consideration to $k = 16$. We plot $t(i, j, n, k = 16)$ for rule 110 for $j = 1$ and -1, and the summed profile $t_s(i, n, k = 16)$, in Fig. 2. Information transfer is highlighted almost exclusively now in the direction of the macroscopic glider motion; this is even more closely aligned with our expectations than was seen for $k = 6$. Much less of the gliders are highlighted than for $k = 6$ or other techniques, with larger values of information transfer concentrated around the leading time-edges of the gliders: this indicates that much of the dynamics following the leading glider edges appear to have comprised non-traveling information. The small, non-zero transfer in the domain remains, but these results provide quantitative evidence that gliders are the dominant information transfer elements here. This is an important distinction to previous filtering work: while the filtered results may appear similar, it is only *local information transfer style filtering that provides quantitative evidence that gliders are the dominant information transfer agents.*

Similarly, note that stationary (i.e. vertical) gliders are not highlighted by this local transfer entropy method, while they are highlighted by other filtering methods (e.g. [8]). With reference to the figures included here, in Fig. 2b this is noticeable in a somewhat similar fashion to the leading glider edges only being highlighted. Stationary gliders are not highlighted as significant information transfer because the next states of their component cells are (almost) completely predictable from their pasts, as is the case in the regular domain: there is no *independent* transfer from the neighbors of those cells.

We also investigated rule 146, which contains domains walls against the background domain. Application of local transfer entropy to the sample run in Fig. 3a highlights the domain walls as containing strong information transfer in each direction of measurement (see Fig. 3c and Fig. 3d). A complete picture is given by its summed $t_s(i, n, k = 16)$ profile in Fig. 3b: the domain walls clearly contain much stronger information transfer than the domain. This highlighting of the domain walls is similar to that produced by other filtering techniques (e.g. [14]), but again quantitatively confirms the domain walls as the dominant information transfer agents.

6   Conclusion

In characterizing the information transfer into each spatiotemporal point in a complex system, the local transfer entropy presents insights that cannot be obtained using the averaged measure alone. Local transfer entropy is a valid filter for coherent structure in CAs, and quantitatively supported the long-held conjecture that particles, gliders and domain walls are the information transfer agents in CAs. It is novel in comparison to other filtering methods: it provides views of information transfer in each generic channel or direction, and highlights subtly different parts of emergent structure (i.e. those which facilitate the information transfer, being the leading glider edges).

The localization of transfer entropy is also a useful tool for investigating parameters of the transfer entropy itself. Here, our results underlined the importance of appropriate destination conditioning lengths, and we intend to use the local metric to investigate other variants of the transfer entropy (e.g. conditioning not only on the past history of the destination but on other information sources) in future work. We intend to provide a more complete analysis of the local transfer entropy metric, its variants, and their application to CAs in the near future.

Separately, we are examining the manner in which the local measure of information transfer introduced here combines with information storage and modification to form three axes of complexity as individual elements of computation [20].

In providing evidence that particles are the dominant information transfer agents in CAs, this result also provides the reverse evidence that transfer entropy is the appropriate measure for information transfer in complex systems. That being said, a comparison should be made with a localization of "information flow" [18] in future work. Local transfer entropy is ready to be applied to more complex systems (e.g. microtubules [6]), for which it may prove similar conjectures about information transfer.

References

1. Klyubin, A.S., Polani, D., Nehaniv, C.L.: All else being equal be empowered. In: Capcarrère, M.S., Freitas, A.A., Bentley, P.J., Johnson, C.G., Timmis, J. (eds.) ECAL 2005. LNCS (LNAI), vol. 3630, pp. 744–753. Springer, Heidelberg (2005)
2. Prokopenko, M., Gerasimov, V., Tanev, I.: Evolving spatiotemporal coordination in a modular robotic system. In: Nolfi, S., Baldassarre, G., Calabretta, R., Hallam, J.C.T., Marocco, D., Meyer, J.-A., Miglino, O., Parisi, D. (eds.) SAB 2006. LNCS (LNAI), vol. 4095, pp. 548–559. Springer, Heidelberg (2006)
3. Lungarella, M., Sporns, O.: Mapping information flow in sensorimotor networks. PLoS Computational Biology 2(10), e144 (2006)
4. Mitchell, M.: Computation in cellular automata: A selected review. In: Gramss, T., Bornholdt, S., Gross, M., Mitchell, M., Pellizzari, T. (eds.) Non-Standard Computation, pp. 95–140. VCH Verlagsgesellschaft, Weinheim (1998)
5. Langton, C.G.: Computation at the edge of chaos: phase transitions and emergent computation. Physica (Amsterdam) 42D(1-3), 12–37 (1990)
6. Brown, J.A., Tuszynski, J.A.: A review of the ferroelectric model of microtubules. Ferroelectrics 220, 141–156 (1999)
7. Shalizi, C.R., Shalizi, K.L., Haslinger, R.: Quantifying self-organization with optimal predictors. Phys. Rev. Lett. 93(11), 118701 (2004)
8. Wuensche, A.: Classifying cellular automata automatically: Finding gliders, filtering, and relating space-time patterns, attractor basins, and the z parameter. Complexity 4(3), 47–66 (1999)
9. Schreiber, T.: Measuring information transfer. Phys. Rev. Lett. 85(2), 461–464 (2000)
10. Wolfram, S.: A New Kind of Science. Wolfram Media, Champaign, IL, USA (2002)
11. Hanson, J.E., Crutchfield, J.P.: The attractor-basin portait of a cellular automaton. J. Stat. Phys. 66, 1415–1462 (1992)
12. Mitchell, M., Crutchfield, J.P., Hraber, P.T.: Evolving cellular automata to perform computations: Mechanisms and impediments. Physica (Amsterdam) 75D, 361–391 (1994)
13. Hordijk, W., Shalizi, C.R., Crutchfield, J.P.: Upper bound on the products of particle interactions in cellular automata. Physica (Amsterdam) 154D(3-4), 240–258 (2001)
14. Shalizi, C.R., Haslinger, R., Rouquier, J.B., Klinkner, K.L., Moore, C.: Automatic filters for the detection of coherent structure in spatiotemporal systems. Phys. Rev. E 73(3), 36104 (2006)

15. Helvik, T., Lindgren, K., Nordahl, M.G.: Local information in one-dimensional cellular automata. In: Sloot, P.M.A., Chopard, B., Hoekstra, A.G. (eds.) ACRI 2004. LNCS, vol. 3305, pp. 121–130. Springer, Heidelberg (2004)
16. MacKay, D.J.: Information Theory, Inference, and Learning Algorithms. Cambridge University Press, Cambridge (2003)
17. Prokopenko, M., Boschetti, F., Ryan, A.J.: An information-theoretic primer on complexity, self-organisation and emergence. Adv. Comp. Sys (submitted)
18. Ay, N., Polani, D.: Information flows in causal networks. Adv. Comp. Sys. (to be published)
19. Wójtowicz, M.: Java Cellebration v.1.50 (2002), online software http://psoup.math.wisc.edu/mcell/mjcell/mjcell.html
20. Lizier, J.T., Prokopenko, M., Zomaya, A.Y.: Detecting non-trivial computation in complex dynamics. In: Almeida e Costa, F., Rocha, L.M., Costa, E., Harvey, I., Coutinho, A. (eds.) ECAL 2007. LNCS (LNAI), vol. 4648, pp. 895–904. Springer, Heidelberg (2007)

An Artificial Development Model for Cell Pattern Generation

Arturo Chavoya[1,3] and Yves Duthen[2,3]

[1] Universidad de Guadalajara, Periférico Norte 799, Zapopan, Jal., Mexico CP 45000
[2] Université de Toulouse 1, 1 Place Anatole France, 31042 Toulouse, France
[3] Institut de Recherche en Informatique et Télécommunications, 118 Route de Narbonne, 31062 Toulouse, France
achavoya@cucea.udg.mx
yves.duthen@univ-tlse1.fr

Abstract. Cell pattern formation has a crucial role in both artificial and natural development. This paper presents an artificial development model for cell pattern generation based on the cellular automata (CA) paradigm. Cellular growth is controlled by a genome consisting of an artificial regulatory network (ARN) and a series of structural genes. The genome was evolved by a Genetic Algorithm (GA) in order to produce 2D cell patterns through the selective activation and inhibition of genes. Morphogenetic gradients were used to provide cells with positional information that constrained cellular replication. After a genome was evolved, a single cell in the middle of the CA lattice was allowed to reproduce until a cell pattern was formed. The model was applied to the problem of growing a French flag pattern.

Keywords: Artificial Development, Cell Pattern, French Flag Problem, Genetic Algorithm, Artificial Regulatory Network, Cellular Automata.

1 Introduction

In biological systems, development is a fascinating and very complex process that involves following an extremely intricate program coded in the organism's genome. One of the crucial stages in the development of an organism is that of pattern formation, where the fundamental body plans of the individual are outlined. It is now evident that gene regulatory networks play a central role in the development and metabolism of living organisms [1]. It has been discovered in recent years that the diverse cell patterns created during the developmental stages are mainly due to the selective activation and inhibition of very specific regulatory genes.

Over the years, artificial models of cellular development have been proposed with the objective of understanding how complex structures and patterns can emerge from one or a small group of initial undifferentiated cells [2][3][4][5][6]. In this paper we propose an artificial development model that generates 2D patterns by means of the selective activation and inhibition of development genes under

M. Randall, H.A. Abbass, and J. Wiles (Eds.): ACAL 2007, LNAI 4828, pp. 61–71, 2007.
© Springer-Verlag Berlin Heidelberg 2007

the constraints of morphogenetic gradients. Cellular growth is achieved through the expression of structural genes, which are in turn controlled by an Artificial Regulatory Network (ARN) evolved by a Genetic Algorithm (GA). The ARN determines when cells are allowed to grow and which gene to use for reproduction, while morphogenetic gradients constrain the position at which cells can replicate. Both the ARN and the structural genes constitute the artificial cell's genome. In order to test the functionality of the development program found by the GA, we applied the evolved genome to a cellular growth model that we have successfully used in the past to develop simple 2D and 3D geometrical shapes [7].

The paper starts with a section describing the cellular growth model, followed by a section presenting the morphogenetic gradients used. The artificial cell's genome is presented next, followed by a section describing the GA and how it was applied to evolve the genome. Results are presented next, followed by a section of conclusions.

2   Cellular Growth Model

Cellular automata (CA) were chosen as models of cellular growth, as they provide a simple mathematical model that can be used to study self-organizing features of complex systems [8]. CA are characterized by a regular lattice of N identical cells, an interaction neighborhood template η, a finite set of cell states Σ, and a space- and time-independent transition rule ϕ which is applied to every cell in the lattice at each time step.

In the cellular growth model presented in this work, a 33×33 regular lattice with non-periodic boundaries was used. The set of cell states was defined as $\Sigma = \{0, 1\}$, where 0 can be interpreted as an empty cell and 1 as an occupied or active cell. The interaction template η used was an outer Moore neighborhood. The CA's rule ϕ was defined as a lookup table that determined, for each local neighborhood, the state (empty or occupied) of the objective cell at the next time step. For a 2-state CA, these update states are termed the rule table's "output bits". The lookup table input was defined by the binary state value of cells in the local interaction neighborhood, where 0 meant an empty cell and 1 meant an occupied cell [9]. A cell can become active only if there is already an active cell in the interaction neighborhood. Starting with an active cell in the middle of the lattice, the CA algorithm is applied allowing active cells to reproduce for 100 time steps according to the rule table. During an iteration of the CA algorithm, the sequence of reproduction of active cells is randomly selected in order to avoid artifacts caused by a deterministic order of cell reproduction. Finally, cell death is not considered in the present model for the sake of simplicity.

3   Morphogenetic Gradients

Since Turing's seminal article on the theoretical influence of diffusing chemical substances on an organism's pattern development [10], the role of these molecules has been confirmed in a number of biological systems. These organizing

substances have been termed *morphogens* due to their role in driving morphogenetic processes. In our proposed development model, morphogenetic gradients
were generated similar to those found in the eggs of the fruit fly *Drosophila*, where
orthogonal gradients offer a sort of Cartesian coordinate system [11]. These gradients provide reproducing cells with positional information in order to facilitate
the spatial generation of patterns. The artificial morphogenetic gradients were
set up as suggested in [3], where morphogens diffuse from a source towards a
sink, with uniform morphogen degradation throughout the gradient.

Before cells were allowed to reproduce in the cellular growth model, morphogenetic gradients were generated by diffusing the morphogens from one of the
CA boundaries for 1000 time steps. Initial morphogen concentration level was
set at 255 arbitrary units, and the source was replenished to the same level at
the beginning of each cycle. The sink was set up at the opposite boundary of
the lattice, where the morphogen level was always set to zero. At the end of
each time step, morphogens were degraded at a rate of 0.005 througout the CA
lattice. We defined two orthogonal gradients in the CA lattice, one generated
from left to right and the other from top to bottom (Fig. 1).

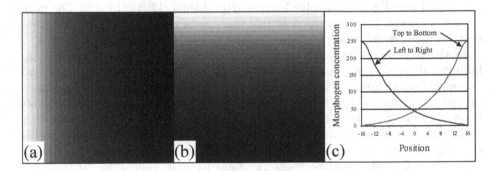

Fig. 1. Morphogenetic gradients (a) Left to Right; (b) Top to Bottom; (c) Morphogen
concentration graph

4   Genome

Genomes are the repository of genetic information in living organisms. They
are encoded as one or more chains of DNA, and they regularly interact with
other macromolecules, such as RNA and proteins. Artificial genomes are typically
coded as strings of discrete data types. The genome used in this model was
defined as a binary string starting with a series of ten regulatory genes, followed
by a number of structural genes (see Fig. 2).

4.1   Regulatory Genes

The series of regulatory genes at the beginning of the genome constitutes an Artificial Regulatory Network (ARN). ARNs are computer models whose objective

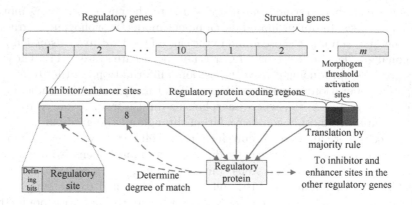

Fig. 2. Genome structure and regulatory gene detail

is to mimic to some extent the gene regulatory networks found in nature. ARNs have previously been used to study differential gene expression either as a computational paradigm or to solve particular problems [12][13][14][15]. The gene regulatory network implemented in this work is an extension of the ARN presented in [16], which in turn is based on the model proposed by Banzhaf [14].

In the present model, each regulatory gene consists of a series of eight inhibitor/enhancer sites, a series of five regulatory protein coding regions, and two morphogen threshold activation sites that determine the allowed positions for cell reproduction (Fig. 2). Inhibitor/enhancer sites are composed of a 12-bit function defining region and a regulatory site. Regulatory sites can behave either as an enhancer or an inhibitor, depending on the configuration of the function defining bits associated with them. If there are more 1's than 0's in the defining bits region, then the regulatory site functions as an enhancer, but if there are more 0's than 1's, then the site behaves as an inhibitor. Finally, if there is an equal number of 1's and 0's, then the regulatory site is turned off [17].

Regulatory protein coding regions "translate" a protein using the majority rule, i.e. for each bit position in these regions, the number of 1's and 0's is counted and the bit that is in majority is translated into the regulatory protein. The regulatory sites and the individual protein coding regions all have the same size of 32 bits. Thus the protein translated from the coding regions can be compared on a bit by bit basis with the regulatory site of the inhibitor and enhancer sites, and the degree of matching can be measured. As in [14], the comparison was implemented by an XOR operation, which results in a "1" if the corresponding bits are complementary. Each translated protein is compared with the inhibitor and enhancer sites of all the regulatory genes in order to determine the degree of interaction in the regulatory network. The influence of a protein on an enhancer or inhibitor site is exponential with the number of matching bits. The strength of excitation en or inhibition in for gene i with $i = 1, ..., n$ is defined as

$$en_i = \frac{1}{v} \sum_{j=1}^{v} c_j e^{\beta \left(u_{ij}^+ - u_{\max}^+\right)} \text{ and} \tag{1}$$

$$in_i = \frac{1}{w} \sum_{j=1}^{w} c_j e^{\beta \left(u_{ij}^- - u_{\max}^-\right)}, \tag{2}$$

where n is the total number of regulatory genes, v and w are the total number of active enhancer and inhibitor sites, respectively, c_j is the concentration of protein j, β is a constant that fine-tunes the strength of matching, u_{ij}^+ and u_{ij}^- are the number of matches between protein j and the enhancer and inhibitor sites of gene i, respectively, and $u_{\max}^+$ and $u_{\max}^-$ are the maximum matches achievable (32 bits) between a protein and an enhancer or inhibitor site, respectively [14].

Once the *en* and *in* values are obtained for all regulatory genes, the corresponding change in concentration c for protein i in one time step is calculated using

$$\frac{dc_i}{dt} = \delta \left(en_i - in_i\right) c_i, \tag{3}$$

where δ is a constant that regulates the degree of protein concentration change.

Protein concentrations are updated and if a new protein concentration results in a negative value, the protein concentration is set to zero. Protein concentrations are then normalized so that total protein concentration is always the unity. Parameters β and δ were set to 1.0 and 1.0×10^6, respectively, as previously reported [16].

The morphogen threshold activation sites provide reproducing cells with positional information as to where they are allowed to grow in the CA lattice. There is one site for each of the two orthogonal morphogenetic gradients described in Section 3. These sites are 9 bits in length, where the first bit defines the allowed direction (above or below the threshold) of cellular growth, and the next 8 bits code for the morphogen threshold activation level, which ranges from 0 to $2^8 - 1 = 255$. If the site's high order bit is 0, then cells are allowed to replicate below the morphogen threshold level coded in the lower order eight bits; if the value is 1, then cells are allowed to reproduce above the threshold level. Since in a regulatory gene there is one site for each of the two orthogonal morphogenetic gradients, for each pair of morphogen threshold activation levels, the pair of high order bits defines in which of the four relative quadrants cells expressing the associated structural gene can reproduce. Quadrants can have irregular edges because morphogenetic gradients are not perfectly generated due to local morphogen accumulation close to the non-periodic boundaries of the CA lattice.

4.2 Structural Genes

Structural genes code for the particular shape grown by the reproducing cells and were obtained using the methodology presented in [9]. Briefly, the CA rule table's output bits from the cellular growth model described in Section 2 were evolved by a GA in order to produce predefined 2D shapes.

Structural genes are always associated to the corresponding regulatory genes, that is, structural gene number 1 is associated to regulatory gene number 1 and its related translated protein, and so on. A structural gene was defined as being active if and only if the regulatory protein translated by the associated regulatory gene was above a certain concentration threshold. The value chosen for the threshold was 0.5, since the sum of all protein concentrations is always 1.0, and there can only be a protein at a time with a concentration above 0.5. As a result, only one structural gene can be expressed at a particular time step in a cell. If a structural gene is active, then the CA lookup table coded in it is used to control cell reproduction. Given that the outer Moore neighborhood used in the cellular growth model consists of the eight cells surrounding the central cell, structural genes are all 256 bits in length ($2^8 = 256$) [9]. The number of structural genes used in the genome depended on the particular pattern grown, as described in Section 6. Structural gene expression is visualized in the cellular growth model as a distinct external color for the cell.

5   Genetic Algorithm

Genetic algorithms (GAs) are search and optimization methods based on ideas borrowed from natural genetics and evolution [18]. A GA starts with a population of chromosomes representing vectors in search space. Each chromosome is evaluated according to a fitness function and the best individuals are selected. A new generation of chromosomes is created by applying genetic operators on selected individuals from the previous generation. The process is repeated until the desired number of generations is reached or until the desired individual is found.

The GA in this paper uses tournament selection with single-point crossover and mutation as genetic operators. As in a previous report, we used the following parameter values [16]. The initial population consisted of 1000 binary chromosomes chosen at random. Tournaments were run with sets of 3 individuals randomly selected from the population. Crossover and mutation rates were 0.60 and 0.15, respectively. Finally, the number of generations was set at 50, as there was no significant improvement after this number of generations.

The fitness function used by the GA was defined as

$$Fitness = \frac{1}{k} \sum_{i=1}^{k} \frac{ins_i - \frac{1}{2}outs_i}{des_i}, \tag{4}$$

where k is the number of different colored shapes, each corresponding to an expressed structural gene, ins_i is the number of active cells inside the desired shape i with the correct color, $outs_i$ is the number of active cells outside the desired shape i, but with the correct color, and des_i is the total number of cells inside the desired shape i. Thus, a fitness value of 1 represents a perfect match.

6 Results

The GA described in Section 5 was used in all cases to evolve the genome for the desired colored patterns, where each color represented a different structural gene being expressed. After a genome was obtained, an initial active cell was placed in the middle of the CA lattice and was allowed to reproduce controlled by the gene activation sequence found by the GA and under the restrictions imposed by the morphogenetic fields. In order to grow the desired pattern with a predefined color and position for each cell, the regulatory genes in the ARN had to evolve to be activated in a precise sequence and for a specific number of iterations inside the allowed space defined by the morphogenetic fields. Not all GA experiments produced a genome capable of generating the desired pattern.

The artificial development model was applied to what is known as the *French flag problem*. The problem of generating a French flag pattern was first introduced by Wolpert in the late 1960s when trying to formulate the problem of cell pattern development and regulation in living organisms [19], and it has been used since then by some authors to study the problem of artificial pattern development [20]. In order to grow the French flag pattern, three different structural genes were used. The first gene drove the creation of the central white square, while the next two genes extended the central square to the left and to the right, expressing the blue and the red color, respectively. The last two structural genes do not code specifically for a square, instead they extend a vertical line of cells to the left or to the right for as many time steps as they are activated.

Figure 3 shows a 27 × 9 French flag grown from the expression of the three structural genes mentioned above. The graph of the corresponding regulatory protein concentration change over time is shown in 3(e). Starting with an initial white cell (a), a white central square is formed from the expression of gene number 1 (b), the left blue square is then grown (c), followed by the right red square (d). The evolved morphogenetic fields are shown for each of the three structural genes. Since the pattern obtained was exactly as desired, the fitness value assigned to the corresponding genome was 1.

In order to increase the complexity of the pattern generated, four different structural genes were used to grow a French flag with a flagpole pattern. The first three structural genes are the same as those used to grow the simple French flag pattern. A fourth gene was added to create the brown flagpole by growing a single line of cells downward from the lower left corner of a rectangle. However, when trying to evolve a genome to produce the French flag with a flagpole pattern, it was found that the GA could not easily evolve an activation sequence that produced the desired pattern. Using the same approach as in [16], in order to increase the likelihood for the GA to find an appropriate genome, instead of using one series of four structural genes, a tandem of two identical series of four structural genes was used, for a total of eight structural genes. In that manner, for creating the central white square, the genome could express either structural

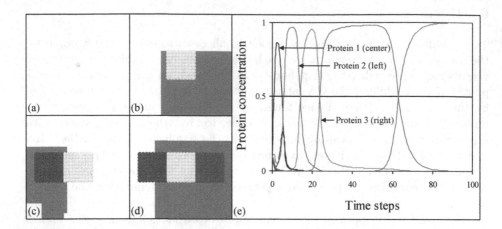

Fig. 3. Growth of a French flag pattern. (a) Initial cell; (b) Central white square with morphogenetic field for gene 1 (square); (c) White central square and left blue square with morphogenetic field for gene 2 (extend to left); (d) Finished flag pattern with morphogenetic field for gene 3 (extend to right); (e) Graph of the protein concentration change from the genome expressing the French flag pattern.

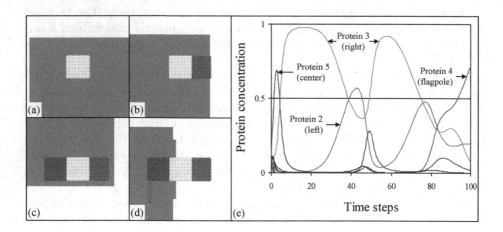

Fig. 4. Growth of a French flag with a flagpole pattern. (a) Central white square with morphogenetic field for gene 5 (square); (b) White central square and right red pattern with morphogenetic field for gene 3 (extend to right); (c) White central square, right red pattern and left blue square with morphogenetic field for gene 2 (extend to left); (d) Finished flag with a flagpole pattern with morphogenetic field for gene 4 (flagpole); (e) Graph of the protein concentration change from the genome expressing the French flag with a flagpole pattern.

gene number 1 or gene number 5, for the left blue and right red squares it could use genes 2 or 6, or genes 3 or 7, respectively, and finally for the flagpole it could make use of structural genes 4 or 8. Thus, the probability of finding an ARN that could express a French flag with a flagpole pattern was significantly increased.

The 21 × 7 French flag with a flagpole pattern produced by the expression of the configuration of structural genes mentioned above is shown in Fig. 4. The graph for the corresponding regulatory protein concentration change is shown in 4(e). After the white central square is formed (a), a right red pattern (b) and the left blue square (c) are sequentially grown, followed by the creation of the flagpole (d). The evolved morphogenetic fields are shown for each of the four structural genes expressed. Note that the white central square is formed from the activation of the first gene from the second series of structural genes, while the other three genes are expressed from the first series of the tandem. It should also be noted that the last column of cells is missing from the red right square, since the morphogenetic field for the gene that extends the red cells to the right precluded growth from that point on (Fig. 4(b)). On the other hand, from the protein concentration graph in 4(e), it is clear that this morphogenetic field prevented the growth of red cells all the way to the right boundary, as gene 3 was active for more time steps than those required to grow the appropriate red square pattern. The fitness value assigned to this pattern was 0.96, which corresponded to the most successful simulation we obtained when trying to grow this particular pattern.

7 Conclusion

The results presented in this paper show that the model proposed can give consistent results when evolving a genome that controls growth of predefined 2D cell patterns starting with a single cell. In particular, it was found that using this model it was relatively easy to generate a French flag pattern from the expression of three structural genes, although some problems were encountered when trying to obtain a slightly more complex pattern that involved the expression of four genes.

In general, the model proved to be suitable for obtaining simple patterns involving the activation of up to four genes, but more work is needed in order to explore pattern formation of more complex forms, both in 2D and 3D. It is also desirable to search for a development model that can reliably synchronize the activation of more than four genes. Furthermore, in order to increase the usefulness of the model, interaction with the environment and other artificial entities may be necessary. Until now our work has been devoted to achieving predefined patterns in a kind of directed evolution. However, it would be desirable to let cells evolve into a functional pattern under environmental constraints without any preconceived notion of the final pattern. The long-term goal of our work is to study the emergent properties of the artificial development process. It is conceivable that highly complex structures could one day be built through the interaction of myriads of simpler entities controlled by a development program.

References

1. Davidson, E.H.: The Regulatory Genome: Gene Regulatory Networks in Development And Evolution, 1st edn. Academic Press, London (2006)
2. Lindenmayer, A.: Mathematical models for cellular interaction in development, Parts I and II. Journal of Theoretical Biology 18, 280–315 (1968)
3. Meinhardt, H.: Models of Biological Pattern Formation. Academic Press, London (1982)
4. Fleischer, K., Barr, A.H.: A simulation testbed for the study of multicellular development: The multiple mechanisms of morphogenesis. In: Langdon, C. (ed.) Proceedings of the Workshop on Artificial Life ALIFE 1992, pp. 389–416. Addison-Wesley, Reading (1992)
5. Kitano, H.: A simple model of neurogenesis and cell differentiation based on evolutionary large-scale chaos. Artificial Life 2(1), 79–99 (1994)
6. Kumar, S., Bentley, P.J.: On Growth, Form and Computers. Academic Press, London (2003)
7. Chavoya, A., Duthen, Y.: Using a genetic algorithm to evolve cellular automata for 2D/3D computational development. In: GECCO 2006. Proceedings of the 8th annual conference on Genetic and evolutionary computation, pp. 231–232. ACM Press, New York (2006)
8. Wolfram, S.: Statistical mechanics of cellular automata. Reviews of Modern Physics 55, 601–644 (1983)
9. Chavoya, A., Duthen, Y.: Evolving cellular automata for 2D form generation. In: Proceedings of the Ninth International Conference on Computer Graphics and Artificial Intelligence 3IA 2006, pp. 129–137 (2006)
10. Turing, A.M.: The chemical basis of morphogenesis. Philosophical Transactions of the Royal Society of London. Series B, Biological Sciences 237(641), 37–72 (1952)
11. Carroll, S.B., Grenier, J.K., Weatherbee, S.D.: From DNA to Diversity: Molecular Genetics and the Evolution of Animal Design, 2nd edn. Blackwell Science, Oxford (2004)
12. Eggenberger, P.: Evolving morphologies of simulated 3D organisms based on differential gene expression. In: Harvey, I., Husbands, P. (eds.) Proceedings of the 4th European Conference on Artificial Life, pp. 205–213. Springer, Heidelberg (1997)
13. Reil, T.: Dynamics of gene expression in an artificial genome - implications for biological and artificial ontogeny. In: Floreano, D., Mondada, F. (eds.) ECAL 1999. LNCS, vol. 1674, pp. 457–466. Springer, Heidelberg (1999)
14. Banzhaf, W.: Artificial regulatory networks and genetic programming. In: Riolo, R.L., Worzel, B. (eds.) Genetic Programming Theory and Practice, pp. 43–62. Kluwer, Dordrecht (2003)
15. Knabe, J.F., Nehaniv, C.L., Schilstra, M.J., Quick, T.: Evolving biological clocks using genetic regulatory networks. In: Rocha, L.M., Yaeger, L.S., Bedau, M.A., Floreano, D., Goldstone, R.L., Vespignani, A. (eds.) Alife 10. Proceedings of the Artificial Life X Conference, pp. 15–21. MIT Press, Cambridge (2006)
16. Chavoya, A., Duthen, Y.: Evolving an artificial regulatory network for 2D cell patterning. In: CI-ALife 2007. Proceedings of the 2007 IEEE Symposium on Artificial Life, pp. 47–53. IEEE Computational Intelligence Society, Los Alamitos (2007)
17. Chavoya, A., Duthen, Y.: Use of a genetic algorithm to evolve an extended artificial regulatory network for cell pattern generation. In: GECCO 2007. Proceedings of the 9th annual conference on Genetic and evolutionary computation, p. 1062. ACM Press, New York (2007)

18. Holland, J.H.: Adaptation in Natural and Artificial Systems: An Introductory Analysis with Applications to Biology, Control and Artificial Intelligence. MIT Press, Cambridge (1992)
19. Wolpert, L.: The French flag problem: a contribution to the discussion on pattern development and regulation. In: Waddington, C. (ed.) Towards a Theoretical Biology, pp. 125–133. Edinburgh University Press, New York (1968)
20. Miller, J.F., Banzhaf, W.: Evolving the program for a cell: from French flags to Boolean circuits. In: Kumar, S., Bentley, P.J. (eds.) On Growth, Form and Computers, pp. 278–301. Academic Press, London (2003)

Rounds Effect in Evolutionary Games

Ayman Ghoneim, Michael Barlow, and Hussein A. Abbass

School of Information Technology and Electrical Engineering,
University of New South Wales,
Australia
{a.ghoneim,m.barlow,h.abbass}@adfa.edu.au

Abstract. Evolutionary games are used to model and understand complex real world situations in economics, defence, and industry. Traditionally, gaming models exhibit interactions among different players or strategies. In the literature, the number of rounds - that a game between different players contains - was treated as an experimental parameter. In this paper, we show for the first time the effect of the number of rounds on the strategic interactions in the Iterated Prisoner's Dilemma. We show that there is a cyclic behavior between the strategies and that the number of rounds per game has a significant affect on the strategies' payoffs, thus the evolutionary process.

1 Introduction

Real life situations exhibit complex behaviors that affect the decisions of all parties involved. Simple games with rich dynamics have been used to understand emergent behaviors in complex situations. The Prisoner's Dilemma (PD) game, despite its simplicity, has been used extensively in modeling several complex real-world problems such as in international politics, economics and social systems [3].

Many biological systems are organized around cooperative interactions [12,13], although natural selection is assumed to favor selfish behavior. Games have proved to be a powerful tool to model and analyze how cooperation can evolve in a population of selfish players, using the iterated version of games [3], or structured populations [13] and investigating some interesting phenomenon such as indirect reciprocity [12]. In evolutionary game theory, players are not assumed to be rational but successful strategies (that have high utility) spread in the population by being inherited or imitated [12]. The rationality in this context is reflected by the player's utility function. Thus, the utility gained by the strategies has a great affect on their spread or elimination from the population.

In many studies [1,2,5,8,9,7,11,15], the PD game was used in modeling and investigating several key aspects (i.e. cooperation evolution, history effect, number of players and information sharing). In spite of the diverse aspects that were investigated, all these studies have a common issue. The number of rounds per game between players was fixed to some experimental value - that differs from one study to another - neglecting its potentially significant effect on the payoffs

M. Randall, H.A. Abbass, and J. Wiles (Eds.): ACAL 2007, LNAI 4828, pp. 72–83, 2007.

gained by the strategies and thus, the evolutionary dynamics of their models. In this paper, we empirically explore the relationships between strategies and show that the number of rounds in any interaction has a great affect on the utility gained by the players. These conclusions are not an artifact of the PD game but apply generally to other evolutionary games.

The rest of the paper is organized as follows: in the following section, we introduce the Iterated Prisoner's Dilemma. Sections III and IV illustrate our experimental setup and results, respectively. Conclusions and future work then follow.

2   Iterated Prisoner's Dilemma

The PD game is a non-zero sum and non-cooperative game. The basic form of the PD game is a two-player game where there are two available choices to each player: to cooperate or to defect. The payoff matrix of the PD game (figure 1) must satisfy two conditions related to the players' preferences [8,15]: $T > R > P > S$ and $2 \times R > (S + T)$.

The PD game models the conflict between self interest (being selfish) and the group interest; hence the dilemma. An individual rationality alone leads to a poor outcome because of the existence of a Pareto optimal solution if both actors cooperate. Iterated Prisoner's Dilemma (IPD) is a series of repeated rounds of the PD game. This feature makes the PD game more capable of modeling complex situations where future interaction between the actors is influenced by their history during playing the game [1,2,3]. For a sufficiently large weight (discount factor) for future interactions, cooperation can emerge spontaneously. This is a very interesting characteristic to observe how cooperation may evolve among a group of potentially selfish players [15]. In many real world situations, the evolution of cooperation is considered the best solution for the long run because it represents the maximum benefit for the group or society. As such, numerous studies have been conducted of the dynamics of the IPD game in order to discover under what conditions cooperation evolves.

	C	D
C	R, R	S, T
D	T, S	P, P

Fig. 1. The payoff matrix of the 2-player PD Game

Understanding the properties of successful strategies in IPD is vital to our understanding of the dynamics of the game. Axelrod [1,3] attempted to discover the properties of successful strategies in 2-players PD game through the formation of a computer tournament of 14 strategies that were submitted by different researchers. The tournament was held in a round robin form (each strategy plays with each other strategy including itself and the RANDOM strategy). Axelrod discovered that properties like "to be nice" (not to be the first to defect), "to be forgiving" (have propensity to cooperate after other's defection, avoiding defection echo that will lead to unending mutual punishment) and "to be provocative" (not to be exploited) existed in the top ranked strategies. The winner TIT FOR TAT (TFT) strategy (start by cooperation and then do whatever the other player does) depends on reciprocity. Axelrod held a second tournament [2] after announcing the results and analysis of the first one, 62 strategies participated and the winner was again TFT. The results of the second tournament were very surprising because all participants knew the results of the first tournament but no one could get a better performing strategy than TFT.

A more sophisticated way was needed to investigate the conditions of cooperation. Axelrod [1,5] proposed the idea of using genetic algorithms to evolve more complex strategies. These strategies co-evolve in a population of competitive strategies. Lindgren [11] started with very simple strategies and used Genetic Algorithms (GA) to evolve them to more complex ones. Axelrod [4,5] used GA for evolving strategies where the strategy representation contains a history portion which is used in remembering the players' actions for the previous l history steps. If there were 3 players and two history steps, then the history portion will consist of 6 bits (2 bits for each player indicating his own previous actions and 4 bits indicating the other players' actions). The rest of the strategy representation will be a lookup table of size 2^{nl} where n is the number of players. Each possible combination of a history has a corresponding action.

Yao and Darwen [15] proposed another representation that is more space-effective than Axelrod's representation in n player games. In their representation, the history portion in the strategy representation will hold the player's own history and the number of players who cooperated in each of the considered historical steps. This representation overcomes the drawbacks of Axelrod's representation like keeping unnecessary information about each player's action and the chromosome length that is significantly affected by the number of players [15]. The rest of the strategy chromosome is also a lookup table. Different ways for evaluating the fitness of the evolved strategies were suggested. Axelrod [4,5] used 8 representative strategies from his second tournament, similarly in [10], six fixed strategies (ALLC, ALLD, TFT, TFTT, PAVLOV and RANDOM) were used in evaluating the fitness, where these six strategies provide a good mix of cooperators, defectors and strategies utilizing memory. Darwen and Yao [7,8,15] used co-evolution for evaluating the fitness, where each strategy in the population plays against every other strategy in the population, causing the environment to continuously evolve. Darwen and Yao [7] used a GA to investigate the time needed for the population to converge (in terms that the population bias will be greater than 85%) and the effect of seeding the initial population with well known strategies such as TFT. Also in [15] the effect of the number of players and the number of history steps taken into account on the evolution of cooperation were discussed. Yao [14] studied evolutionary stable strategies (Collective Stability [3]), where strategies are called stable if they can't be invaded by other strategies.

Vital features were neglected in the PD abstraction formulation like the possibility of communication between players and uncertainty about the other players' previous actions [3]. Introducing new features to the PD game and considering different scenarios for the game were very helpful to move the PD game closer to modeling complex real world situations. Introducing different levels of cooperation in the PD game and investigating their influence on the emergence of full mutual cooperation was investigated in [8]. The introduction of multiple levels of cooperation into IPD helps in studying the dynamics of real-world situations that offer intermediate responses between full cooperation and full defection. Chang and Yao [6] introduced noise to the IPD game, investigated the effect of different (low and high) noise levels and how modeling mistakes in the players' decisions influence the evolution of cooperation and the behavioral diversity in the multiple levels of cooperation (how different the played choices are in the game). Also studied was the effect of reputation on the dynamics of the game [16] where information about players' past actions are available for future opponents. Information sharing between IPD players was introduced in [9], where an extra bit was added to the history portion, this bit holds the value of 0 if the decisions to cooperate were greater than the decisions to defect in the previous generation. The Addition of this extra bit doubles the chromosome length and alters the dynamics of the game.

3   Experimental Design

Our aim in this paper is to investigate the effect of number of rounds in evolutionary games (IPD as an example). We first investigated the number of rounds in a non-evolutionary sense, then using conclusions drawn from our first experiment, we conducted another set of experiments using evolutionary model. Axelrod's representation [4,5] is used (sufficient for 2-players IPD game) and the simplest case where the players remember only one history step (remember his own and opponent's previous action) is considered. A lookup table in this case is represented by a chromosome of six bits where the first two bits represent the history portion in the chromosome and the other 4 bits for the strategy itself as shown in figure 2, the history portion is used in determining the first action for the strategy in any game. We carried out two types of experiments. For the first experiment, we considered all the possible lookup representations for one history step, a total of 64 lookup tables (2^6, where 6 is the total length). Originally, we have 16 different strategies, and each of these 16 strategies has four variants according to the different histories that can be associated with it ('00', '01', '10' and '11'). For any strategy, the history portion is mapped from a

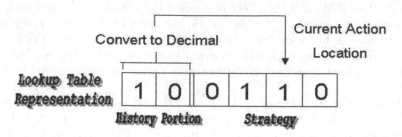

Fig. 2. The lookup table representation

binary representation to a real one in order to determine the location of the first action. For a strategy, history portions that will lead to the same initial actions will not affect the strategy behavior, and are considered to be redundant. We start encapsulating the lookup representation space by considering two history portion variants for each strategy, one that will lead to a 'C' initial action and another that will lead to a 'D' initial action. The special cases, always cooperate '0000' and always defect '1111' strategies were considered once because their actions are fixed and independent from the history portion. After the encapsulation, we have in total 30 unique lookup representations. We then established a tournament between all the lookup representations and investigated the results of the tournament after each round.

In the second experimental setup, We used GA to investigate the cooperation level that evolves in a population of IPD players using conclusions from our first experiment. A population of $N = 100$ strategies is initialized randomly. Each player is evaluated by playing against each other player in the population; hence, each player plays $(N - 1)$ 2-players IPD game. Each game lasts for a certain pre-defined number of rounds (varied to serve analytical purposes). After all players finish playing against each other, each player is awarded a cumulative payoff from the played (IPD) games. The fitness is calculated by dividing a strategy's cumulative payoff by the number of games it participated in multiplied by the number of rounds in each IPD game to obtain the average payoff per round for this strategy. Proportional (roulette wheel) selection is used where the probability of selecting a strategy for mating is proportional to the ratio between its payoff and the cumulative payoff of the whole population. We then apply a one-point crossover and bit mutation for generating new offspring with probabilities of 0.6 and 0.001 respectively. These parameters settings were used by Yao and Darwen [15]. We used the following values for the IPD payoff matrix for both experiments where $R = 3$, $S = 0$, $P = 1$ and $T = 5$.

4   Results and Discussions

Using our first experimental setup, we tried to investigate the interplay of the interactions between different lookup tables (strategies with different initial actions). Considering the unique lookup tables, we established a pair-wise tournament between all the lookup tables (each two play against each other an IPD game). We monitored the results of the tournament across the rounds, categorizing the result of a game into three types: Win, Draw and Loss. A winner strategy is the one that scores higher cumulative payoff (across the rounds) than its opponent, while a tie happens when both competing strategies score the same cumulative payoff. We found that after what we call, a "warming-up" period, each lookup table starts repeating its results against all the other lookup tables. For example, as shown in table 1, TFT strategy '0101' after the first two rounds (the warming-up period) starts repeating its results in a cyclic behavior of length 12 rounds. At rounds 3 and initial action 'C', it does not win any game, has a draw with 18 lookup tables and loses to 11 lookup tables. This behavior is repeated exactly the same at round 15, as shown in table 1, it does not win any game, has a draw and loses to exactly the same strategies as in round 3. Comparing the TFT strategy results from rounds 3 to 14 with its results from 15 to 26, we will find that the strategies' behavior is exactly repeated.

The explanation for this cyclic behavior is straight forward. Given that any two strategies are fixed and deterministic, they will reproduce the same sequence of actions if the history possessed by them is repeated. Knowing that the possible histories between any two strategies are finite, the same history will be repeated in an opposed game after a certain number of rounds. The warming-up period is to overcome the effect of the pre-assumed initial action, that is independent from the opponent that the strategy will face. The pre-assumed initial action

Table 1. Games' results for the TFT strategy with 'C' and 'D' initial actions

Rounds	(0)'0101'			(0)'0101'		
	Win	Draw	Loss	Win	Draw	Loss
1	0	14	15	15	14	0
2	0	14	15	15	14	0
3	0	18	11	11	18	0
4	0	18	11	11	18	0
5	0	14	15	15	14	0
6	0	19	10	10	19	0
7	0	15	14	14	15	0
8	0	17	12	12	17	0
9	0	18	11	11	18	0
10	0	16	13	13	16	0
11	0	14	15	15	14	0
12	0	21	8	8	21	0
13	0	15	14	14	15	0
14	0	15	14	14	15	0
15	0	18	11	11	18	0
16	0	18	11	11	18	0
17	0	14	15	15	14	0
18	0	19	10	10	19	0
19	0	15	14	14	15	0
20	0	17	12	12	17	0
21	0	18	11	11	18	0
22	0	16	13	13	16	0
23	0	14	15	15	14	0
24	0	21	8	8	21	0
25	0	15	14	14	15	0
26	0	15	14	14	15	0

significantly affects the strategy performance against its opponents. For example, the TFT strategy with 'C' initial action does not win any game and losses a considerable number of games. The same strategy with 'D' initial action does not lose any game and wins a considerable number of games. This behavior is consistent with varying the number of rounds as is clear from table 1.

Not all the lookup tables share the same warming-up period or the same cycle length. Table 2 shows the warming-up periods and the cycle lengths for all the 30 unique lookup tables. Some lookup tables (i.e. S1 - S7, S14, S15, S22, S23 and S30) have a cycle length of 1, others have a cycle length of 3 (i.e. S16, S17, S28 and S29), 4 (i.e. S20, S21, S24 and S25), 6 (i.e. S26 and S27) and 12 (i.e. S10 - S13, S18 and S19). Different initial actions do not affect the cycle length of a strategy. But, different initial actions do affect the length of the warming-up period, as shown in table 2. Until now, we were discussing the properties of individual lookup tables. Considering the whole strategy space (all the 30 lookup tables), what will be the warming-up period and the cycle length for the whole system?. The warming-up period for the whole system will be the maximum warming-up period possessed by all the strategies which is 6 rounds. Thus, after a maximum of 6 rounds, all the strategies will enter their cycles. The cycle length of the whole system is 12 rounds, which is the lowest (least) common multiple (LCM) for all the strategies cycle lengths. Thus, after a warming-up period of 6 rounds, the results of games between strategies in the whole system will be repeated every 12 rounds. In other words, assuming a population of strategies in an evolutionary context, where each strategy plays against all other strategies in the population an iterated IPD game that consists of a pre-defined number of rounds. The results (in term of win, draw or loss) of all games played in the population, if the game consists of 7 rounds or of 19 rounds or of 31 rounds (any number of rounds that is more than 6 - warming-up period, and increased continuously by 12 rounds - cycle length) will be the same.

But in an evolutionary context, the evolutionary process does not care much about the results of games between strategies in term of win, draw and loss. It cares only for the average payoff accumulated by a strategy from playing against other strategies that exist in the population. A strategy's average payoff (fitness) is used to determine if it is a successful strategy that will be inherited and imitated, compared to other strategies in the population. The question now becomes: Do strategies accumulate the same payoff across different cycles, knowing that the games' results are fixed? For addressing such a question, using our experimental first setup, we reported the average accumulated payoff by a lookup table playing against all the other lookup tables - 15 complete cycles for the whole system after the warming-up period - for 186 rounds per game. These 15 complete cycles are sufficient that at least each strategy experienced 15 individual cycles. Figure 3 shows the average accumulated payoff by some strategies across their different cycles. The dashed line represents a strategy's average payoff in its first cycle. For lookup tables such as S1 and S2, the average payoff is fixed and does not change from one cycle to another, and this represents one type of lookup tables. But for the rest of the strategies in figure 3, this

Table 2. The "warming-up" period and cycle length for each strategy

S	Initial Action	Strategy	Warming-up	Cycle Length
S1	#	0000	1	1
S2	0	0001	1	1
S3	1	0001	5	1
S4	0	0010	1	1
S5	1	0010	5	1
S6	0	0011	1	1
S7	1	0011	1	1
S8	0	0100	1	6
S9	1	0100	5	6
S10	0	0101	2	12
S11	1	0101	2	12
S12	0	0110	3	12
S13	1	0110	6	12
S14	0	0111	5	1
S15	1	0111	1	1
S16	0	1000	4	3
S17	1	1000	6	3
S18	0	1001	6	12
S19	1	1001	3	12
S20	0	1010	5	4
S21	1	1010	5	4
S22	0	1011	5	1
S23	1	1011	1	1
S24	0	1100	5	4
S25	1	1100	5	4
S26	0	1101	5	6
S27	1	1101	1	6
S28	0	1110	6	3
S29	1	1110	4	3
S30	#	1111	1	1

is not the case. The average payoff keeps changing from one cycle to the next in an increasing (i.e. S8, S10 and S13) or decreasing (i.e. S3, S9, S11, S17 and S19) pattern (except S20 and S21S). The initial action showed a great effect on average payoff that a strategy accumulates across cycles, although it does not affect the cycle length. For example, S2 and S3 are the same strategy '0001' but with different initial actions. It is clear from figure 3 that the average payoff of S2 is fixed across cycles but the average payoff of S3 is changing in a decreasing pattern. It is clear also that the average payoff that a strategy accumulates in its first cycle (the dashed line) varies dramatically depending on the position in the cycle. The differences between a strategy's average payoff across cycles is large in early cycles. But these differences start vanishing after a considerable number of cycles, and also the average payoff becomes stable and approximately a straight line, which indicates that the position in the cycle does not matter any more.

After showing the effects of number of rounds in a non-evolutionary sense, we will illustrate these effects in an evolutionary environment using our second experimental setup. Figure 4 shows the cooperation level that evolves if we set the number of rounds per game to be within the warming-up period (from 1 to 6 rounds per game). The cooperation level keeps increasing by the increase in the number of rounds. But the cooperation level is very low, and approximately the nash equilibrium of the PD game in very early rounds (1 and 2 rounds

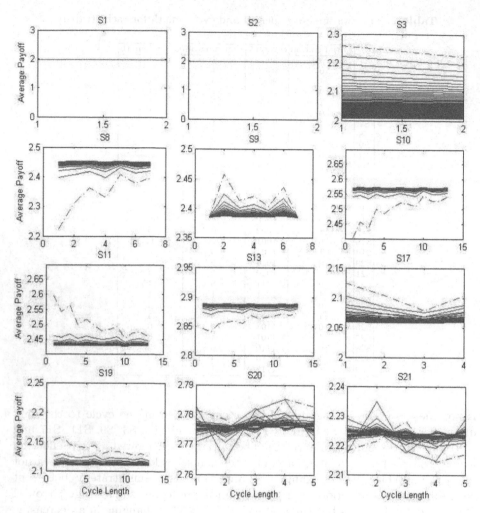

Fig. 3. The average payoff accumulated by different strategies (i.e. S1-S3, S8-S11, S13, S17 & S19 - S21) by playing against all other strategies, the dashed line indicates the average payoff of the strategy's first cycle.

per game) which means that cooperation does not evolve at all (as if we are in the zero history case). Thus, setting the number of rounds per game to be in the warming-up period will evolve a very low level of cooperation and the model's results will be misleading. We then investigated the cooperation levels that evolve inside the cycles. We set the number of rounds to be in the first cycle (from 7 to 18 rounds per game). Figure 5 shows the average payoff of 30 runs at the last generation (generation 500), it is clear that the average payoff changes dramatically depending on your positions in the cycle. The difference between the maximum average payoff obtained (at 13 rounds per game) and the minimum one (at 8 rounds per game) in the first cycle is 0.804, which is considered relatively

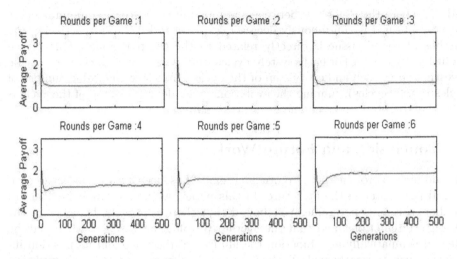

Fig. 4. The average payoff in 30 runs for the warming-up period, from 1 to 6 rounds per game

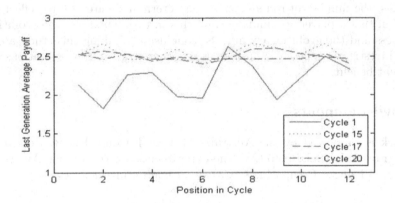

Fig. 5. The average payoff of the last generation in 30 runs for system cycles 1, 15, 17 & 20.

a high difference when compared to the payoff matrix cardinalities we are using. Comparing the average payoffs obtained in both cycle 1 and cycle 15 (shown in figure 5), it is clear that the variance of average payoffs in cycle 15 (from 175 to 186 rounds per game) is much smaller where the difference between the maximum average payoff and the minimum is 0.257. This indicates that the dynamics of the evolutionary model is almost stable after a considerable number of cycles. We can observe in cycle 15 that the average payoff is even repeated in different number of rounds (i.e. the same in 176, 182 and 185 rounds per game, and in 177 and 178 rounds per game, also in 175, 181 and 184 rounds per game), this is not the case in cycle 1. Figure 5 shows also the average payoffs obtained in cycle 17 and 20, where the difference between the maximum and the minimum average payoffs obtained is continuously decreasing, 0.161 and 0.135, respectively. The number of rounds issue is directly related to the computational effort of the evolutionary model. For early system cycles, the dynamics of the game must be investigated at each and every step of the cycle. After a considerable number of cycles (i.e. 20 cycles), running the evolutionary model at any step of the cycle is sufficient to obtain the true and accurate dynamics.

5   Conclusion and Future Work

The number of rounds per evolutionary game has been treated as an experimental parameter in the literature. In this paper, we investigated the effect of number of rounds per game. We showed that there is a cyclic behavior in the strategic interactions, and that the average payoff gained by a strategy will be affected dramatically as a function the number of the cycle and its position inside the cycle (for early cycles). Moving to evolutionary models, we showed that a very low cooperation level evolves within the warming-up period. Setting the number of rounds to be in the first system cycle will give a misleading indicator for the cooperation level. In order to determine the true dynamics of evolutionary games, the number of rounds has to be determined carefully to reflect the true payoff that a player accumulates, and this in turn will affect the evolutionary process and the evolutionary models' conclusions. Taking this study a step further, investigating these conclusions using other strategy representations will be of great benefit.

Acknowledgements

This work is supported by the Australian Research Council (ARC) discovery scheme grant number DP0667123. The experiments are run on the Australian Center for Advanced Computing (AC3) super computing facilities.

References

1. Axelrod, R.M.: Effective choice in the prisoner's dilemma. Journal of Conflict Resolution 24(1), 3–25 (1980a)
2. Axelrod, R.M.: More effective choice in the prisoner's dilemma. Journal of Conflict Resolution 24(3), 379–403 (1980b)
3. Axelrod, R.M.: The Evolution of Cooperation. Basic Books, New York (1984)
4. Axelrod, R.M.: The Evolution of Strategies in the Iterated Prisoner's Dilemma. In: Genetic Algorithms and Simulated Annealing, Morgan Kaufmann, San Francisco (1987)
5. Axelrod, R.M.: The Complexity of Cooperation. Princeton University Press, New Jersey (1997)
6. Chong, S.Y., Yao, X.: Behavioral diversity, choices, and noise in the iterated prisoner's dilemma. IEEE Transactions on Evolutionary Computation 9(6), 540–551 (2005)

7. Darwen, P.J., Yao, X.: On evolving robust strategies for iterated prisoner's dilemma. In: Yao, X. (ed.) Progress in Evolutionary Computation. LNCS, vol. 956, pp. 276–292. Springer, Heidelberg (1995)

8. Darwen, P.J., Yao, X.: Why more choices cause less cooperation in iterated prisoner's dilemma. In: Proceedings of the 2001 Congress on Evolutionary Computation, Piscataway, NJ, USA, May 2001, pp. 987–994. IEEE Press, Los Alamitos (2001)

9. Ghoneim, A., Abbass, H., Barlow, M.: Information sharing in the iterated prisoner's dilemma. In: 2007 IEEE Symposium on Computational Intelligence and Games, Honolulu, Hawaii, USA, April 2007, vol. 8, pp. 56–62 (2007)

10. Goh, C.K., Quek, H.Y., Teoh, E.J., Tan, K.C.: Evolution and incremental learning in the iterative prisoner's dilemma. In: The 2005 IEEE Congress on, vol. 3, pp. 2629–2636. IEEE Press, Los Alamitos (2005)

11. Lindgren, K.: Evolutionary phenomena in simple dynamics. Artificial Life II 10, 295–312 (1991)

12. Nowak, M.A., Sigmund, K.: Evolution of indirect reciprocity. Nature 437(27), 1291–1298 (2005)

13. Ohtsuki, H., Hauert, C., Liberman, E., Nowak, M.A.: A simple rule for the evolution of cooperation on graphs and social networks. Nature 441(25), 502–505 (2006)

14. Yao, X.: Evolutionary stability in the n-person iterated prisoner's dilemma. BioSystems 37(3), 189–197 (1996)

15. Yao, X., Darwen, P.J.: An experimental study of n-person iterated prisoner's dilemma games. Informatica 18(4), 435–450 (1994)

16. Yao, X., Darwen, P.J.: How important is your reputation in a multi-agent environment. In: Proc. of the 1999 IEEE Conference on Systems, Man, and Cybernetics, Piscataway, NJ, USA, October 1999, pp. 575–580. IEEE Press, Los Alamitos (1999)

Modelling Architectural Visual Experience Using Non-linear Dimensionality Reduction

Stephan K. Chalup, Riley Clement, Chris Tucker, and Michael J. Ostwald

Faculty of Engineering and Built Environment, The University of Newcastle,
Callaghan 2308, Australia
Stephan.Chalup@newcastle.edu.au
Tel.: +61 2 492 16080

Abstract. This paper addresses the topic of how architectural visual experience can be represented and utilised by a software system. The long-term aim is to equip an artificial agent with the ability to make sensible decisions about aesthetics and proportions. The focus of the investigation is on the feature of line distributions extracted from digital images of house façades. It is demonstrated how the dimensionality reduction method isomap can be applied to calculate non-linear "streetmanifolds" where each point on the manifold corresponds to a house façade. Through interpolation between manifold points and the application of an inverse Hough transform, basic structure plans for new house façades are obtained. If the interpolated points are close to the manifold it can be argued that the new plans reflect the character of the surrounding streetscape. The method is also demonstrated using basic examples which can be represented by circles.

1 Introduction

Aesthetical perception is an important factor in understanding the interaction of a living individual with its environment. The discipline of environmental aesthetics argues that the environment is fully integrated with the individual [1] and that "aesthetic values pervade the entire range of human culture" [2] which includes environmental and architectural design of gardens, landscapes, cities, and virtual space.

The concept of streetmanifolds was introduced in previous papers [5,6] to provide a holistic geometrical representation of the visual experience which can be gained through evaluation of a large set of house façades. Navigation in the streetmanifold would correspond to continuous morphing and interpolating between façade designs represented by the data set of images of house façades. The concepts of holism, continuity, and clustering are associated with manifold learning [26,29] but can also be found in Gestalt psychology [18,30] which has close links to the concepts of visual neuroscience [7].

The hypothesis of the present study is that the visual experience gained by an architect through visual perception of thousands of house façades during his education and professional life may be captured in a structure which corresponds to some form of streetmanifold.

M. Randall, H.A. Abbass, and J. Wiles (Eds.): ACAL 2007, LNAI 4828, pp. 84–95, 2007.

The house façades along a street contribute to the character of a streetscape [9,27,28]. This is an important factor for architects who design a new house for an empty spot between the other houses of a street such that the new house harmoniously relates to the neighbourhood [12].

The present study's streetmanifolds are based on the calculation of pairwise distances between digital images of house façades. At the current stage of the project the focus is on an important feature in the visual perception of houses which is the distribution of lines determined by the edges of the main components of a house façade. Typically most of the lines have horizontal or vertical direction with a few approximately diagonally oriented lines along the roof or gable. Figure 1 shows examples of house façades with virtual lines along edges extracted using a Hough transform [14,24].

The approach to take line directions as the central feature for the calculation of streetmanifolds is supported by research in visual neuroscience which found that detection of edge directions is a key component of the human visual system [15,19]. It also was found that the visual system has specialised areas for representation of different entities such as buildings [11].

The main new contribution of the present article is to utilise the streetmanifolds calculated from our dataset of house façades in [5] to generate basic plans for new house façades. The new plans are distributions of lines which are obtained through linear interpolation of points on the streetmanifold and application of an inverse Hough transform.

Previous related work which addresses how artificial life methods can be applied in architecture include philosophical discussions [21] or software development associated with the area of emergent design [13,20,22]. Reich [21] addressed the topic of how aesthetic judgment can be incorporated in computational design. He claimed that aesthetic criteria are embedded in designers' expertise and their use is manifested in existing designs. Reich discussed how rationalistic and romanticistic aesthetic criteria can synergistically be applied to design. A practical example was presented in a system for the design of cable-stayed brides. Frazer [10] proposed a generative design tool for architects based on cellular automata. An artificial life based emergent design software system was developed by Ross et al. [22] which allows architects to endow elements of an architectural scenario with agency and dynamic spatial interaction. Hemberg et al. [13] developed computational generative design software for architects which can generate three dimensional forms and surfaces. Their system used evolutionary algorithms and L-systems grammars with the aim of being able to grow and evolve organic forms.

The remaining sections of this paper address the topic of manifold learning (section 2), some basic examples of learning circle manifolds (section 3), the procedure required to extract a streetmanifold from a set of digital images of

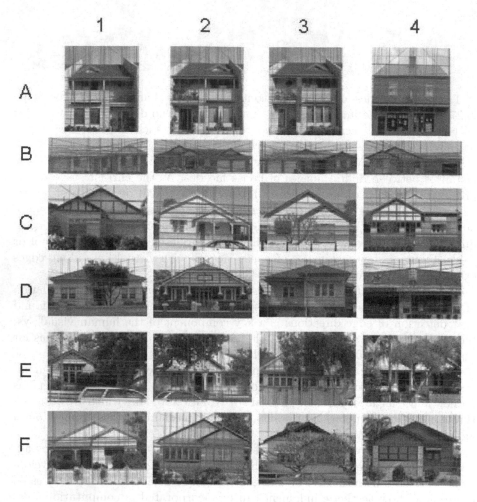

Fig. 1. Six clusters of houses (A-F) found in the streetmanifolds

house façades (section 4), and a description of how to generate plans for new houses through interpolation of manifold points (section 5). In section 6 a brief discussion and summary of the results is provided.

2   Manifold Learning

Manifolds are locally Euclidean spaces with some additional very general mathematical properties [25]. In dimension one they appear as continuous deformations of lines and circles and in dimension two they are surfaces derived from spheres, tori, pretzel surfaces, or similar objects. The manifold concept generalises to higher dimensions.

Manifold learning describes algorithms for non-linear dimensionality reduction [4,23]. The aim of manifold learning algorithms is to detect the essential underlying geometric structure of a high-dimensional data set, to extract it as a low-dimensional manifold and to embed it faithfully into a low-dimensional space.

In contrast to the relatively new manifold learning techniques, traditional methods for dimensionality reduction such as principal component analysis (PCA) [16] or multidimensional scaling (MDS) [8] were designed for reducing the dimensionality of data when the underlying structure was linear.

Two manifold learning methods, isomap [26] and maximum variance unfolding (MVU) [29], have been employed in the present project to calculate streetmanifolds [5]. Both methods can be applied by first calculating a distance matrix based on a weighted k-nearest neighbour graph of the data points.

In isomap [26] these pairwise distances, which can be regarded as approximations to geodesic distances on the manifold, are fed into MDS. That is, isomap can be regarded as a modification of MDS where instead of the Euclidean distances approximations to geodesic distances are used. MDS then maps the data into a lower dimensional space while preserving the pairwise distances [8].

The aim of MVU [29,23] is to maximise the sum of pairwise distances of all data points, i.e. $\sum_{ij} \left(\|y_i - y_j\|^2 \cdot \delta_{NN}(x_i, x_j) \right)$, where $\delta_{NN}(x_i, x_j)$ is 1 if x_i and x_j are nearest neighbours and 0 otherwise; the maximisation is subject to two conditions which postulate that: (I) distances between nearest neighbour inputs should be the same as between the associated outputs, i.e. $\|y_i - y_j\|^2 = \|x_i - x_j\|^2$ and (II) the outputs should be centered at the origin, i.e. $\sum_i y_i = 0$.

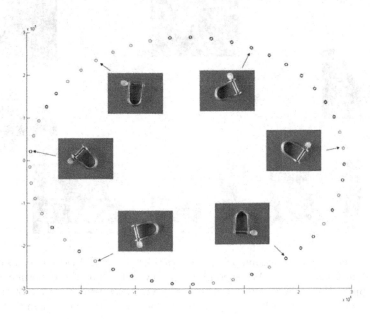

Fig. 2. Circle extracted from images of a rotating shackle using 4-isomap

3   Extracting Circle Manifolds

A simple example of how manifold learning works is shown in figure 2. A sequence of digital images of a rotating shackle was taken, i.e. the underlying dynamics of the data set was a rotation. The dimension of the space of digital images is the number of pixels in each image, i.e. 192×292. Isomap with $k = 4$ was able to extract a 1-dimensional circle embedded in $\mathbf{R}^2$ from the rotating shackle data.

In a second experiment, instead of taking pictures of a rotating object, we rotated the camera at the center point of a circle. Figure 3 shows that the result 4-isomap extracted from an image sequence taken by an HD video camera while rotated in the middle of Wheeler place in Newcastle is, as expected, again a circle. The data consisted of about 200 overlapping frames sampled from the video sequence.

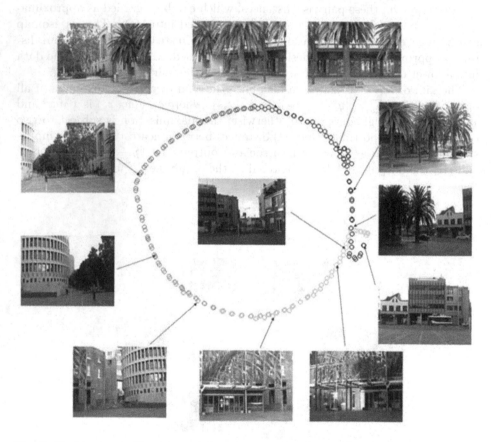

Fig. 3. Circle extracted by 4-isomap from images taken by an HD video camera rotating about $360°$ in the middle of Wheeler place in Newcastle. For the experiment about 200 overlapping frames were extracted from the video sequence. Twelve of them are displayed above together with the corresponding points on the circle manifold.

4 Calculating Streetmanifolds

The calculation of the streetmanifolds was based on a dataset of several hundred digital images of house façades which were taken in Newcastle and selected by a team of researchers from architecture. Some example images are shown in figure 1.

A line can be regarded as a set of points $\mathbf{x} = (x_1, x_2) \in \mathbf{R}^2$ and can be determined by using the Hessian normal form $\{ \mathbf{x} \in \mathbf{R}^2 ; \ [\cos\varphi, \sin\varphi] \cdot \mathbf{x} - b = 0 \}$, where $\varphi \in [0, 360°[$ controls the slope of the line's normal vector and $b \in \mathbf{R}$ is its perpendicular distance from the origin. Using the Hough transform [14,24] each image was associated with an array of discrete parameters $(\varphi, b) \in [0, 360°[\times\mathbf{R}$— the Hough array—where each point corresponds to a line in the image.

For the application of isomap and MVU the distance between each pair of Hough arrays was calculated. The discrete set of point values in the Hough arrays was smoothed by multiplying each point in the array with a Gaussian function. Then for each pair of arrays $A = (a_{ij})_{\substack{i=1,\ldots,m \\ j=1,\ldots,n}}$ and $B = (b_{ij})_{\substack{i=1,\ldots,m \\ j=1,\ldots,n}}$ their Euclidean distance was calculated using $d_2(A, B) = (\sum_{\substack{i=1,\ldots,m \\ j=1,\ldots,n}} (a_{ij} - b_{ij})^2)^{1/2}$. As an alternative a distance based on the Bhattacharyya distance measure [3,17] was applied after normalisation of the arrays: $d_{Bhat}(A, B) = 1 - \sum_{\substack{i=1,\ldots,m \\ j=1,\ldots,n}} \sqrt{a_{ij}} \sqrt{b_{ij}}$. Application of isomap or MVU allowed to embed the manifold of Hough arrays into two or three-dimensional space (figures 4, 5, and 6).

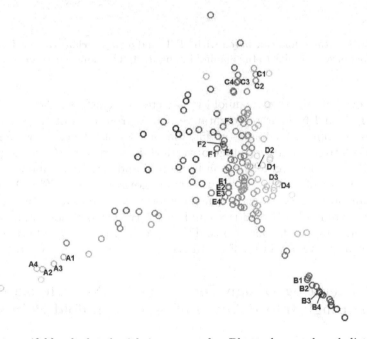

Fig. 4. Streetmanifold calculated with isomap and a Bhattacharyya based distance. Greylevel encodes the third dimension.

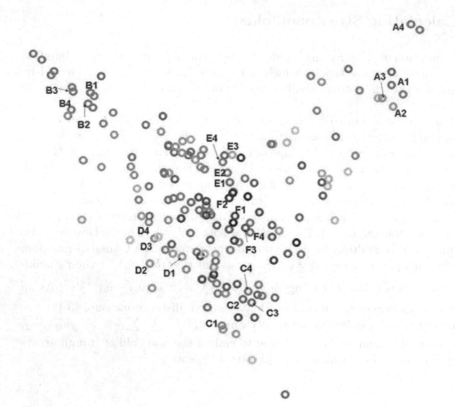

Fig. 5. Streetmanifold calculated with MVU and a Bhattacharyya based distance. The manifold is very similar to the manifold in figure 4. The same clusters can be identified.

To evaluate the streetmanifold we selected six clusters of houses (A-F) in figures 4, 5, and 6. Four representative houses from each of the six clusters are shown in figure 1. We found (cf. [5]) that houses of category A were narrow and had a relatively high percentage of vertical lines. In contrast the houses of category B were wide and had strong horizontal and vertical components. Category C had houses of medium width with many horizontal lines. The D category was very similar to the C category but the houses were wider in D. In the E category houses were hidden behind trees and the distribution of horizontal and vertical lines tended to be homogeneous. The associated cluster was located at a close to central position. Cluster F contained houses with average characteristics.

5   Generating Design Templates for New House Façades Through Interpolation of Streetmanifold Points

The geometry of the streetmanifold is determined by the distances between all records of the dataset. Therefore the streetmanifold calculated from the image

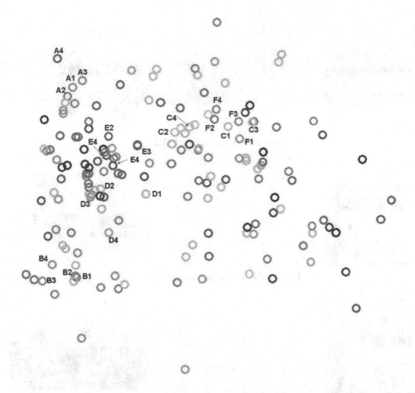

Fig. 6. Streetmanifold calculated with MVU and Euclidean distance appears to have a different shape but shows similar clusters as the manifolds in figures 4 and 5.

dataset of the houses of a street or neighbourhood can be regarded as a representation of the aesthetical character of the streetscape.

Points on or close to the streetmanifold represent Hough arrays of façades which have similar features as those of the images which were used to generate the manifold. Through application of an inverse Hough transform it is possible to generate for each manifold point a line distribution as shown in the middle column of figure 7. These patterns of lines may be used as plans for architects to outline basic proportions of a house which should fit into the streetscape.

In the present study several pairs of house façades were selected and for each pair a linear interpolation of the associated Hough arrays was calculated. Then an inverse Hough transform was applied to the result of the interpolation. Before interpolation the Hough arrays were smoothed by multiplying each peak with a Gaussian function. The inverse Hough transform was calculated by selecting the 30 highest local maxima of the sum of the two smoothed Hough Arrays.

The middle column of figure 7 shows the resulting plans obtained by the procedure of interpolation between Hough arrays of the pairs (A2, B2), (F4, B1), (A1, A2), (A1, D2), and (A1, F1), respectively. The house façades were selected from the data used to calculate the streetmanifolds and are also displayed in figure 1. The format and size of the plans in the middle column of figure 7 was

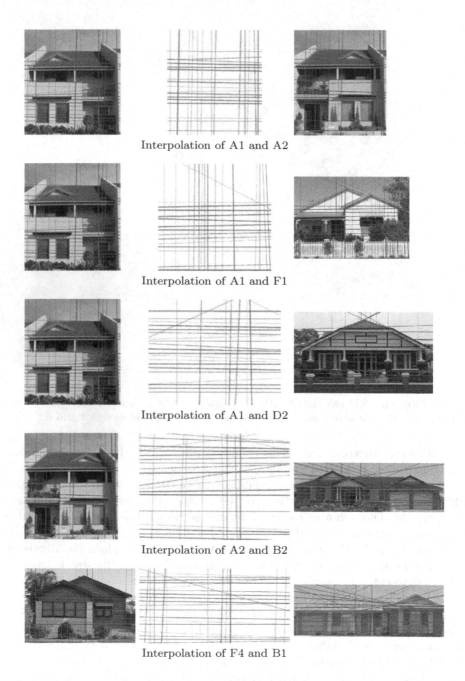

Interpolation of A1 and A2

Interpolation of A1 and F1

Interpolation of A1 and D2

Interpolation of A2 and B2

Interpolation of F4 and B1

Fig. 7. The middle column shows the inverse Hough transforms of interpolations between Hough arrays corresponding to five pairs of house façades. The outcome indicates that interpolation between A1 and A2 or A1 and F1 led to sensible results in contrast to interpolation between distant points such as A2 and B2.

determined by taking the maximum of heights and the maximum of widths of the two images of the house façades which were used in the interpolation process.

6   Discussion and Summary

The streetmanifolds in figures 5 and 6 show a comparable structure of clusters to the streetmanifolds of figure 4 and our previous results [5]. The resulting clusters suggest that the streetmanifolds have captured and smoothly organised a variety of line-based features of the whole data set in one object.

Although streetmanifolds are non-linear we have employed linear interpolations of smoothed Hough arrays to generate plans for new houses. That means that for points close to each other such as A1 and A2 the interpolation result is likely to be close to the manifold. But for points distant to each other such as A2 and B2 the interpolation result may lie far outside the manifold and may hence not be representative for the character of the streetscape.

In some cases we added the interpolation result to the initial data set of Hough arrays and recalculated the streetmanifolds. For arrays resulting from interpolation of nearby points the manifold did not change much. However, for some arrays resulting from interpolation of distant points, the newly calculated streetmanifolds changed significantly compared to the original.

These results seem to support the hypothesis that local interpolation (e.g. between A1 and A2), or interpolation on the streetmanifold, may lead to plans which are conform with the character of the streetscape.

Future research may investigate alternative options of interpolation on street-manifolds and their use in software systems for generative design in virtual worlds or application software for architects.

References

1. Berleant, A.: The Aesthetics of Environment. Temple University Press (1995)
2. Berleant, A.: Aesthetics and Environment: Theme and Variations on Art and Culture. Ashgate Publishing, Limited (2005)
3. Bhattacharyya, A.: On a measure of divergence between two statistical populations defined by their probability distributions. Bulletin of the Calcutta Mathematical Society 35, 99–109 (1943)
4. Burges, C.J.C.: Geometric Methods for Feature Extraction and Dimensional Reduction. In: Data Mining and Knowledge Discovery Handbook: A Complete Guide for Researchers and Practitioners, Kluwer Academic Publishers, Dordrecht (2005)
5. Chalup, S.K., Clement, R., Marshall, J., Tucker, C., Ostwald, M.J.: Representations of streetscape perceptions through manifold learning in the space of hough arrays. In: 2007 IEEE Symposium on Artificial Life, April 1-5, 2007, IEEE Computer Society Press, Los Alamitos (2007)
6. Chalup, S.K., Clement, R., Ostwald, M.J., Tucker, C.: Applications of manifold learning in architectural façade and streetscape analysis. In: Workshop on Novel Applications of Dimensionality Reduction at NIPS, Whistler CN 2006 (2006)
7. Chalupa, L.M., Werner, J.S. (eds.): The Visual Neurosciences. MIT Press, Cambridge (2004)

8. Cox, T.F., Cox, M.A.A.: Multidimensional Scaling. 2nd edn. Chapman & Hall/CRC (2001)
9. DIPNR: Neighbourhood Character. NSW Department of Infrastructure Planning & Natural Resources, Sydney (2004)
10. Frazer, J.: An Evolutionary Architecture. Architectural Association, London (1995)
11. Grill-Spector, K., Malach, R.: The human visual cortex. Annual Reviews Neuroscience 27, 649–677 (2004)
12. Groat, L.: Contextual compatibility in architecture: An issue of personal taste? In: Nasar, J. (ed.) Environmental aesthetics: Theory, research, and applications, pp. 228–253. Cambridge University Press, Cambridge (1988)
13. Hemberg, M., O'Reilly, U.M., Menges, A., Jonas, K., Goncalves, M., Fuchs, S.: Exploring generative growth and evolutionary computation for architectural design. In: Machado, P., Morelo, J.J. (eds.) Art of Artificial Evolution, Springer, Heidelberg (2006)
14. Hough, P.V.C.: Methods and means for recognizing complex patterns. U.S. Patent 3,069,654 (1962)
15. Hubel, D.H., Wiesel, T.N.: Receptive fields, binocular interaction, and functional architecture in the cat's visual cortex. Journal of Physiology (London) 160, 106–154 (1962)
16. Jolliffe, I.T.: Principal Component Analysis. Springer, New York (1986)
17. Kailath, T.: The divergence and bhattacharyya distance measures in signal selection. IEEE Transactions on Communication Technology 15, 52–60 (1967)
18. Koffka, K.: Principles of Gestalt Psychology. Harcourt Brace, New York (1935)
19. Larsson, J., Landy, M.S., Heeger, D.J.: Orientation-selective adaptation to first- and second-order patterns in human visual cortex. Journal of Neurophysiology 95, 862–881 (2005)
20. O'Reilly, U.M., Hemberg, M., Menges, A.: Evolutionary computation and artificial life in architecture: Exploring the potential of generative and genetic algorithms as operative design tools. Architectural Design, Special Issue on Emergence 74, 48–53 (2004)
21. Reich, Y.: A model of aesthetic judgment in design. Artificial Intelligence in Engineering 8, 141–153 (1993)
22. Ross, I., O'Reilly, U.M., Testa, P.: Emergent design: Artificial life for architecture design (2000)
23. Saul, L.K., Weinberger, K.Q., Sha, F., Ham, J., Lee, D.D.: Spectral methods for dimensionality reduction. In: Chapelle, O., Schölkopf, B., Zien, A. (eds.) Semi-Supervised Learning, pp. 293–308. MIT Press, Cambridge, MA (2006)
24. Shapiro, L.G., Stockman, G.C.: Computer Vision. Prentice-Hall, Englewood Cliffs (2001)
25. Spivac, M.: A Comprehensive Introduction to Differential Geometry, 2nd edn. Publish or Perish, Inc. (1979)
26. Tenenbaum, J.B., de Silva, V., Langford, J.C.: A global geometric framework for nonlinear dimensionality reduction. Science 290, 2319–2323 (2000)
27. Tucker, C., Ostwald, M.J., Chalup, S.K.: A method for the visual analysis of streetscape character using digital image processing. In: Bromberek, Z. (ed.) Contexts of Architecture: Proceedings of the 38th Annual Conference of the Architectural Science Association ANZAScA and the International Building Performance Simulation Association, Launceston, Tasmania: Australia and New Zealand Architectural Science Association, pp. 134–140 (2004)

28. Tucker, C., Ostwald, M.J., Chalup, S.K., Marshall, J.: Sustaining residential social space: a visual and spatial analysis of the nearly urban. In: ANZAScA 40th Annual Conference of the Architectural Science Association, "Challenges for architectural science in changing climates", November 22-25, 2006, The University of Adelaide Adelaide, South Australia (2006)
29. Weinberger, K.Q., Saul, L.K.: An introduction to nonlinear dimensionality reduction by maximum variance unfolding. In: Proceedings of the National Conference on Artificial Intelligence (AAAI), Nectar paper, Boston, MA (2006)
30. Wertheimer, M.: Untersuchungen zur Lehre von der Gestalt II. Psychologische Forschung 4, 301–350 (1923)

An Evolutionary Benefit from Misperception in Foraging Behaviour

Lachlan Brumley, Kevin B. Korb, and Carlo Kopp

Clayton School of Information Technology, Monash University, Australia
lbrumley@infotech.monash.edu.au, korb@infotech.monash.edu.au,
carlo@infotech.monash.edu.au

Abstract. Misperception is a common cause of error for individuals and organisations. Conventional wisdom suggests that its effects are detrimental to the misperceiver or its society as a whole. However, in some circumstances misperception can provide a benefit either by diversifying the behaviour of a population or by discouraging behaviour that has a negative impact on the population. In such cases adaptive pressures will drive the population to evolve a probability of misperception that is optimal for that environment. We explore this hypothesis using an evolutionary artificial life simulation.

Keywords: Artificial Life, Misperception, Evolutionary Simulation.

1 Introduction

Misperception can be said to occur when an entity gathers information from its environment and uses that information to produce an internal model of the world that may or may not accurately represent the surrounding physical environment [1]. Misperception may be caused unintentionally by flaws within the misperceiver or intentionally by other entities performing Information Warfare attacks [2]. Any information sensor that is used to gather information from the environment can be affected by misperception. Entities may misperceive any element of their environment – such as the existence or non-existence of other entities, their attributes or the relationships between entities in the environment.

Russell and Norvig [3] describe the basic cycle of a simple intelligent agent as consisting of information collection, orientation relative to the environment, decision-making and action execution. This cycle describes a feedback loop between the agent and its environment, as the agent's actions will affect the state of its environment, which can be observed in the future. Similar models of the decision-making cycle are also discussed in psychology [4] and military science [5].

The first opportunity for misperception in an agent's decision-making cycle occurs while it gathers information from its environment. There are many possible causes of a dysfunction here, including sensor limitations, natural deterioration and external attack. In all of these cases the agent is unable to correctly perceive the environment and unknowingly gathers incorrect information, which it uses

M. Randall, H.A. Abbass, and J. Wiles (Eds.): ACAL 2007, LNAI 4828, pp. 96–106, 2007.

to update the its internal representation of the world. The agent's future actions can be affected by the incorrect information.

The second opportunity for misperception in the agent's decision-making process occurs when the agent incorrectly processes the gathered information and then updates its internal representation of the world with the incorrectly processed information. This introduces inaccuracies into the agent's representation of the world.

The various causes of misperception introduce some form of singular or repeated error into an agent's representation of their environment. Typically, considerations of misperception assume that it is due to an unintentional error of the misperceiver and that it reoccurs with some frequency. Unintentional repeated misperception can therefore be modelled as a random error that occurs with a certain probability.

1.1 Artificial Life Simulations of Misperception

Misperception is an everyday occurrence in the real world, yet it is rarely found in artificial life simulations. Presumably this is because of the common belief that misperception is always detrimental. However, this is not true in some cases.

Doran [6] demonstrated two similar cases where agents may hold incorrect beliefs without the individual agents or their society suffering a detrimental effect. In both of these cases the agent's misbeliefs discourage detrimental behaviour. Doran simulated an environment where agents could move around a two-dimensional space, harvest resources and asexually reproduce. Agents were able to misbelieve in both cases, with their misbeliefs spread by communication or inherited from their parent.

The first simulated environment contained a fatal zone, which killed any agents who entered immediately. Agents could only harvest a resource if they believed that they were the nearest agent to that resource. The agents were able to misbelieve the existence of pseudo-agents where none actually existed. Agents are deterred from harvesting a resource whenever they believe a pseudo-agent is closer to that resource. Most of the agents in this society developed the belief that pseudo-agents existed in the fatal zone, which deterred them from entering the fatal zone to harvest resources. The misbelieving population was fitter than one without misbelief: individual agents benefited by avoiding the fatal zone, while the agent's society benefited by increasing in size.

A second experiment looked at the formation of cults. In this experiment (absent a fatal zone) the agents were able to form friendships, which allowed them to exchange information about resource locations. Agents were also able to kill other agents. However, agents were prevented from killing any agents with whom they shared a common friend. The misbelief that could form in this society was the belief that resources were actually agents, called "resource agents". Agents could also decide that resource agents were their friends. These rules allowed the agents to construct long-lasting "cults", where many agents shared a common misbelief in a resource agent who was their friend. For a cult to outlast its individual member agents its figurehead must be a resource agent, as resource agents only "die" when no agents believe in them. The misbelief in resource agents allowed cults to survive as long as there were followers, and the restriction on killing fellow cult members allowed the agents' society to grow in size.

More directly relevant to our work, Akaishi and Arita [7,8] hypothesised that misperception could prove to be adaptive in cases where it increases the diversity of a population's collective beliefs and thereby increases the diversity of the population's collective behaviour. Increased behavioural diversity should help reduce direct competition between agents for access to popular locations or resources.

This hypothesis was tested with a simulation of a two-dimensional grid world populated by agents and resource nodes. As agents traversed their environment, they would gather resources from stable resource nodes and maintain an internal map of where they believed resource nodes existed. The fitness of individual agents was a function of the quantity of resources gathered, while the population's fitness was determined by the average resources gathered. Agents could misperceive every time they viewed the environment, with the probability being a constant for the entire population. Misperception only affected the perception of resources, either their existence or their location. Misperception of existence could cause the appearance of either a resource where none existed or an empty location where there was a resource. Misperception of location occurred when an agent correctly perceived the contents of a location but stored the information in a random location in its world map.

The results of their simulation demonstrated that a population with a misperception probability up to 10% collected more resources on average than a population with no misperception. Optimal resource gathering occurred when the misperception probability was 1%. The fact that a misperception probability of up to 10% is better than no misperception is counter-intuitive. Their results support their hypothesis that behavioural diversity caused by misperception is beneficial. Consider an agent in a region that is densely populated with other agents and so is constantly competing for access to resources. If this agent misperceives, it may convince itself to head into a less populated, less competitive area. In this example, misperception would have provided a benefit.

One of Akaishi and Arita's [7] claims was that misperception would provide an evolutionary benefit. However, their simulated system implemented no evolutionary mechanisms. Although they found a "fitness" benefit for misperception, fitness was defined strictly in terms of resource gathering. It is easy to believe that this can translate into an evolutionary benefit, but they failed to demonstrate it. Some of our work here was inspired by the idea of making such a demonstration. If there is evolutionary value to misperception, it should be possible for a population of foraging agents to evolve to a stable state with a misperception probability that is significantly above 0%.

2   Methodology

We now describe our simulation technique. As in Akaishi and Arita's simulation, our agents inhabit a two-dimensional square grid world containing resource nodes. The agents move about this world gathering resources from these nodes. While exploring, the agents maintain an internal map of where they believe resources are located. In this simulation all misperceptions are caused by sensor failures, which result in either location or existence misperception with an equal probability. Cells may only be occupied by one agent at a time.

Evolutionary simulation requires a population of agents that evolve over the duration of the simulation. Existing agents reproduce to produce periodically, while agents die from old age or starvation. A population cap is used to prevent overcrowding.

Each turn agents must metabolise resources at the basic metabolic rate (BMR), or else starve. Agents also require a predetermined quantity of resources in order to reproduce, which equals one half of the parental investment in the health of their offspring. Once an agent can afford to have offspring it can reproduce with any other agent it encounters who also has sufficient resources; in other words, reproduction is sexual but genderless.

Each agent has its own inherited misperception probability that determines how likely it is to misperceive any object it observes. This misperception probability derives from one of its parents and may be altered by mutation.

During each turn of the simulation, every agent is activated (in a random order) and proceeds through its action cycle, consisting of perception, decision-making, movement and, possibly, mating and gathering. First, the agent perceives its surroundings, with its misperception probability determining whether or not it misperceives what is in each location. The agent then updates its resource map. Next, the agent decides which resource node is closest and adopts this resource node as its intended destination. The agent then moves toward its destination. If two agents who may reproduce meet, they will reproduce if there is room in the simulation for a new agent. Once the agent has finished moving, if its location contains a resource node it will gather any available resources. Finally, the agent consumes the amount of resources determined by its basic metabolic rate. If the agent has insufficient resources, then it dies of starvation and is removed from the simulation.

The major simulation parameters are the maximum agent density, the resource density, the basic metabolic rate and the parental investment cost. These parameters determine how much competition there will be for resources and how many resources are needed both to stay alive and to reproduce.

As there are too many potential combinations of these parameters to investigate, we have paired some, investigating them in combination. In this way, agent density and resource density were paired (Table 1) as they main determiners of resource competition. Basic metabolic rate and offspring cost were also paired together (Table 2), as jointly affecting the cost of living. Combining these pairs produced 36 different parameter sets to explore. For these simulations we used EnFuzion, a commercialised derivative of Nimrod [9], a software tool that supports efficient parallel search through the parametric space on a computer cluster.

Table 1. Agent Density and Resource Density pairings

Agent Density	Resource Density
30%	25%
30%	10%
30%	5%
25%	15%
20%	10%
15%	10%
15%	5%
10%	5%
5%	5%

Table 2. Basic Metabolic Rate and Offspring Cost pairings

Basic Metabolic Rate	Offspring Cost
0.2	25
0.15	100
0.1	50
0.05	500

2.1 Experiment 1

The first experiment was aimed at demonstrating that the previously claimed benefit of misperception in a foraging environment exists, by showing that an evolved misperception probability could be greater than zero. This simulation was performed for each of the 36 parameter sets.

2.2 Experiment 2

Our second experiment tested Akaishi and Arita's hypothesis that misperception provides a benefit specifically by increasing the agents' behavioural diversity. If this claim is true, then any mechanism (misperception or otherwise) that introduces relevant diversity into the behaviour of the agent population should provide a noticeable benefit. To test this hypothesis, four different foraging behaviours were implemented in the simulation and compared against each other. These foraging behaviours were: misperception-affected foraging (as in Experiment 1), misaction-affected foraging, reflexive foraging and perfect-perception foraging.

Misperception-affected foraging is the standard foraging method used in the simulation. Behavioural diversity is introduced by the agents' differing beliefs about resource node locations.

Misaction-affected foraging is similar to misperception-affected foraging, except that the random errors occur during the movement stage instead of the perception stage. A misaction causes the agent to move in an unintended direction.

Reflexive foraging replaces the agent's resource node discovery and path planning with random movement. Agents move randomly about the world until they observe a resource node within their perception range. They then move to the node's location and gather its resources. Clearly, the randomness of agents' movement introduces substantial diversity to their behaviour.

Perfect-perception foraging agents are agents who utilise the same decision-making methods as misperception- or misaction-affected foraging agents, but are unaffected by misperception or misaction. These agents will have very little behavioural diversity and make a baseline for comparison. If behavioural diversity provides a benefit to the population, then perfect-perception foraging should perform the worst of all the foraging methods.

3   Results

3.1   Experiment 1

If misperception has evolutionary value then the population of foraging agents should evolve to a stable state with a misperception probability that is significantly above zero. When agents reproduce, the new offspring may have its misperception probability mutated by adding a small normally distributed delta value with a mean of 0.0 and a controllable standard deviation (σ). Hence, 95% of the mutated misperception probabilities will be within $\pm 1.96\sigma$ of the original misperception probability. Call this range of misperception probabilities the mutation range. For all simulations the standard deviation was 0.02, so the mutation range was 0.0392 (or 3.92%). If the difference between the two agents' misperception probabilities is greater than the mutation range, then the two agents are unlikely to be parent and child. We report a population's average misperception probability as significantly different from 0% if that probability is greater than 3.92%. (Because of the very large number of agents sampled — around 50 million per estimated probability — the sample variance in estimating the average misperception probability was ignored.)

The average misperception probabilities for all the parameter sets are shown in Figure 1. There are seven identified parameter sets (numbered 1-7) where the average misperception probability is significantly different from zero, especially for parameter sets 1 and 2. There are two further points (8 and 9) identified in Figure 1, because they contain substantial subpopulations of misperceivers.

The majority of these numbered points occur when the agent density is 5% (relatively low density) or when the offspring cost is 500 resources (at its highest). In the latter case, the need for resources is highest, so the selection pressure in favour of successful foragers is greatest. Following the interpretation of [7,8], misperception benefit occurs in this simulation when clusters of agents gather around resource nodes and form a "traffic jam". Occasional misperception aids by sending some agents away from the food source allowing their relatives the opportunity to collect the resource, supporting a inclusive fitness (kin selection) advantage for the misperception. So long as the misperception rate is low, the misdirected agents may well also then locate a new, uncongested resource node. In any case, were misperception to provide no benefit at all, we would see a completely flat plot with the average misperception probabilities not significantly different from 0%. The results show that there is an evolutionary benefit to misperception in many of our simulations.

We also divided the agent population into three groups based on their misperception probability and compared their total population size and fitness. One group contains agents whose misperception probability is 0%, another contains

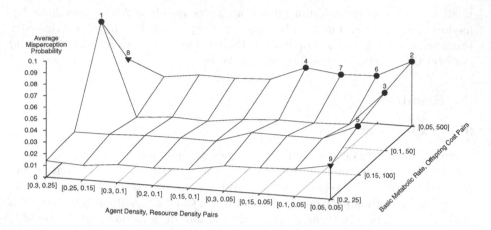

Fig. 1. Average Misperception Probabilities

agents whose misperception probability not significantly different from 0%, while the last contains agents whose misperception probability significantly different from 0%. If misperception is beneficial, then the agents who are significantly different from 0% should be more numerous and fitter. The percentage of the total agent population of each distinct agent group is shown in Figure 2 for all 36 parameter sets. The numbered parameter sets have a substantial percentage of the population that is significantly different from the 0% misperception probability; also, less than 10% of their populations has a misperception probability of 0%.

To measure agent's fitness we used its potential offspring, that is, how many offspring it could afford to parent from its surplus resources. A measure of actual offspring is not suitable for our simulation, as agents are prevented from reproducing whenever the environment is full. The potential number of offspring is only calculated after the agent has died, as the calculation (1) requires the agent's total resources gathered and its age.

$$\text{Potential Offspring} = \frac{\text{Total Resources Gathered} - (\text{Age} \times \text{BMR})}{\text{Offspring Cost}} \tag{1}$$

The average potential offspring of the 36 parameter sets is shown in Figure 3, again divided into three groups based on their misperception probability. In the majority of the parameter sets there is no difference in the average potential offspring between agents with different misperception probabilities. From this we can argue that misperception is not providing a noticeable benefit through increased potential offspring. However, some parameter sets show subpopulations with very substantially larger fitness corresponding to subpopulations with higher misperception probabilities, especially sets 1, 3 and 9.

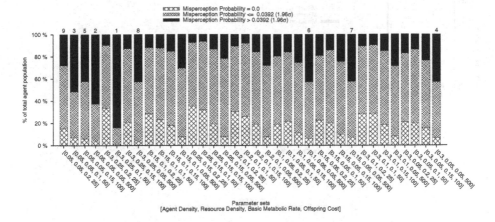

Fig. 2. Proportions of misperception probabilities

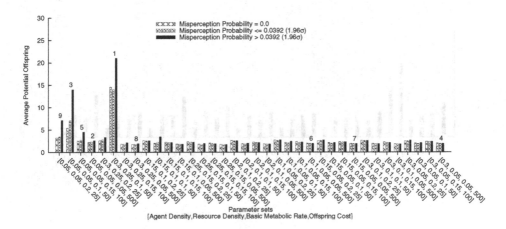

Fig. 3. Average potential offspring

3.2   Experiment 2

Any measure of the misperception probability is meaningless for populations that use either reflexive foraging or perfect-perception foraging, as both foraging methods lack misperception. The only value worth comparing is the average potential offspring due to different foraging behaviours.

Misaction-affected foraging (Figure 4) has several parameter sets where the agents whose misaction probability is significantly different from 0% have more potential offspring than their competitors with lower misaction probabilities. There are many more parameter sets where this occurs than compared to misperception-affected foraging, which implies that misaction as implemented in this simulation offers a greater benefit than misperception. This benefit is

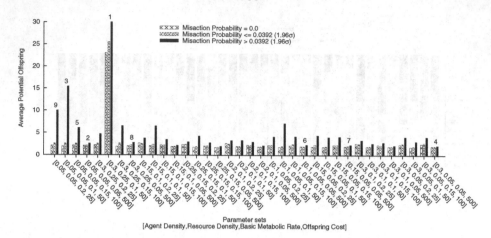

Fig. 4. Average potential offspring (misaction-affected foraging)

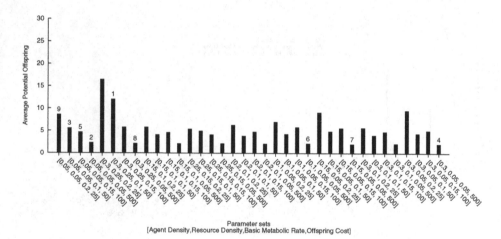

Fig. 5. Average potential offspring (reflexive foraging)

expressed as more effective foraging, which allows an agent potentially to produce more offspring. While these benefits are more observable in the parameter sets where offspring cost less, in all cases agents whose misaction probability is significantly different from 0% have more potential offspring.

As the misperception probability is meaningless for reflexive foraging agents, the total population was combined together to determine their average potential offspring (Figure 5). The agent populations had more potential offspring than any of the other foraging methods for many of the parameter sets. This may have been due to a higher level of behavioural diversity arising from their random foraging.

The average potential offspring measure for perfect-perception foraging is calculated together, as all the agents have a misperception probability of 0%.

Populations of agents with perfect-perception foraging behaviours (Figure 6) only outperformed the reflexive foraging behaviour for a few parameter sets. When compared against misperception-affected foraging and misaction-affected foraging, the perfect-perception foraging behaviour had either slightly less or slightly more potential offspring. When the four different foraging methods are compared, perfect-perception foraging has the least potential offspring in the majority of cases. Following Akaishi and Arita's hypothesis, this would be due to the lack of behavioural diversity in its agent populations. While conventional wisdom suggests perfect-perception foraging will be fitter than the three alternatives, this does not always occur.

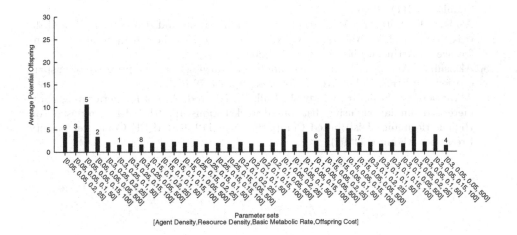

Fig. 6. Average potential offspring (perfect-perception foraging)

4   Conclusion

Our results support and extend earlier work that showed a general benefit from misperception. In particular, this benefit is demonstrated in an evolutionary environment, with misperception achieving evolutionary stability. As in prior work, misperception is only beneficial when it is infrequent. Our results also directly support prior speculation that the benefit works through introducing behavioural diversity. Furthermore, in at least some circumstances, we have shown that a more direct introduction of behavioural diversity can have greater benefit than misperception itself.

Contemplated future work includes identifying other situations where misperception could be adaptive, such as cases where individuals misperceive the value of various attributes of objects they can perceive in their environment. Also, the evolution of self-deception in social simulation is a likely extension of these ideas. Finally, another potential area to explore is to focus on the link between misperception and altruism and on kin selection as a driving force in the evolution of misperception.

References

1. Brumley, L., Kopp, C., Korb, K.: Causes and effects of perception errors. Journal of Information Warfare 5(3), 41–53 (2006)
2. Kopp, C., Mills, B.: Information Warfare and Evolution. In: Hutchinson, W. (ed.) IWAR 2002. Proceedings of the 3rd Australian Information Warfare & Security Conference 2002, pp. 352–360 (2002)
3. Russell, S.J., Norvig, P.: Artificial Intelligence: A modern approach. Prentice Hall, Englewod Cliffs, New Jersey (1995)
4. Neisser, U.: Cognition and Reality. W. H. Freeman, San Francisco (1976)
5. Boyd, J.R.: The essence of winning and losing. Slideshow (January 1996)
6. Doran, J.: Simulating collective misbelief. Journal of Artificial Societies and Social Simulation 1(1) (1998)
7. Akaishi, J., Arita, T.: Misperception, communication and diversity. In: Standish, R.K., Bedau, M.A., Abbass, H.A. (eds.) Proceedings of the 8th International Conference on Artificial Life, pp. 350–357 (2002)
8. Akaishi, J., Arita, T.: Multi-agent simulation showing adaptive property of misperception. In: FIRA Robot World Congress, pp. 74–79 (2002)
9. Abramson, D., Sosic, R., Giddy, J., Hall, B.: Nimrod: A tool for performing parameterised simulations using distributed workstations. In: 4th IEEE Symposium on High Performance Distributed Computing, pp. 112–121. IEEE Computer Society Press, Los Alamitos (1995)

Simulated Evolution of Discourse with Coupled Recurrent Networks

Kazutoshi Sasahara[1], Bjorn Merker[2], and Kazuo Okanoya[1]

[1] Laboratory for Biolinguistics, RIKEN Brain Science Institute (BSI),
2-1 Hirosawa, Wako, Saitama 351-0198, Japan
[2] Gamla Kyrkv. 44, SE-14171 Segeltorp, Sweden
sasahara@brain.riken.jp

Abstract. The origin and evolution of language have been the subjects of numerous debates and hypotheses. Nevertheless, they remain difficult to study in a scientific manner. In this paper, we focus on the string-context mutual segmentation hypothesis proposed by Merker and Okanoya, which is based on experimental findings related to animal songs. As a first step in formally exploring this hypothesis, we model the evolution of agent discourse using coupled recurrent networks (RNNs). This model is a simplified representation of this hypothesis; that is, agents are situated in a single context (e.g., behavioral, social, or environmental) and they mutually learn their utterance strings from the prediction dynamics of their RNNs. Our simulation demonstrates the emergence of shared utterance patterns, which are culturally transmitted from one generation to the next. Furthermore, the distribution of the shared patterns changes over the course of this evolution. These findings demonstrate an important aspect of language evolution: "language shaped by society."

1 Introduction

1.1 Approaches to Language Evolution

How did modern human language evolve into a complex system? While this issue has attracted the attention of many scholars and has been the subjects of numerous debates and hypotheses, the origin and evolution of language are still unknown. Language is not a single faculty but a complex system comprising many cognitive sub-faculties, some of which are shared by non-human animals. As suggested by Hauser et al. [4], to understand the origin and evolution of language, it is important to compare the prelinguistic ability of animals and human language in terms of not only homologies but also differences . This is a plausible approach to clarifying the origin and evolution of language.

Another plausible approach is computational modeling. The iterated learning model (ILM) proposed by Kirby shows that, even without natural selection, cultural transmission from generation to generation can produce the syntactical structure of language such as compositionality and recursion [5]. Many other

M. Randall, H.A. Abbass, and J. Wiles (Eds.): ACAL 2007, LNAI 4828, pp. 107–118, 2007.

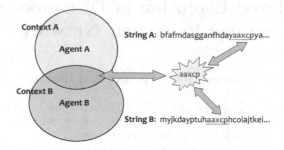

Fig. 1. Schematic illustration of mutual segmentation hypothesis: When agents with segmentation ability collaborate, the common parts of the context they face and the sound strings they utter can be mutually segmented. Small portions of the sound strings can be linked to specific contexts.

computational models have been studied, each of which models a different aspect of language evolution, such as the evolution of a signal system [19], the emergence of linguistic communication [8,18], the evolution of syntax [3,9], and the origin of meaning [15]. Computational models provide good testbeds for the systematic exploration of hypotheses on language evolution. They can also be used to identify novel linguistic phenomena resulting from the interactions of agents.

Both approaches are promising for clarifying the origin and evolution of language in a scientific manner. Moreover, the interplay between them may be beneficial [13].

1.2   String-Context Mutual Segmentation Hypothesis

Here, we briefly review the string-context mutual segmentation hypothesis, which is a language evolution hypothesis proposed by Merker and Okanoya [6,10]. This hypothesis was motivated by the biological evidences of animal songs. Figure 1 is a schematic illustration of this hypothesis. Let us consider a society without language in which agents make utterances (like songs) that are specific to behavioral, social, and/or environmental contexts. Further, let us suppose that each agent has segmentation ability, which is the prelinguistic ability to find discrete patterns in contexts and utterance strings. It should be noted that primitive segmentation ability can be found in non-human animals, particularly in songbirds. Juvenile songbirds learn sound patterns (called chunks) found in the tutor birds' songs by using statistical cues such as the transition probability of sound elements [17].

When agents with segmentation ability collaborate, the existence of shared substrings embedded in strings uttered in contexts that also have some features in common, the possibility exists of extracting the substring as a marker for the shared contextual aspect by mutual segmentation on the basis of statistical learning. Small segmented parts of sound strings can be linked to ever more

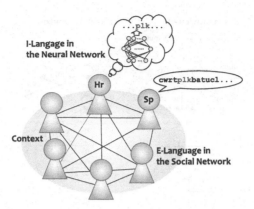

Fig. 2. Schematic illustration of simplified model: As evolution proceeds, the neural network of each agent and the social network of agents coevolve, and shared utterance patterns emerge

specific contexts, as shown in Fig.1. As a consequence, a meaningful word could come into existence, and its iterated usage could popularize it among the agents.

The key concept of such a "holistic " mechanism was proposed independently by Kirby [5] and Wray [20], and can also be found in the work of Mithen [7]. In the publications [6, 10], Merker and Okanoya show how this key concept allows a path to be completed from unsemanticized song to language on the basis of attested behavioral biology and neural mechanisms, emphasizing the importance of vocal leaning as a driving force of language evolution. Experimental findings for non-human animals provide a considerable amount of information on which to model the string-context mutual segmentation hypothesis. In this paper, we explore this hypothesis by using a simple artificial life model.

2   Model

We begin by modeling the evolution of agent discourse, which is a simplified representation of the string-context mutual segmentation hypothesis with only a single context. That is, agents are situated in the same context and mutually learn utterance strings, as shown in Fig.2.

To model the segmentation ability, we use a simple recurrent network (RNN). As many previous studies have shown, RNNs can learn the sequential structures in a self-supervised manner [11, 2, 1]. Furthermore, language evolution has been simulated by using the population of agents modeled by RNNs [8, 18]. These studies revealed that the prediction dynamics of RNNs is essential to learning the structure of temporal sequences like language.

While previous models focused on the accuracy of a sequential leaning task, we focus on an emergent property – how shared utterance patterns emerge from the prediction dynamics of agents with RNNs. In this section, we describe our model architecture and simulation setup.

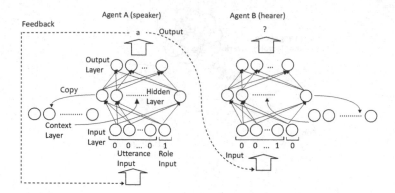

Fig. 3. Discourse of coupled recurrent networks: The first ten input neurons of RNNs receive utterance information, in which alphabetic letters are coded with ten basis vectors. An additional input neuron is used for assigning an agent's role: a speaker (1) or a hearer (0).

2.1   Agents

Each agent is modeled as a simple recurrent network (RNN), used for both speech and recognition during discourse. The RNN has a layered structure with an additional input neuron, as shown in Fig.3. The input layer receives two types of inputs. The first ten input neurons receive utterance information, in which alphabetic letters (from "a" to "j", in this model) are coded with ten basis vectors; e.g., a→[000000001], b→[001000000], for example. The other input neuron is used for assigning an agent's role as a speaker (1) or a hearer (0). If a hearer gets a letter "a", his input vector is expressed as [00000000010] (see Fig.3). This additional input is important for making different states in the hidden layer, related to the agent's role. The outputs of the RNN are translated into an alphabetic letter by a winner-take-all process, in which the maximum neural output becomes one and the others become zero.

The dynamics of the RNN is expressed as the following equations:

$$y_j(t) = g(\sum_i w_{ij}x_i(t) + \sum_l w_{lj}y_l(t-1) + b_j) \tag{1}$$

$$z_k(t) = g(\sum_j w_{jk}y_j(t) + b_k) \tag{2}$$

$$g(x) = \frac{1}{1+e^{-x}} \tag{3}$$

where g is the sigmoid function, w is the neural connection weights, and y and z are the neural outputs of the hidden and output layers, respectively. In this paper, the size of the input layer is set to 11, the hidden and context layers to 22, and the output layer to 10.

2.2   Discourse

Agent discourse is simulated by coupling two RNNs, as shown in Fig.3 [16]. During a discourse, two agents are randomly chosen from the population of agents; one as a speaker and the other as a hearer. The speaker begins an utterance by setting the initial inputs [00000000001](i.e., no utterance information with the speaker's identifier (1)) and continues the utterance by feeding back the outputs to the input layer. When the speaker utters a string, the hearer receives it and predicts the speaker's next utterance. During the discourse, the hearer learns the sequential structure of an utterance string in a self-supervised manner; in this model, the RNN of a hearer is trained using the error back propagation (BP) algorithm [12].

In the BP algorithm, the connection weights of the RNN are updated as the following equation:

$$\mathbf{w}(t+1) = \mathbf{w}(t) - \eta \frac{\partial E(\mathbf{w})}{\partial \mathbf{w}}\bigg|_{\mathbf{w}=\mathbf{w}(t)} + \alpha \varDelta \mathbf{w}(t), \tag{4}$$

where $\mathbf{w}(t)$ is the connection weight vector of the RNN at learning step t, and $E(t)$ is the error function. The constants η and α are the coefficients of learning and inertia, respectively. In this paper, these values are set as $\eta = 0.1$ and $\alpha = 0.8$.

The discourse is evaluated in terms of its predictability (i.e., the number of shared patterns) and the complexity of the strings that agents utter or predict. A pattern is defined as a substring found in both an utterance and a prediction string and that consist of more than two types of letters (i.e., a one-letter repetition like "aaa" is not a pattern). For example, for the discourse

- a speaker (utterances): uuukcwruk*plk*batuclaat...
- a hearer (predictions): _xtxtoutrudixtx*plk*bq...,

there is one shared pattern: *plk*. After the discourse, we perform a simple pattern matching procedure, in which the number of patterns shared between the speaker and hearer is calculated.

The scores of the speaker (S_{sp}) and hearer (S_{hr}) in a discourse are calculated using the following equations:

$$S_{sp} = \sum_i N_i^{ltr} N_i^{ptn} \times H_{sp} \tag{5}$$

$$S_{hr} = \sum_i N_i^{ltr} N_i^{ptn} \times H_{hr} \tag{6}$$

where N_i^{ltr} denotes the number of letter types per shared pattern-i and N_i^{ptn} denotes the number of pattern-i. Furthermore, H_{sp} and H_{hr} denote the information entropies of the speaker's utterance string and the hearer's prediction one. Using these H_{sp} and H_{hr}, we can consider the endogenous trend in string complexity in this model.

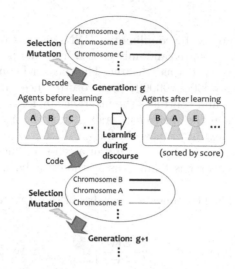

Fig. 4. Evolutionary time course of model. The connection weights of an RNN are coded by a chromosome. Agents are created by decoding the inherited chromosomes. After all discourses, the agents are sorted by score and the top ten agents leave offspring (i.e., encode their innate chromosomes with mutations).

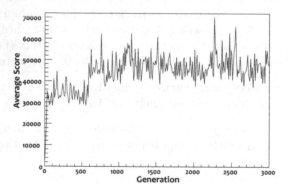

Fig. 5. Average score across 3000 generations: Step-wise evolution is observed at about generation 550

2.3   Evolution

To evolve the agents, a simple genetic algorithm (GA) with ranking selection and point mutation is used after a certain number of discourses. Figure 4 shows the evolutionary time course. In this model, the connection weights of an RNN are encoded by an artificial chromosome, which crosses from one generation to another.

The agents leave offspring in accordance with their total scores across all discourses. Only the top ten agents can leave offspring: each of them leaves one copy offspring without mutations; furthermore, the best agent can leave 4 mutant offspring, the second can leave 3, the third can leave 2, the fourth can leave 1. The other agents are removed from the population.

A mutant offspring inherits a parent chromosome with point mutations, where a small amount of noise is added (at most 20% per chromosome). As shown in Fig.4, the "Darwinian mechanism" is used in this model; that is, offspring inherits a parent's innate chromosome, which encodes the initial connection weights of the RNN, NOT the one that encodes the learned connection weights, which is used in the "Lamarckian mechanism." We use the Darwinian mechanism because it is more adaptive in a dynamical environment [14] and more natural even in our abstracted evolution.

3   Simulation

At the initial state (generation zero) of the simulation, we make 20 agents, each of which has an RNN with random connection weights. The only differences between the agents are the connection weights. The other simulation parameters are set as mentioned in the previous sections. Every agent makes an utterance or a prediction sting with the length $L_{str} = 30$ per discourse, and the usable alphabetic letters are 10 (from "a" to "j"). In each generation, 1000 discourses are carried out for two randomly selected agents. The Darwinian evolution (Fig.4) proceeds for 3000 generations.

3.1   Evolution of Discourse

We observe a step-wise evolution, as shown in Fig.5. The score increases rapidly at about generation 550; it then remains approximately the same score up to generation 3000. These findings indicate that there is a transition through which the agents become more communicative in their discourses, as we will see in the next section. The evolution of the information entropies of the utterance strings (H_{hr}) and prediction strings (H_{sp}) are shown in Fig.6. These entropies also exhibits a step-wise evolution; both H_{sp} and H_{hr} suddenly increase at about generation 550 and then decrease slightly over the subsequent generations.

Comparing the information entropies between utterances and predictions in Fig.6, we see that the entropy of predictions (H_{hr}) always larger than that of utterances (H_{sp}). This suggests that the prediction ability of agents preceded their utterance ability. When agents play a hearer role, they predict and learn utterance patterns by updating the connection weights of their RNNs. On the other hand, when agents play a speaker role, they make utterances on the basis of their RNNs, which have already been structured and do not change during a discourse. Playing a hearer role is the only chance agents get to modify their innate RNNs during their lifetime and thereby increasing their prediction ability. This secondarily affects the utterance ability.

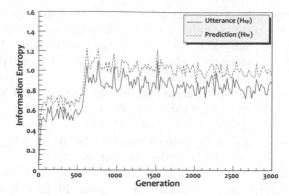

Fig. 6. Information entropies of utterance and prediction strings: The information entropy of prediction strings are larger than that of utterance strings

```
Generation 0                                    Generation 100
agent[17]  aaaaaaaaaaaaaaaaaaaaaaaaaaaaaa        agent[14]  aaaaccacaacaaaccacaccccaacccccc
agent[16]  _jjjjjaaaaaaaaaaaaaaaaaaaaaaaa        agent[00]  _accccacaaaccacaacaaaccaaaacc

agent[16]  bbbbbbbbbbbbbbbbbbbbbbbbbbbbbb        agent[00]  cacacccaacacccaaccaaaacccccccc
agent[13]  _aaaaaaabbbbbbbbbbbbbbbbbbbbbb        agent[01]  _ccccccccccccccccccccaaacccccc

agent[00]  acacaaaccaacaccaccacaccaacaaaa        agent[06]  acaaacaacccaccaaccacacacccaaaca
agent[16]  _iiiiiiiiiiiiiiiacccccccccacaaa       agent[07]  _cccaccccaaaacaacaacaacaacaaca


Generation 500                                  Generation 600
agent[14]  iaiaaaaaciciaaaaccaicaciaiaiaa        agent[12]  hhichiiihiaccahaccchhciiiahhih
agent[03]  _iiiiiiiiiiiiiiiiiiiaiiaaiaiaia       agent[18]  _cchiaiaahhaiaiciaacichcaihhah

agent[13]  aaaacaacccaaaaccaaacaccaacacc         agent[10]  iaaaccihchhciaiahcaicahahhaicc
agent[02]  _iiaaicaicaiaacaaiaccaaciiciai        agent[19]  _accccahcacaccahhachhhchhccaac

agent[02]  cicaiiccaiaiiiacaiciaiccciiacc        agent[07]  cacacahcaacahaacaachaiahahcaca
agent[10]  _chchchccfcfhcchhchhcccahaaac         agent[14]  _cccccccccccccccacaaacaaaaaaaaa


Generation 3000
agent[04]  chchhaaaahcchchachhachahchach
agent[08]  _cacccacahhccaacaaahhhaahhaach

agent[01]  ahacachahacacaaccacchccccchhhha
agent[04]  _haachcchccaachcacccccahahhahh

agent[15]  ccccccccccccccccccccccccccccccc
agent[04]  _aahacchcahchhhachhchahchhhaac

agent[01]  acaaccacaaahchcahhachhchchchhc
agent[11]  _acaaacacaaaaaacaaaaaahchchchh
```

Fig. 7. Examples of agent discourses: At the initial stage, the utterance and prediction strings are simple; after generation 550, they become more complex

Figure 7 shows examples of agent discourses in each generation. In generation zero, the utterance and prediction strings are simple. Almost all the speakers repeat a single letter like "aaa..." or "bbb..."; only a few utter multiple letters like "acac..." It should be noted that agents with randomly connected RNNs are not random speakers (i.e., random letter generators) and that most of them have poor utterance ability. In this generation, we can identify adaptive behaviors of hearers, who easily change their predictions through learning. As a consequence, the information entropy of hearers (H_{hr}) does not become zero. For example, agent 16 can recall a pattern "ac" after a sequence of wrong predictions with "i". In this way, agents who make better predictions obtain higher scores and leave more offspring in the early stages of evolution. We can also see in Fig.7 that agent 16 behaves differently depending on his role, which is an important finding. When he is a speaker, he utters only "i", while he recall a several letters when he is a hearer. This shows that the additional input neuron we introduced works as intended.

As the evolution proceeds, the strings of both utterances and predictions evolve and become more complex, and several shared patterns emerge. In generation 100, the utterance strings consist of only two letters, "a" and "c", which have been in steadily use in the agent society. In generation 500, a novel letter "h" that has never been observed before begins to be used, and then it becomes used more and more. Interestingly, in generation 3000, we can find stereotyped discourses in which agent 4, who is the best agent at the time, produces similar strings despite his role. Even when an utterance string is a simple repetition of "c", agent 4 robustly predicts different patterns such as "ac" and "ah," and thus obtains the highest scores on the whole. This is in contrast to the adaptive behaviors of initial agents in the early stages of evolution. This interesting phenomenon may be related to the statistical property of shared utterance patterns in the society of agents. As shown in the next section, this is because "ah" and "ac " are dominant in the frequency of shared patterns, so the BP learning for the rare repetition of "c" has little effect.

3.2   Statistical Properties of Shared Patterns

Figure 8 shows the statistical properties of the shared utterance patterns in the agent society; the number for each pattern and their rank are plotted on a log-log scale, which is called a Zipf plot. In human language, the frequency of any word is inversely proportional to its rank in the frequency table, and the slope is -1 in a log-log plot, which is called "Zipf's law" [21]. It should be noted that the shared patterns in this model are not strikingly parallels with actual words and that they have many overlaps; for example, given a pattern set {abab,ababab}, "ac" is repeatedly counted and the resulting counts are ab(5), abab(2), and ababab(1). Only statistical cues such as frequency are available to segment such patterns – the model has no semantics to serve as cues for segmenting.

Different characteristics in the Zipf plots (Fig.8) are evident in the shared utterance patterns. In generation zero, the most common pattern is "ac," and the top five shared patterns are combinations of "a" and "c." The slope is close to -1, but this is not a case of human language. In generation 100, we observe the same top five patterns, except " aaac." In generation 600, a new letter, "h", appears in the top five patterns, resulting more diversity in the shared patterns. This causes the information entropies to rapidly increase between generations

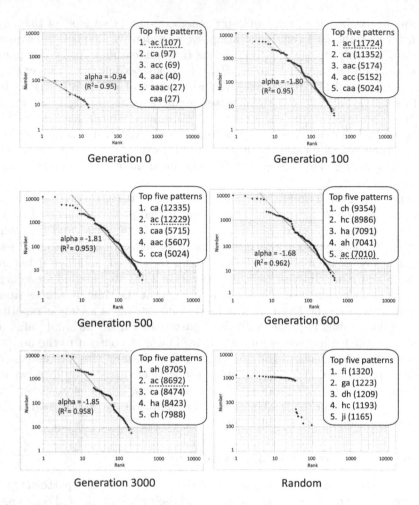

Fig. 8. Statistics for shared utterance patterns in agent society: The building block pattern "ac" is culturally transmitted from generation to generation. Zipf slope varies from −1.85 to −0.94. When the same pattern matching method is applied to random dummy data, the slope is flat, unlike the other slopes.

500 and 600, as shown in Fig.6, resulting in the step-wise evolution of average scores shown in Fig.5.

Furthermore, we can find pattern "ac" in the top five patterns not only in the early stages of the evolution but also in the later stages. This suggests that this pattern originates in the initial agent society because it is more popular than the other patterns. It is thus culturally transmitted through agent discourse and becomes frequently used in the shared patterns. Through the evolution, pattern "ac" functioned as a building block for longer patterns.

Through the evolution, the Zipf slope varied from −1.85 to −0.94. As the evolution proceeds, the Zipf plots exhibits terraced slopes, and each terrace corresponds to a certain group of patterns; for example, in the generation 3000, the top five patterns have almost equal frequency, and they are building blocks for larger patterns, such as "ahaccahach." In addition, the slopes are different from that for the random dummy data, as shown in Fig.8.

4 Discussion and Conclusion

We have demonstrated the evolution of agent discourse by using coupled recurrent networks (RNNs). This model is a simplified version of the string-context mutual segmentation hypothesis, in which agents are situated in a single context and mutually learn their utterance patterns suitable for the context. As a result, we observe the emergence of shared utterance patterns, which are culturally transmitted from one generation to the next. Furthermore, the distribution of shared patterns changes over the course of evolution. These findings demonstrate an important aspect of language evolution; namely, "language shaped by society."

In the emergence of shared patterns, "the prediction chain reaction" (PCR) of the agents with RNNs is an important driving force; that is, agents learn the verbal behaviors of others that learn. The introduction of information entropy into the scores (i.e., eqs.(5) and (6)) models an endogenous driving force for the string complexity. To predict the utterance patterns precisely, then simple ones are more effective for this task, but there exists a trend for the string complexity in this model. Thus, two driving forces balance between predictability and complexity, affecting the emergence of shared patterns.

Our model at present has only a single context, and the context affect neither the utterances nor prediction strings. utterance and prediction strings. As mentioned in Section1.2, contexts may help differentiate the usage of utterance patterns by agents, and the strings and contexts may interplay with one another. We plan to introduce more contexts into this model and use it to explore the string-context mutual segmentation hypothesis in greater depth.

Acknowledgments

This work was partly supported by a Grant-in Aid (No.18800083) from the Ministry of Education, Culture, Sports, Science and Technology of Japan.

References

1. Christiansen, M., Allen, J., Seidenberg, M.: Learning to Segment Speech Using Multiple Cues: A Connectionist Model. Language and Cognitive Processes 13, 221–268 (1998)
2. Elman, J.L.: Language as a Dynamical System, pp. 195–223. MIT Press, Cambridge (1995)
3. Hashimoto, T., Ikegami, T.: Emergence of net-grammar in communicating agents. Biosystems 38, 1–14 (1996)

4. Hauser, M.D., Chomsky, N., Fitch, W.T.: The Faculty of Language: What Is It, Who Has It, and How Did It Evolve? Science 298, 1569–1589 (2002)
5. Kirby, S.: Natural Language from Artificial Life. Artificial Life 8(2), 185–215 (2002)
6. Merker, B., Okanoya, K.: The Natural History of Human Language: Bridging the Gaps without Magic. In: Emergence of Communication and Language, pp. 403–420. Springer, Heidelberg (2006)
7. Mithen, S.: The Singing Neanderthals: The Origins of Music, Language, Mind and Body. Weidenfeld & Nicolson (2005)
8. Munroe, S., Cangelosi, A.: Learning and the Evolution of Language: The Role of Cultural Variation and Learning Costs in the Baldwin Effect. Artificial Life 8(4), 311–339 (2002)
9. Nowak, M.A., Komarova, N.L., Niyogi, P.: Evolution of Universal Grammar. Science 291, 114–118 (2001)
10. Okanoya, K., Merker, B.: Neural Substrates for String-Context Mutual Segmentation: A Path to Human Language. In: Lyon, C., Nehaniv, C.L., Cangelosi, A. (eds.) Emergence of Communication and Language, pp. 421–434. Springer, Heidelberg (2006)
11. Pollack, J.B.: The Induction of Dynamical Recognizers. Machine Learning 7(2), 227–252 (1991)
12. Rumelhart, D.E., McClelland, J.L.: Parallel Distributed Processing: Explorations in the Microstructure of Cognition. MIT Press, Cambridge (1993)
13. Sasahara, K., Ikegami, T.: Evolution of Birdsong Syntax by Interjection Communication. Artificial Life 13(3), 259–277 (2007)
14. Sasaki, T., Tokoro, M.: Adaptation toward Changing Environments: Why Darwinian in Nature? In: Husbands, P., Harvey, I. (eds.) Fourth European Conference on Artificial Life, MIT Press, Cambridge (1997)
15. Steels, L.: Self-organizing Vocabularies. In: Langton, C.G., Shimohara, K. (eds.) Artificial Life V, pp. 179–184 (1996)
16. Taiji, M., Ikegami, T.: Dynamics of internal models in game players. Physica D 134(2), 253–266 (1999)
17. Takahasi, M., Okanoya, K.: (in preparation)
18. Tonkes, B., Wiles, J.: Methodological Issues in Simulating the Emergence of Language. In: Wray, A. (ed.) The Transition to Language, pp. 226–251. Oxford University Press, Oxford (2002)
19. Werner, G.M., Dyer, M.: Evolution of communication in artificial organisms. In: Langton, C., Taylor, C., Farmer, D., Rasmussen, S. (eds.) Artificial Life II, pp. 659–687. Addison-Wesley, Reading (1992)
20. Wray, A.: Dual Processing in Protolanguage: Performance Without Competence. In: Wray, A. (ed.) The Transition to Language, pp. 113–137. Oxford University Press, Oxford (2002)
21. Zipf, G.K.: Human Behavior and the Principle of Least Effort. Addison-Wesley, Reading (1949)

How Different Hierarchical Relationships Impact Evolution

Susan Khor

Concordia University, Canada, H3G 2W1
slc_khor@cse.concordia.ca

Abstract. The evolutionary behavior of three hierarchical relationships, HIFF-C, HIFF-II and HIFF-M is studied in the context of two computational models, J and JGA. In J, entities are composed from other entities in the population. JGA is a panmictic genetic algorithm. Results from our experiments indicate that *specificity* in a relationship enhances convergence to a global optimum in both models. When there is little specificity in the relationship, external conditions such as join rate, crossover rate, agitation type or selection mechanism need to be set appropriately. Our results also suggest that cooperation is neither necessary nor sufficient for the evolution of higher level entities. We found that cooperation was evolutionary advantages in J only for relationships with little to no *top-down inter-level conflict*.

Keywords: major evolutionary transition, hierarchical relationships, multi-level selection, inter-level conflict, specificity, genetic algorithms, population diversity.

1 Introduction

A major theme in evolutionary biology is the formation of higher level entities from lower level entities. This formation is also known as a major evolutionary transition (MET) [6]. A difficulty with the MET theory is, understanding why higher level entities can be stable and replicated as wholes in the face of selection forces at play amongst their self-interested lower level entities. Reeve and Keller speak of the need for attractive forces to exceed the repulsive and centrifugal forces for there to be stability within a collective [2, p.7]. They define these forces in terms of absolute inclusive fitness. Michod [7] stresses the necessity of cooperative interactions among lower level units to form emergent higher level groups, and conflict mediation in favor of the higher-level unit for groups to transition to new evolutionary individuals. Further, how the units of a group are reorganized and conflict mediated in a transition to individuality, can influence the individual's evolvability in the future [7]. The problem of *stability*, that is keeping autonomous units together as cooperating wholes, is not limited to the biological realm but applicable also to the evolution of complexity from simplicity, in general.

In reference [5], we explored this problem of stability from a logical (vs. physical-chemical) point of view. In that study, we assumed that two necessary conditions for a MET, as suggested by existing theory, namely multi-level selection in favour of

M. Randall, H.A. Abbass, and J. Wiles (Eds.): ACAL 2007, LNAI 4828, pp. 119–130, 2007.
© Springer-Verlag Berlin Heidelberg 2007

higher level adaptation (cooperation) and protective barriers against disintegration such as membrane enveloped compartments are in place, and focused instead on how relationships between parts affect the formation and stability of composite entities. By relationship we mean how parts of a whole interact with each other within the whole. A composite entity is an entity formed from previously existing entities. The study in [5] was made with a model called J. In that study we found that composite entities formed under a relationship with high specificity (section 2.3) were more stable, and their stability were impervious to conditions created by different selection mechanisms and agitation type (section 3.1).

In this paper, we use the same J model used in our previous experiments [5], but on a different relationship (HIFF-II). This relationship differs from those studied in [5] because it has high top-down inter-level conflict (TDILC) (section 2.2). TDILC makes the problem of bottom-up evolution via a MET approach more interesting because *short-term mutualism can degenerate into long term behavior that is detrimental to the whole*. Relationships between biological entities are dynamic and on a continuum [1]. We report in this paper that evolving higher level entities under the HIFF-II relationship with J is more challenging than the previous two relationships (HIFF-C and HIFF-M). Enforcing cooperation and using protective barriers actually made it more difficult to evolve HIFF-II entities in the J model. This result is not too surprising given that HIFF-II does not evolve easily under the two selection mechanisms available to J [3]. However, even when we adapted J to use SM3 (RMHC3 in [3]), a selection strategy that is known to be successful for HIFF-II, the success rate was less than 100% given the parameters of the experiment. SM3 involves cooperation at all levels, not just between lower level entities for the interest of higher level entities as in SM2, but also between higher level entities for the interest of lower level entities. From our experience with J and HIFF-II, we conclude that it is more difficult for entities with high top-down inter-level conflict to evolve with the MET approach because (i) more negotiation of interests between parts at different levels is required, (ii) the evolution is more sensitive to external conditions, and (iii) the evolution takes a greater length of time.

The difficulties we experienced with J and HIFF-II led us to design the JGA model which we introduce in this paper. JGA combines aspects of the J model into a genetic algorithm (GA). Instead of incrementally evolving larger entities from smaller ones, entities in JGA start out at the target size but with randomly chosen parts. As in our experiments with the J model [5], we found HIFF-M entities least particular about the parameter settings of JGA. HIFF-C and HIFF-II entities were more particular. In addition, HIFF-C entities performed slightly worse than HIFF-II entities under JGA. JGA is a panmictic genetic algorithm and so is susceptible to loss of population diversity. A population losses its diversity when all individuals in the population carry the same value for one or more genes. From our JGA experiments, we hypothesize that the high level of specificity in the HIFF-M relationship helps to decelerate population diversity loss. By our definition in section 2.3, specificity is lowest in HIFF-C and highest in HIFF-M. We plan to test this hypothesis and study population diversity under the different relationships. As with any computer simulation study, it remains to be seen how dependent this conclusion is on the models and the parameter settings we used, and also the characteristics of the problem.

2 Background

This section defines the three relationships used in our experiments and reviews previous work to compare and understand the behavior of these relationships. The concepts of *top-down inter-level conflict* and *specificity* are defined.

2.1 The Relationships (HIFF-C, HIFF-II, HIFF-M)

A *relationship* is a set of weighted links that defines how variables (genes) of an entity's genotype interact with each other. We experiment with three hierarchical relationships: HIFF-C, HIFF-II and HIFF-M. These relationships are variants of the Hierarchical-If-And-Only-If (HIFF) problem [8]. The three relationships lend entities the same hierarchical structure. An entity's size refers to the length of its genotype, a $\{0, 1\}^N$ string. Discussion in this paper assumes the binary alphabet $\{0, 1\}$ is used. An entity of size $N=2^n$ where $n \in \mathbf{Z}^+$, is decomposed to $\log_2 N$ levels. The levels of the hierarchy are labeled 1 ... n from the bottom-up and there is a total ordering on the set of levels. At level λ, the N variables are partitioned into $N/2^\lambda$ non-overlapping modules of consecutively located genes. Every variable belongs to exactly one module of level λ. Each module at level λ has 2^λ variables. The minimum module size is 2 and each module of size 2^i where $i \in \mathbf{Z}^+$ and $i > 1$ consists of exactly two other distinct sub-modules. Figure 1 illustrates how a size 8 entity is decomposed into levels and modules.

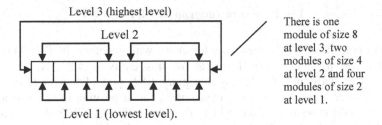

Fig. 1. Hierarchical decomposition for an entity of size 8

The maximum fitness contribution of a module at any level is 1. Therefore the optimal fitness value for level λ is $N/2^\lambda$. The total fitness of an entity is the sum of all level fitness values in its hierarchy. Therefore the optimal total fitness is N-1. When necessary, we write the level fitness values of an entity in level descending order and call this structure the *phenotype*. The optimal phenotype is thus $\langle 2^0, 2^1, ..., 2^{n-2}, 2^{n-1} \rangle$.

The weight of a link between genes i and j of an entity defines the contribution made by i and j to the entity's fitness when i and j satisfy the constraint associated with the link. In this paper, all constraints are IFFs. Therefore optimal solutions are maximally similar. Since the values (alleles) we use are $\{0, 1\}$, there are two optimal genotypes for each relationship, the all ones genotype **1** and the all zeroes genotype **0**. The structure (which variables interact) and the weights of the links for each of the relationships are defined next.

HIFF-C. This is the continuous version of the HIFF problem [8]. Every variable interacts with every other variable in an entity. Fitness of a HIFF-C module is given by $(p \times q) + (1 - p) \times (1 - q)$ where p and q are the proportion of ones in the first and second halves of the module respectively. For example, the HIFF-C phenotype for entity with genotype 1000 1100 is $\langle$ 0.5, 0.5, 3.0 $\rangle$ and the total HIFF-C fitness for this entity is 4.0. The level 3 fitness is 0.5 because p =1/4, q = 2/4, and $(0.25 \times 0.5) + (1 - 0.25) \times (1 - 0.5) = 0.5$. At level 2, the two modules are 1000 and 1100. Fitness of the 1000 module is $(0.5 \times 0) + (0.5 \times 1) = 0.5$. Fitness of the 1100 module is $(1.0 \times 0) + (0 \times 1) = 0$. Thus, fitness at level 2 is $(0.5 + 0) = 0.5$.

HIFF-II. [3] Every variable interacts with n other distinct variables in an entity. n is the number of levels in the hierarchy for the entity. Fitness of a HIFF-II module at level λ is calculated by doing a pair-wise comparison of genes in the first half of a module with genes in the second half of a module (Figure 2):

(i) gene i is compared with gene $2^{\lambda-1} + i$ of a module for $i = 0, ..., 2^{\lambda-1} -1$,
(ii) the number of matches is divided by half the module size, $2^{\lambda-1}$.

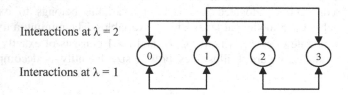

Fig. 2. Interaction diagram for HIFF-II, N = 4

For example, the HIFF-II phenotype for entity with genotype 1000 1100 is $\langle$ 0.75, 0.5, 3.0 $\rangle$ and the total HIFF-II fitness for this entity is 4.25. At level 3, the interactions are between the following four pairs: (0-1, 4-1), (1-0, 5-1), (2-0, 6-0), (3-0, 7-0). An interaction pair (i-a, j-b) means the gene at position i has value a, and it interacts with the gene at position j which has value b. For an IFF problem (i-a, j-b) = 1 if a = b, and 0 otherwise. So the interaction pairs at level 3 return the values 1, 0, 1 and 1 respectively. This means there are 3 matches at level 3. Since a module at level 3 has 8 variables, by rule (ii) above, the fitness value of level 3 is 3/4 = 0.75.

HIFF-M. [4] For all modules m at every levels λ, if the first (0) and the middle ($2^{\lambda-1}$) variables of m have the same value (Figure 3), 1 fitness point is awarded to m. For example, the HIFF-M phenotype for entity with genotype 1000 1100 is $\langle$ 1, 0, 3 $\rangle$ and the total HIFF-M fitness for this entity is 4. At level 2, the interaction pairs are (0-1, 2-0) and (4-1, 6-0). Since there are no matches, level 2 fitness is 0.

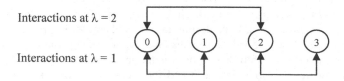

Fig. 3. Interaction diagram for HIFF-M, N = 4

A general principle for all three relationships is intra-module interactions outweigh inter-module interactions. This accounts for the more rapid optimization of lower level modules under normal conditions; that is when total fitness is used by selection (section 2.2). Further, all three relationships are non-linear; the fitness contribution of a pair of interacting variables may be higher than the sum of their individual fitness contributions.

2.2 Top-Down Inter-Level Conflict (TDILC)

The combination of (i) lower level modules adapting quicker than higher level modules, (ii) the existence of two optimal solutions for every module, and (iii) the requirement that all modules adapt to the same optimal solution if a globally optimal solution is to be found, creates conflict between levels. What is evolutionary advantages to lower level modules need not be beneficial to the whole entity in the long run. This bottom-up conflict presents itself when the selection mechanism compares entities by their *total* fitness values because by default, lower level modules make a larger fitness contribution than higher level modules in all three relationships. We refer to this selection mechanism as *SM1*. With SM1, a variant entity replaces its parent entity if its total fitness is equal to or greater than the total fitness of its parent entity. Otherwise, the variant is discarded.

Another selection mechanism that we will use in our experiments is a multi-level selection scheme that prioritizes the optimization of higher levels. This selection mechanism is called *SM2* (also known as RMHC2 in [4]) and it works by comparing entities using their phenotype values in level descending order. SM2 replaces a parent entity with its variant entity if the variant is fitter than its parent at level λ and is as fit as its parent at all levels above λ. A variant also replaces its parent entity if it is as fit as its parent at all levels. Otherwise, the variant is discarded.

Because SM2 favours higher level modules, inter-level conflict arises when adaptations that are good for higher levels prevent optimization of lower level modules. Such top-down inter-level conflict (TDILC) exists for the HIFF-II and HIFF-M relationships. HIFF-C does not have TDILC under SM2 because the optimal set of genotypes at the highest level consists of global optima only (Table 1). Thus optimization of higher level modules also optimizes lower level modules.

Table 1. Optimal genotypes (local optima) by level, N = 4

Level	HIFF-C	HIFF-II	HIFF-M
2	{ 0000, 1111 }	{ 0000, 0101, 1010, 1111 }	{ 0*0*, 1*1* }
1		{ 0000, 0011, 1100, 1111 }	
TDILC	None	High	Low

* is a placeholder and may take either a '0' or '1' value.

Top-down inter-level conflict is higher in HIFF-II than HIFF-M because HIFF-II defines more distinct constraints per level (N/2) than HIFF-M ($N/2^{\lambda}$) at levels above level 1. Further, the set of constraints for any level in HIFF-II involves all variables. Therefore, fit higher level HIFF-II modules lock in more genes per level and leave fewer degrees of freedom for lower level modules than HIFF-M.

2.3 Specificity

Specificity refers to how difficult it is for independent entities to become part of a composite entity. Specificity at level λ refers to the number of genotype configurations with 0 fitness value at level λ.

For HIFF-C, there are only two situations when a module's fitness is 0: (i) $p = 0$ and $q = 1$ and (ii) $p = 1$ and $q = 0$. Since fitness of a level is the sum of the fitness of its modules, and there are $N/2^\lambda$ modules at level λ, specificity for HIFF-C at level λ is $2^{N/2^\lambda}$. Specificity at the highest level ($\lambda = n$) for HIFF-C is therefore 2. $N=2^n$ (section 2.1).

HIFF-II has $N/2$ distinct constraints per level and these constraints involve all variables. Since there are 2 configurations per constraint that contributes 0 to fitness, specificity per level for HIFF-II is the same for all levels and is $2^{N/2}$.

HIFF-M has $N/2^\lambda$ distinct constraints per level and each constraint is applied to the first and middle variables of each module at level λ. This leaves $(2^\lambda - 2) \cdot N/2^\lambda$ variables free to take on any value without affecting the fitness of level λ. Therefore specificity at level λ for HIFF-M is $2^{N/2^\lambda} \cdot 2^{(2^\lambda - 2) N / 2^\lambda}$ which simplifies to $2^{N(1 - 1/2^\lambda)}$. At the highest level, specificity for HIFF-M is 2^{N-1}.

The calculations in this section serve to show that for $\lambda > 1$, HIFF-C < HIFF-II < HIFF-M where '<' means is less specific than.

3 Models

The objective of this paper is to study how the relationships described in section 2 influence evolution in two computational models: J and JGA.

The J model adopts the major evolutionary transition (MET) [6] approach to evolution; entities are recursively composed from smaller entities until they reach the target size. A difficulty with the MET approach as we mentioned in section 1 is keeping parts of a composite entity together. We explored this stability problem for HIFF-C and HIFF-M using J in [5]. In that study, runs were made under different selection mechanisms (SM1, SM2) and agitation types (R, NR). These two dimensions represent enforced cooperation and enforced cohesion respectively. With SM1 (section 2.2), all parts "act for their own self-interest", there is no cooperation between parts. With SM2 (section 2.2), cooperation between parts is enforced in the sense that the interest of the whole takes precedence over the interests of the parts. NR (non-random) agitation type (section 3.1) enforces cohesion by limiting the number of parts a composite entity can disintegrate into. In this paper, we focus on the stability of HIFF-II entities evolving with the J model. Stability of HIFF-C and HIFF-M entities was discussed in [5].

This paper also introduces JGA. In JGA, entities start out at the target size and evolution is via crossover, mutation and selection. JGA adopts the approach common in genetic algorithms. The replacement strategy for the crossover operator in JGA is similar to the decision condition for the join and exchange operators in J, a new or child entity replaces its donor or parent entities in the population only if it is strictly fitter than both of its donor or parent entities. JGA uses the same mutation operator as

J with SM1. Part of the motivation for JGA was the difficulty we experienced trying to evolve HIFF-II entities with *J* (section 4.1).

3.1 The *J* Model

The specific *J* algorithm used in our experiments is outlined in this section. Evolution in *J* starts with a population of entities each having a size 2 random genotype. Evolution in *J* proceeds with either a join, exchange or mutation operation in each iteration, until either an optimal entity of the size desired is created or the maximum number of iterations is reached. The total amount of genetic material (512 × 2 bits) stays constant throughout a run.

Algorithm for *J*
create 512 entities each with a size 2 random genotype
while number of iterations < 50,000
increment number of iterations by 1
if number of iterations is divisible by 50
if fittest entity is also globally optimal and of size 128, stop.
record statistics   —   *0.5 is the join rate*
if random real number in [0.0, 1.0] < 0.5
chose 2 random distinct entities, e1 and e2
if e1.size = e2.size and e1.size × 2 ≤ 128 and random integer in [1, 2] = 1
join e2 to e1
else
exchange parts of e1 and e2
else
select an entity, e0, using fitness-proportionate (roulette-wheel) selection
mutate e0

A **join** event enables two random distinct entities (e1, e2) of the same size to concatenate their genotypes to form a new entity (e3) and see whether there is additional benefit to exist as one instead of two entities. If e3 is strictly fitter than the sum fitness of e1 and e2, e3 is added to the population and e1 and e2 are removed from the population. A successful join reduces population size by 1.

An **exchange** event enables parts (modules) of two random distinct entities (the donor entities), which may be of different sizes, to try out a different configuration. This new configuration succeeds if it provides a better context for the parts involved. If this is the case, the two donor entities disintegrate.

The first step in an exchange between two donor entities (e1 and e2) is to split e1 and e2 into their constituent parts. The granularity of parts from this split depends on the *agitation type* (AT), which is a parameter of *J*. If AT is random (R), then e1 and e2 may be split into parts of any size 2^i where $1 \leq i \leq \log_2$ of the smaller entity size. If AT is non-random (NR), then part size is either the smaller entity size or when entities are of equal size, half the size of a donor entity.

These parts are then assembled into a new entity e3 of size equal to the larger of the donor entities. If a part in e3 is strictly fitter on average than a part in e1 and a part in e2, then we add e3 and the remaining unused parts to the population, and remove e1 and e2 from the population. Otherwise, e1 and e2 are left intact in the population[1].

Fitness of a part in an entity e is the total fitness of e divided by the number of parts used to create e or the number of parts e is split into. Suppose e1 is 1111 1001, e2 is 11, the relationship is HIFF-C and AT is NR. Then e1 is split into 4 equal sized parts, i.e. 11, 11, 10 and 01. Fitness of a part in e1 is 4/4 = 1. Fitness of a part in e2 is 1/1 = 1. Let e3 = 1111 1101. This exchange succeeds because fitness of a part in e3 is 5.25/4 = 1.3125, which is greater than 1. Simple average is used so that all parts in an entity have the same fitness value. This makes the decision whether to keep e3 or discard it straightforward.

A **mutate** operation on e0 makes a clone entity e3 of e0 and then complements 1 to k genes randomly chosen with replacement from e3's genotype where k is $P_m \times$ e3.size. e3 competes with e0 for a place in the population using either the SM1 or SM2 replacement strategy (section 2). P_m is the mutation rate. In the experiments (section 4), P_m is 0.03125 or 4/128.

3.2 JGA

The specific JGA algorithm used in our experiments is outlined in this section. The population size is constant (steady-state) throughout a run.

Algorithm for JGA
create 128 entities each with random genotype of length 128
while number of iterations < 100,000
increment number of iterations by 1
if number of iterations is divisible by 50
if fittest entity is also globally optimal, stop.
record statistics
chose 2 random distinct entities, e1 and e2
if random real number in [0.0, 1.0] < 0.5
produce c1 and c2 by doing a 2-point crossover between e1 and e2
if c1 is strictly fitter than both e1 and e2, replace e1 with c1
if c2 if strictly fitter than both e1 and e2, replace e2 with c2
else
produce c1 by mutating e1
if c1 is as fit as or fitter than e1, replace e1 with c1
produce c2 by mutating e2
if c2 is as fit as or fitter than e2, replace e2 with c2

0.5 is the crossover rate

A crossover between a pair of entities e1 and e2 is made by randomly choosing two locations x and y in a genotype where x < y. The genotype of e1 (e2)'s child entity, c1 (c2), inherits e1 (e2)'s genes at all locations except those between x and y which it inherits from e2 (e1). A child entity c1 produced by mutating an entity e1 inherits all

---

[1] In [5], we mistakenly said that a successful exchange increases population size by at least 1. A successful exchange does not decrease population size.

of e1's genes except at 1 to k random locations where the complement of e1's genes are inherited instead. k is $P_m \times$ e1.size = 4 since P_m is 0.03125 or 4/128 and all entities in JGA are of the same size which in our experiments is 128.

4 Experiments and Discussion of Results

In the experiments with J, we gauge stability by monitoring the weighted average age (WAA) of a population throughout a run. A steady rise in WAA indicates stability. In JGA, the emphasis is on the success rate and speed of evolution.

4.1 The J Model

Table 2 summarizes the results of our experiments with J. Except for the HIFF-M relationship, introducing enforced cooperation (via SM2) and/or enforced cohesion (via non-random agitation type) either lengthens average time to optimum or reduces success rate.

Table 2. Number of times a globally optimal entity (N=128) is found out of 30 random runs and the average number of iterations successful runs took

Agitation Type	HIFF	Selection Scheme			
		SM1 (compares total fitness)		SM2 (does multi-level selection)	
		Success	Iterations	Success	Iterations
Non-random (NR)	C	5	14,960 (9,447)	30	22,480 (2,827)
	H	5	21,960 (9,941)	0	-
	M	30	10,820 (805)	30	10,700 (683)
Random (R)	C	30	13,690 (2,100)	30	19,380 (5,939)
	H	30	15,040 (2,617)	0	-
	M	30	11,530 (820)	30	11,490 (784)

Standard deviation is reported in parentheses.

In [5], we reported that entities evolving with J under the HIFF-M relationship were more stable than entities evolving with J under HIFF-C because the HIFF-M relationship is more specific. We explained in [5] how specificity (section 2.3) helps to improve stability of composite entities by making it more difficult for joins and exchanges to succeed, thus giving more time for entities to optimize themselves before becoming parts of composite entities. Successful joins and exchanges exert downward pressure on WAA while unsuccessful exchanges push WAA up. Therefore, the WAA graph for HIFF-M show steady rise because there is less downward than upward pressure on WAA for HIFF-M. Another effect of specificity is slower decline in population size for HIFF-M than HIFF-C, which in turn affects the granularity of exchanged parts when non-random agitation type is used.

We also reported in [5] that lowering the join rate from 0.5 to 0.25 for HIFF-C (NR, SM2) shortened evolutionary time to optimum and improved stability. Here we provide further evidence to support this (Figure 4).

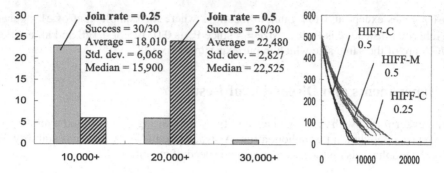

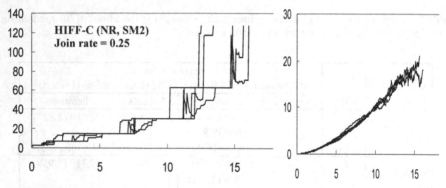

Left: Number of HIFF-C (NR, SM2) runs that took x iterations on average to complete successfully. A join rate of 0.25 brings average evolutionary time required for HIFF-C (NR, SM2) closer to the average (13,690) found with HIFF-C (R, SM1) runs. **Right:** Population size over iterations for (NR, SM2) with different join rates. HIFF-C (NR, SM2) benefits from a lower join rate which results in a slower decline in population size because population diversity in terms of entity size is maintained for a longer period of time. Diverse entity sizes means more configuration possibilities for exchanges with NR agitation type.

Left: Best fitness over iterations in 000's for 5 HIFF-C (NR, SM2) random runs which completed in less than 20,000 iterations with J at join rate 0.25. The step pattern of these graphs shows evidence of optimal modules combining to form larger optimal modules. Recall that optimal fitness for a module of size 2^n is $2^n - 1$. Hence we see vertical jumps around fitness values 15, 31 and 63. **Right:** WAA over iterations in 000's for runs plotted in the graph on the left. Steady rise in WAA indicates exchanges are on the whole not successful, thus composite entities are stable. The volatile nature of the WAA graph for HIFF-C (NR, SM2) when join rate is 0.5 can be seen in Fig. 3B of reference [5]. Space constraint prevents us from reproducing it here.

Fig. 4. Lowering join rate improves stability and performance for HIFF-C entities evolving with J under (NR, SM2) conditions

HIFF-II

Stability of the successful HIFF-II runs reported in Table 2 is poorest compared with HIFF-C and especially HIFF-M runs. Here we try to improve the stability and success rate for HIFF-II entities evolving with the J model. Like HIFF-C, HIFF-II is less specific than HIFF-M. Since lowering the join rate improved stability for HIFF-C, we try to do the same for HIFF-II. Through trial and error, we found that lowering the join rate to 0.125 and increasing the maximum number of iterations to 100,000, improved the success rate for HIFF-II (NR, SM1) to 25/30 (83%) and improved stability (Figure 5). However, the successful runs took much longer to complete, 42,670 iterations on average with a standard deviation of 14,750. Only 20/30 (67%) HIFF-II (NR, SM1) *J* runs completed successfully when the join rate was 0.25.

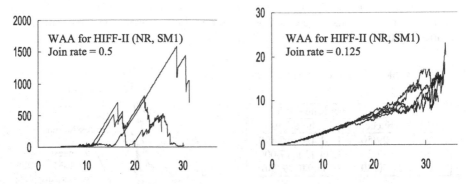

Fig. 5. WAA over evolutionary time for successful HIFF-II (NR, SM1) runs at different join rates. Note the different scales for the y-axis.

We could not find any configuration of parameter values that was successful at least 50% of the time for the SM2 categories. We attribute this to (i) the high level of top-down inter-level conflict in HIFF-II entities under SM2 that prevents further adaptation of entities via mutation once their highest level is optimal, and (ii) to the genotype configurations that SM2 produces for HIFF-II which has a higher number of switches on average than those produced by SM1. A *switch* marks a change in value. For example, the genotype configuration 0111 0011 has 3 switches. The average switch count for a best genotype configuration at the end of 5 randomly sampled unsuccessful HIFF-II (NR, SM1) runs at join rate 0.5 was 2.6 while the same statistic was 52.6 and 60.2 for HIFF-II (NR, SM2) and HIFF-II (R, SM2) runs at join rate 0.5 respectively. It is less probable to produce a genotype with a low number of switches through an exchange of parts with a high than low number of switches. Further, subsequent mutation and replacement under SM2 could increase the number of switches for a genotype.

Previously in [3], we found that HIFF-II entities can evolve to optimality under the SM3 (RMHC3 in [3]) selection (replacement) scheme. With SM3, a variant replaces its parent only if it is not less fit than its parent at any level. With join rate at 0.125, 24/30 (80%) HIFF-II (NR, SM3) *J* runs completed successfully using 47,800 iterations on average with standard deviation of 16,480. The WAA graphs for 5 randomly sampled HIFF-II (NR, SM3) is similar to Fig. 5 (Right). However, even with a selection scheme that removes inter-level conflict (SM3) in the sense that no level adapts at the expense of another higher or lower level, evolution time to optimality is high compared with the average time found for HIFF-II (R, SM1) runs.

Hence we conclude that introducing cooperation (SM2 or SM3) and/or barriers against disintegration (NR) seems to make the evolutionary problem more difficult for

HIFF-II entities. In contrast, the addition of SM2 and NR reduced evolutionary time for HIFF-M entities, while an appropriate join rate could reduce evolutionary time for HIFF-C (NR, SM2). But for HIFF-II, the (R, SM1) combination seems to work the best for bottom-up evolution under *J*.

4.2 JGA

Table 3 reports the results for our experiments with JGA. Figure 6 shows the percentage of genes (bits) where > 90% of the population carries the same value, over evolutionary time.

Table 3. JGA results N=128

HIFF	Crossover Rate = 0.5	
	Success	Iterations
C	29/50	31,450 (15,190)
II	32/50	34,420 (17,000)
M	50/50	34,940 (9,946)

HIFF	Crossover Rate = 0.25	
	Success	Iterations
C	40/50	40,060 (13,210)
II	42/50	45,360 (17,160)
M	50/50	54,320 (11,370)

Standard deviation is reported inparentheses.

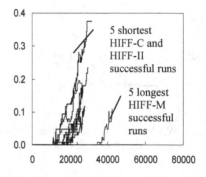

Fig. 6. At crossover rate 0.25, HIFF-C and HIFF-II JGA runs lose diversity faster then HIFF-M.References

References

1. Dimijian, G.G.: Evolving together: the biology of symbiosis, part 2. Baylor University Medical Center Proceedings 13(4) (October 2000)
2. Keller, L. (ed.): Levels of Selection in Evolution. Princeton University Press (1999)
3. Khor, S.: HIFF-II: A Hierarchically Decomposable Problem with Inter-level Interdependency. In: IEEE Symposium on Artificial Life (April 2007)
4. Khor, S.: On Solving Hierarchical Problems with Top-Down Control. In: Proceedings of Genetic and Evolutionary Computation Conference (GECCO) Companion (July 2007)
5. Khor, S.: Specificity Increases Stability of Composite Entities. In: Proceedings of Extending the Darwinian Workshop held at ECAL (September 2007)
6. Maynard Smith, J., Szathmáry, E.: The Major Transitions in Evolution. W.H. Freeman & Co. Ltd. (1995)
7. Michod, R.E., Nedelcu, A.M.: On the Reorganization of Fitness During Evolutionary Transitions in Individuality. Integr. Comp. Biol. 43, 64–73 (2003)
8. Watson, R.A.: Compositional Evolution: The Impact of Sex, Symbiosis and Modularity on the Gradualist Framework of Evolution. The MIT Press, Cambridge (2006)

A Dual Phase Evolution Model of Adaptive Radiation in Landscapes

Greg Paperin, David Green, Suzanne Sadedin, and Tania Leishman

Monash University, Faculty of Information Technology
Clayton Campus, Bldg. 63, Wellington Rd, Clayton, 3800 Vic, Australia
{Gregory.Paperin,David.Green,Suzanne.Sadedin,Tania.Leishman}
@infotech.monash.edu.au

Abstract. In this study, we describe an evolutionary mechanism – Dual Phase Evolution (DPE) – and argue that it plays a key role in the emergence of internal structure in complex adaptive systems (CAS). Our DPE theory proposes that CAS exhibit two well-defined phases – selection and variation – and that shifts from one phase to the other are triggered by external perturbations. We discuss empirical data which demonstrates that DPE processes play a prominent role in species evolution within landscapes and argue that processes governing a wide range of self-organising phenomena are similar in nature. In support, we present a simulation model of adaptive radiation in landscapes. In the model, organisms normally exist within a connected landscape in which selection maintains them in a stable state. Intermittent disturbances (such as fires, commentary impacts) flip the system into a disconnected phase, in which populations become fragmented, freeing up areas of empty space in which selection pressure lessens and genetic variation predominates. The simulation results show that the DPE mechanism may indeed facilitate the appearance of complex diversity in a landscape ecosystem.

Keywords: Dual Phase Evolution, complex systems, speciation, adaptive radiation, simulation.

1 Introduction

An intriguing question motivated by new fields of research, such as artificial life and evolutionary computation, is whether biological evolution can provide insights about self-organisation in complex adaptive systems (CAS). Our recent studies show that deep similarities do exist between biological evolution and adaptive processes in other systems [2, 3]. Based on these similarities, we have proposed a theory of the existence of a family of adaptive processes, which we term Dual Phase Evolution (DPE) [4].

In this study, we further develop the theory of Dual Phase Evolution. We show how DPE operates in several systems and investigate its implications for patterns of species evolution. We begin by reviewing the theory of DPE and present some of the supporting evidence. We then provide a model of adaptive radiation (global speciation dynamics) in landscapes and use it to demonstrate how DPE can facilitate the appearance of perpetual novelty in ecosystems.

M. Randall, H.A. Abbass, and J. Wiles (Eds.): ACAL 2007, LNAI 4828, pp. 131–143, 2007.

2 Dual Phase Evolution

CAS exhibit a sustained diversity of their locally interacting components. In the absence of a global controller CAS exhibit far-from-equilibrium dynamics, and permanent novelty and adaptation [5, 6]. This is facilitated by the complex organisation of the locally interacting systems' components and their interrelations.

There is a large body of evidence that suggests that structure of CAS emerges through self-organisation [5], however, the specific mechanisms governing this process are not well understood. There is a large amount of evidence (see next section) that CAS generally tend to self-organise towards a stable, balanced state. A number of adaptive mechanisms present in CAS cause these systems to exhibit little large-scale variation over long periods of time. Such mechanisms may include lower order dynamics such as feedback loops and higher order dynamics such as evolution driven by selection (in a general sense) [7]. It has been demonstrated in analytical [8, 9] and computational [7] models that lower-order local dynamics are capable of stabilising a system over a large range or external forcing, and that higher order local dynamics (evolutionary dynamics) can greatly increase the stabilising effect.

The same adaptive forces that are responsible for global stability of CAS may work to inhibit novelty and change within such systems. In particular, selection acting on systems' components at various hierarchical levels of organisation, as well as on the topology and types of their interactions, may drive a system as a whole to a local optimum state, thereby preventing innovation [5]. There are two mechanisms that have the potential to work against such long-term stasis.

One such mechanism is co-evolution. Local adaptation of system components driven by selection may affect the selection criteria (the fitness landscape) for other components, which will also adapt as a result. The adaptation of the components affected in such a way may in turn cause changes in the fitness landscapes of other components, including the components which initiated the changes in the first place. The feedback loops which can arise in this way may function as sources of perpetual novelty because the selection acts on random variation and the results of such feedbacks may be different for each loop. However, it is not clear that co-evolution is capable of providing the degree of innovation observed in many natural CAS. For instance, analytical models [10] suggest that selection rather than variation (in this case genetic drift) drives speciation. As a result, co-evolutionary feedback loops are likely to quickly (on evolutionary timescales) lead to stable system states that reside at local optima of the global fitness landscape. Once such a state is reached, selection towards the optimum makes variations that could disturb the stability highly unlikely [10], and therefore rare.

The second mechanism that may function as a source of continual novelty in CAS is external disturbance. It has been shown, that evolutionary innovations in various natural CAS often coincide with external perturbations (e.g. [11], see also next section), some of such examples are discussed in the following sections. External disturbances may affect a system in several ways and move it from the local fitness optimum thereby disturbing the stable configuration.

Once away from a local optimum, the system enters a phase in which it is driven by variation and change. Any chance variation of some local component or substructure may provide a better adaptation to the local constraints and selection will facilitate the proliferation of such change. As long as the disturbed system is far from a fitness optimum, selection will therefore amplify rather than inhibit some local random variations. Over time, components and their interactions on various system levels will be driven towards new local fitness optima and the inhibitory effects of selection on variation will increase again. Eventually, the whole system will develop towards a new stable balance-state. Different perturbations will continue to affect the system causing it to flip between balance phases dominated by stabilising selection and exploration phases dominated by directional selection.

We propose a general mechanism governing many processes in CAS. This mechanism, Dual Phase Evolution (figure 1), can be summarised as follows: CAS develop towards a balanced state. In this state they are stabilised through various processes including selection, and exhibit little large-scale variations (on evolutionary timescales). The balance state is disturbed by external perturbations which unbalance systems and flip them into a phase in which they exhibit variation on all scales. Over time, stabilising processes drive systems into a new balance-state.

While some parts of the system may be completely or partly reorganised during a variation phase following a particular disaster, others will remain stable. These stable parts may form new interactions and assume new roles within the changing system. Such stable sub-systems can act as functional components during a variation phase. We speculate that this may be the mechanism that facilitates the emergence of closed components in complex systems. When a sub-system consisting of several components remains stable during a variation phase, it may act as a functional component in the re-organised system. It is possible that this mechanism is responsible for the emergence of hierarchical levels of organisation found in CAS.

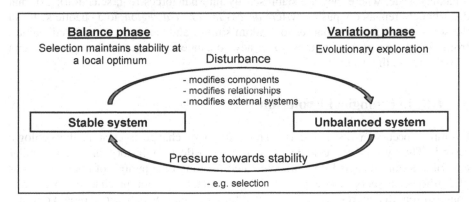

Fig. 1. The mechanism of dual phase evolution. Systems flip between balance and variation phases. External disturbances unbalance stable systems, variation facilitates evolutionary exploration, internal pressures drive the system into a new stable state.

2.1 Contrast Between DPE and SOC

In contrast to DPE, the theory of Self-Organized Criticality (SOC) suggests [12, 13] that CAS self-organise to a critical state, in which the complexity of systems' responses to external stimuli emerges through a propagation of the stimuli through local component interactions with thresholds at each component. These propagations result in avalanches of different sizes. This theory of SOC suggests that CAS evolve to reside at the "edge-of-chaos" [14, 15], a transition state between the general stasis of equilibrium systems and the random behaviour of chaotic systems.

As the response propagation avalanches in SOC systems follow a power law distribution, an observation of this distribution in data is often used as an indication that a system may self-organise to a critical state. A number of models [12] led to suggestions that various complex systems, some of which are adaptive, may exhibit SOC dynamics. For instance, it has been suggested [16] that the self-organisation of the biosphere to a critical state may be an explanation for punctuated equilibria [17], since the sizes of extinction events observed in the fossil record follow a power law distribution. However, the extent to which SOC presents the general form for organisation of CAS remains doubtful. In many cases there are several processes which may lead to power-law distributed data. For instance, [18] demonstrates a non-critical extinction model without any species interactions that yields a power-law with an exponent closer [7] to the empirical punctuated equilibria data. It has been suggested at various occasions (e.g. [19, 20]) that the critical behaviour requires fine-tuning of an order parameter. Furthermore, it remains unclear whether SOC occurs in non-conservative systems [19, 21]. It has been attempted to avoid some of the problems related to SOC by using a notion of nearly-critical behaviour which can be applied to a wider range of systems (e.g. [21]), however, the generality of SOC theory remains inconclusive. Here, we aim to pinpoint the key difference between DPE and SOC, as it is a widely considered theory of self-organisation in CAS.

The SOC view is that CAS self-organise towards a critical region (see above). If we were to describe DPE using the SOC-vocabulary, we would say – CAS develop to a balance-state, where they are stabilised by internal forces (e.g. selection); external disturbances repeatedly push a system *across the critical region*, to a chaotic state (in the sense that systems responses to random stimuli and variations are unpredictable), from which the system returns to a new balance-state, accumulating order and complexity on the way.

3   DPE in Biological Evolution

Evolution occurs in fits and starts. This pattern of change is clear in the geologic record. The system of geologic classification reflects a history in which similar assemblages of fauna and flora predominated for long time periods of time, often tens of millions of years. These periods are punctuated by abrupt changes in species composition. Recent research has revealed that the changes between geologic eras are associated with mass extinction events. Recognising this pattern led Eldredge and Gould [17] to put forward the idea of punctuated equilibrium, in which the general pattern of evolution is constant composition punctuated by mass extinctions, followed by brief periods of rapid speciation.

Alvarez et al. [22] found evidence that the Cretaceous-Tertiary boundary was associated with impact of a large comet. Research since then has produced evidence that asteroid impacts, volcanic activity and climate change are associated with many other geological boundaries as well.

Green et al. [2] proposed a mechanism to explain the above observations. For most of the time, pressure for space or niches within a landscape impose selective pressure. Established populations restrict the spread of invaders. Widespread populations are genetically "connected" and genetic variation is suppressed [2]. When a major disturbance occurs, the above patterns are reversed: vast areas of free space are opened up; suppressed species are free to expand into the new territory; selective pressure becomes negligible; and established populations become fragmented.

There is abundant evidence that Dual Phase Evolution, and processes closely related to it, occur in many contexts. There are striking similarities between species evolution, on a scale of millions of years, and forest change, which occurs on a scale of thousands of years. Forest history, as recorded by preserved pollen, shows that during postglacial (the last 12,000 years), forest composition changed in fits and starts. This pattern is reflected by the systematics used by palaeontologists, who divide the postglacial history into pollen zones. The zones have more or less constant composition, with rapid changes from one zone to another. Studies of the process have shown that major forest fires triggered the rapid changes, with the species composition being determined by climate at the time [11].

In certain regions, fluctuations in landscape connectivity have been linked to the evolutionary radiation of whole groups of animals. In Great: In lakes of east Africa, for instance, the explosive speciation in cichlid fishes has been linked to changes in water level [23, 24]. During periods of high water level, environments are connected, but become fragmented when water levels are low. Similarly, Hewitt [25] argues that repeated glaciations throughout the Quaternary caused species ranges in North America and Europe to fragment, leaving surviving populations in isolated refugia. These isolated populations diverged genetically, but later reunited, creating a complex genetic patchwork in species such as the European hedgehog, *Chorthippus* grasshoppers and bears, and sometimes leading to speciation. Numerous similar parapatric species occur in the mountains of Sulawesi that are thought to have diverged during periods of habitat fragmentation [26]. Taxa include *Chitaura* grasshoppers, macaques, pond-skaters, cicadas, carpenter bees, butterflies, limacodid moths and tiger beetles. Likewise, Amazonian insects are thought to have diversified in response to fluctuating connectivity in forest canopy density [27].

Habitat fragmentation at fine temporal scales does not always lead to speciation: instead, the outcome may be formation of genetic suture zones, where populations that have diverged while separated meet and interbreed. Some authors suggest that fragmentation may even contribute to evolutionary stasis. For example, Bennet [28] argues that Milankovitch climate oscillations, which occur on the order of 10-100ky, cause continual changes in the direction of selection, preventing species from adapting locally and therefore speciating. Similarly, Coope [29] notes the prevalence of stasis among temperate Quaternary insect species despite the appearance of incipient ecological species in modern fragmented fauna. Clearly, there is still much to be discovered about the impact of habitat fragmentation on evolution.

4 Simulation Model of DPE

In order to further investigate the effect of the DPE-process on evolutionary ecosystems, we created a computational simulation model. The model investigates the adaptive radiation exhibited by a population of individuals in a landscape. Adaptive radiation is usually described as the evolution of ecological and phenotypic diversity within a rapidly multiplying lineage leading to utilisation of new ecological niches [30]. Our aim is to investigate the potential consequences of the DPE mechanism on adaptive radiation in landscape ecosystems.

The model consists of a population of haploid individuals situated on a two-dimensional landscape. The model is based on a well-known model of adaptive radiation [30] that did not incorporate DPE. The landscape consists of a 100×100 grid of cells. Each cell has a maximum carrying capacity of up to 4 individuals. (The numbers given here are parameter values for a base scenario. Other scenarios and the sensitivity analysis systematically varied these parameters).

The environment allows 60 possible niches, where each niche is represented by a string of 20 bits; a bit represents the requirement, that a particular trait must be present (1) or absent (0) in an individual in order to be well adapted to that niche. The niches are not location specific. Individuals' genotypes are also represented by bit-strings of length 20; the bits represent the presence (1) or absence (0) of the above traits (note, here genotype equals phenotype; elsewhere [31] we show analytically that this approach is computationally equivalent to the genotype-phenotype setup used in [30] when the number of traits in our model equals to the number of traits × the number of loci per trait in Gavrilets' model). The individuals evolve to adapt to one of the niches. At each time-step an individual is assumed to occupy the niche which best matches the individual's traits.

The *fitness* of an individual is determined in proportion to the hamming distance between the individual's genotype and its niche and is scaled by the niche condition. The *niche condition* is a number between 0 and 1 that describes how appropriate a particular niche is within the current environment (i.e. in a desert environment, the niche "hot & dry sand" may have condition = 1 and individuals well adapted to that niche will have a high fitness; individuals which are well adapted to the niche "cold and wet soil" will have a low fitness as such niche will have a much lower condition in this climate). 30 out of the 60 model niches are called "normal", they have a condition = 1. The other 30 niches describe the environmental conditions short after a disaster. In the base setup, in which there are no disturbances, the condition of these *disaster-niches* is set to 0.2.

The life-cycle of the model organisms is reproduction – selection – dispersal, the generations are non-overlapping. Individuals mate within their occupied cell only. Each individual in a cell is selected once as "mother". A "father" is then randomly selected from the remaining individuals of the cell (regardless of their niche). If there is only one individual in a cell, it will engage in hermaphroditism. The number of offspring each couple produces is Poisson-distributed with $\lambda = 5$. Reproduction is through free recombination; each offspring is subject to a mutation rate of 0.00001 per gene, which corresponds to background mutation rate in nature [32]. Once all individuals have mated, the old population is replaced by the offspring. It is then determined which individuals of the new population will survive to the age of reproduction, by reducing the number of individuals in each cell to its carrying capacity (4). The probability of survival is given directly by an organism's fitness. Finally, the surviving individuals may disperse across the landscape. With probability of 0.1, each individual will migrate to one of the neighbouring cells. After all individuals have migrated, the current generation will engage in reproduction and complete the live-cycle.

The model is initialised with a small population of clones of 2 randomly chosen individuals and thereafter simulated for 40,000 generations.

4.1 Basis Scenario

Initially we ran the model without any disturbances. The results are generally in line with what was observed in [30]. At the beginning of the simulation there is a burst of adaptive radiation which leads to a large number of niches being utilised. After a while (typically 5 to 10 thousand generations), the number of utilised niches begins to decline slowly. Most niche-proportions (proportion of the population occupying a certain niche out of the total model population) fall below 0.1 and then either engage in uncorrelated fluctuations or decline to 0. The proportions of 1 to 3 niches typically remain above the threshold of 0.1. Out of these "dominating" niches, one typically grows in proportion slowly, while the others decline accordingly. Sometimes, two rather than one niches gain a stable proportion while the others decline (figure 2).

In addition, we observed some patterns that were not initially expected from the model, but are known to occur in nature. These patterns include the dominance of lower-fitness populations over closed spatial patches [33] and the occurrence of stable hybrid zones [34] maintained by a balance between dispersal and selection against hybrids. These observation provide an indication that our model correctly captures the relevant landscape dynamics [35].

4.2 Disturbances

We modified the basis scenario such that at each generation a disaster occurred with probability 5×10^{-5}. During each disaster, all individuals in most landscape cells were wiped out. The cells were selected by setting a random point in the landscape as disaster centre and wiping out all cells within a certain radius around the centre; the radius was normally distributed around 30 cells (sensitivity analysis, not shown, shows that if the average radius exceeds a certain threshold, the model behaviour is not sensitive to the impact radius). Then a new impact centre was randomly selected. This process was repeated until 95% of cells were wiped out. Whatever the nature of a disaster (bush fire, volcanic eruption, disease, etc) it will not only wipe out the population in affected areas, but also alter the local environment. To model this, the conditions of normal niches in areas affected by the disaster were reduced to 0.2, while the conditions of the disaster-niches in these areas were raised to 1. In a sensitivity analysis (not shown) we have verified that reducing the normal niche condition to any value below 0.5 does not affect the qualitative system behaviour. This altered environment was maintained at the impact sites for several thousand generations and then stepwise returned to normal using a linear interpolation between the disaster environment and the normal environment. This strategy is similar to a number of cellular landscape models incorporating disturbances (e.g. [33]).

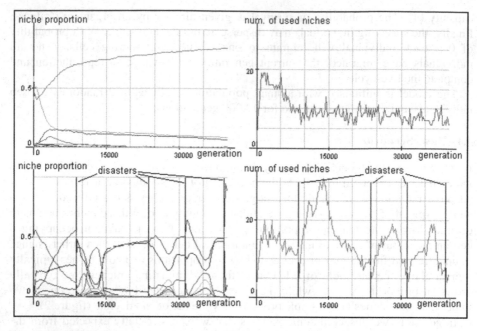

Fig. 2. Typical runs. *Basis case (top)*. *Left*: The population proportions of all niches are displayed. An initial burst of adaptive radiation is followed by populations of 3 niches taking over the entire population. After 40,000 generations it can be expected that a single niche will soon take over the entire landscape. *Right*: the number of utilised niches at each generation from the same simulation run. *Disaster case (bottom)*. *Left*: The population proportions of all niches are displayed. While in long phases between disturbances the dynamics are similar to above, disasters are followed by phases of intense variation. After the effects of a disaster have faded out, new niches appear to be colonised. *Right*: the number of utilised niches drops immediately after a disaster as the populations utilising the niches are wiped out. However, the reduced competition in the freed-up areas allows genetic variations to colonise new niches. This figure was generated using the real-time plotter LiveGraph [1].

The general model behaviour in the disaster scenario can be described as follows: A disaster kills a large part of the population, wiping out entire niche-populations. The number of utilised niches therefore drops. However, the remaining population can now expand into the freed-up areas and colonise the disaster-niches available in the disturbed landscape, which leads to a quick growth of utilised niches. As the disturbed areas recover from the disaster, some of the normal niches are re-populated. In addition, some of the normal niches not used before the disturbance are also utilised. By the time the landscape has fully recovered, the number of utilised niches is typically large. As the landscape is fully populated at this point, selection now slowly reduces the number of utilised niches until the next disturbance (figure 2).

5 Results

The quantities offering the most insights into how disasters affect landscape evolution in the context of DPE are niche turnover and diversity. *Niche turnover* is the number of niches that have been utilised throughout the history of the landscape up to a certain point of time. The continual new adaptation to unused habitat niches is one of the key properties in biological CAS. *Diversity* is a necessary basis on which selection can act in CAS. While continual evolutionary innovation does not necessarily lead to increased diversity, a certain degree of diversity is required for such innovation. There are various ways to measure diversity [36, 37]. As phylogeny based measures are not meaningful within the model, we will apply several information theoretic measures: species richness (taken here as "niches richness") as well as the Simpson, Shannon and Pielou measures of diversity [37].

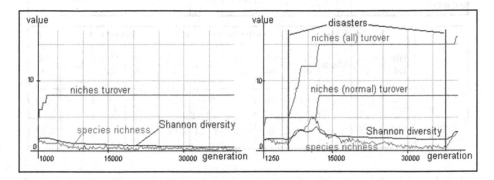

Fig. 3. Measurement results of typical runs. Measurements were taken every 250 genera- tions; measurements taken during variation phases (*crowedness* < 0.9) are removed. Basis scenario (left): Initially, a burst in niches turnover, Shannon diversity as well as species richness (taken here as niches richness) can be observed. After a while selection drives to inhibit diversity and further turnover of niches. Disaster scenario (right): Variation phases induced by disasters are followed by a jump in diversity as well as the discovery of new habitat niches. This figure was generated using the real-time plotter LiveGraph [1].

Sensitivity analysis shows that the model is sensitive to extremely small parameter values. For instance, the disaster dynamics described here could not always be observed on grids of size 30×30 and smaller. This may be because the spatial distribution of genotype differences is too small on grids of this size. Runs involving 8 and less niches also did not always produce the disaster scenario dynamics. This was mostly due to the fact that all niches were utilised and no further diversification was possible. In addition, selection will always drive the genotypes towards the available niches and a very small number of niches will work strongly against diversification. A very small number of traits (< 10) also tended to not produce interesting dynamics. This is probably due to the small number of possible genotypes in such a setup and a small mutation probability per genotype. Once the parameters exceed some threshold, the qualitative dynamics of the model are not sensitive to specific values. In general, the parameters used for both, the basis and the disaster scenario, lie well above these thresholds. Of course, particularly large parameter values lead to unreasonably long simulation run times. Our strategy of choosing the parameter values was therefore to choose values as large as possible, while still achieving a reasonable computation time (one model run on our workstation – Pentium 4 Prescott, 3.2 GHz, 1GB RAM – takes approx. 9 hours).

Table 1. Niche turnover and diversity measured at the end of each run. 20 control runs were performed for each scenario. Given are the minimum and the maximum values, the average, and the standard deviation.

	Simpson diversity				Shannon diversity				Pielou diversity			
	Min val.	Max val.	Mean	Std. div.	Min val.	Max val.	Mean	Std. div.	Min val.	Max val.	Mean	Std. div.
Basis scenario	0.01	0.75	0.42	0.19	0.04	2.20	1.10	0.50	0.03	0.58	0.35	0.13
Disaster scenario	0.18	0.90	0.60	0.17	0.47	3.81	1.87	0.76	0.26	0.69	0.46	0.11

	Species richness				Niches turnover (all)				Niches turnover (normal)			
	Min val.	Max val.	Mean	Std. div.	Min val.	Max val.	Mean	Std. div.	Min val.	Max val.	Mean	Std. div.
Basis scenario	0.09	1.89	0.82	0.52	2	9	5.38	1.88	2	9	5.29	1.79
Disaster scenario	0.09	4.15	1.86	1.06	6	22	14.81	3.7	3	11	8.05	1.88

For both, the basis and the disaster scenario we performed 20 control runs. At intervals of 250 generations we calculated the turnover of utilised niches and the four measures of diversity described above. In order to ensure the comparability of the results, we have calculated the turnover of *all* niches as well as the turnover of *normal niches only*. This is because disaster-niches are expected to be rarely utilised in the basis setup. In addition, the model population during the variation phase is expected to exhibit increased diversity. This can be expected and we are primarily interested whether this diversity will persist into the next balance phase. We therefore removed the data points observed during the variation phases. Selection in the model is due to restricted carrying capacity of the landscape cells. Therefore, selection is inhibited while the landscape is under-populated. We define:

$$crowdness = \frac{current\ population}{maximum\ popolation} = \frac{current\ population}{number\ of\ cells \times cell\ carrying\ capacity}$$

We can now define *variation phases* as phases during which *crowdness* < 0.9. The general behaviour of niche populations in each case is discussed in the previous section. The measured quantities at the end of each run are summarised in table 1.

Typical simulation results are depicted in figure 3. It can be seen that all kinds of diversity as well as the niche turnover are consistently higher in the disaster scenario. Two-sample t-tests confirm with confidence 99% ($\alpha = 0.01$) that the means of all 6 statistics listed in table 1 are higher in the disaster scenario.

6   Discussion and Future Work

The results support our claim that the DPE mechanism contributes to the emergence of sustained diversity and perpetual novelty in biological CAS. As discussed earlier, another mechanism that can lead to such novelty is co-evolution. The present model continuously exhibits novelty, but it does not include any species interactions. This supports the argument from an earlier section that sustained novelty does not necessarily require co-evolution. It is interesting to see this work being extended in the future to incorporate co-evolution in some form. For instance, this could be done by varying the "normal" fitness landscape over time. For instance, this could be done by introducing new normal niches over the run-time of the simulation or by varying the conditions of the existing niches.

In the present model, disturbances facilitate continuous novelty in two ways. By wiping out large areas, the disasters separate the remaining populations in disconnected islands, preventing gene flow between sub-populations. This can lead to divergence of two population islands through genetic drift. In addition, the disasters free up large areas, into which the remaining populations can expand. In an adapted population, random mutations are likely to be selected out. As selection is temporarily inhibited after a disaster (because the freed-up cells are not filled to their carrying capacity), mutations can lead to adaptations to new niches, even if the fitness within those niches it initially low. Such adaptations will be amplified by the high condition of the disaster niches in disturbed areas even after the disturbed areas have been repopulated. The sensitivity analysis of the model has shown that if the impact radii of the disasters are extremely small, the dynamics of the disaster scenarios cannot be observed. A possible reason for this is that such setup is essentially equivalent to selecting the cells affected by disasters in an independently randomly distributed manner. Rather than freeing up large areas, such method leads to feeing up many small patches which will be immediately repopulated by individuals from remaining population sites, which will, on average, be close by. The resulting variation phase of inhibited selection will be very short and not sufficient for sub-populations to diverge. In nature, sites affected by disasters are not independently distributed. If a landscape patch was affected by a disaster, the nearby patches are more likely to be affected than those far away. However, further research of the relationship between particular disaster types (impact distribution, size, etc.) and their effects on ecosystem properties (diversity, niche utilisation, etc.) may provide interesting results.

In summary, the simulation results of both the basis and the disaster scenarios exhibit patterns [35] observed in the corresponding ecosystems [11, 33, 34]. The model also confirms the implications of the DPE theory (see previous sections) on landscape evolution. Clearly, much interesting research remains to be done to investigate the role of DPE in biological and other CAS.

References

1. LiveGraph - a framework for real-time data visualisation, analysis and logging (accessed on 30.08.2007), http://www.live-graph.org
2. Green, D.G., Newth, D., Kirley, M.G.: Connectivity and catastrophe - towards a general theory of evolution. In: Bedau, M., McCaskill, J.S., Packard, N.H., Rasmussen, S., McCaskill, J., Packard, N. (eds.) Artificial Life VII (2000)
3. Green, D.G., Sadedin, S.: Interactions matter- complexity in landscapes and ecosystems. Ecological Complexity 2, 117–130 (2005)

4. Green, D.G., Leishman, T.G., Sadedin, S.: Dual phase evolution: a mechanism for self-organization in complex systems. InterJournal Complex Systems (2006)
5. Holland, J.H.: Hidden Order: How Adaptation Builds Complexity. Perseus Books (1995)
6. Levin, S.A.: Ecosystems and the Biosphere as Complex Adaptive Systems. Ecosystems 1, 431–436 (1998)
7. Lenton, T.M., Van Oijen, M.: Gaia as a Complex Adaptive System. Philosophical Transactions of the Royal Society: Biological Sciences 357, 683–695 (2002)
8. Watson, A.J., Lovelock, J.E.: Biological homeostasis of the global environment: the parable of Daisyworld. Tellus B 35, 284–289 (1983)
9. Weber, S.L.: On Homeostasis in Daisyworld. Climatic Change 48, 465–485 (2001)
10. Gavrilets, S.: Fitness Landscapes and the Origin of Species. Princeton University Press, Princeton / Oxford (2004)
11. Green, D.G.: Fire and Stability in the Postglacial Forests of Southwest Nova Scotia. Journal of Biogeography 9, 29–40 (1982)
12. Bak, P.: How Nature Works: The Science of Self-Organized Criticality. Reprint edn. Springer-Verlag Telos (1999)
13. Bak, P., Tang, C., Weisenfeld, K.: Self-Organized Criticality. Physical Review A 38, 364–374 (1988)
14. Langton, C.G.: Computation at the edge of chaos: Phase transitions and emergent computation. Physica D: Nonlinear Phenomena 42, 13–37 (1990)
15. Langton, C.G.: Life at the Edge of Chaos. In: Artificial Life II, Addison-Wesley, Reading (1991)
16. Bak, P., Sneppen, K.: Punctuated equilibrium and criticality in a simple model of evolution. Physical Review Letters 71, 4083 (1993)
17. Eldredge, N., Gould, S.J.: Punctuated Equilibria: An Alternative to Phyletic Gradualism. Freeman Cooper, San Francisco (1972)
18. Newman, M.E.J.: A model of mass extinction. Journal of Theoretical Biology 189, 235–252 (1997)
19. de Carvalho, J.X., Prado, C.P.C.: Self-Organized Criticality in the Olami-Feder-Christensen Model. Physical Review Letters 84, 4006 (2000)
20. Sornette, D., Johansen, A., Dornic, I.: Mapping Self-Organized Criticality onto Criticality. Journal de Physique I 5, 325–335 (1995)
21. Kinouchi, O., Prado, C.P.C.: Robustness of scale invariance in models with self-organized criticality. Physical Review E 59, 4964 (1999)
22. Alvarez, L.W., Alvarez, W., Asaro, F., Michel, H.V.: Extraterrestrial Cause for the Cretaceous-Tertiary Extinction. Science 208, 1095 (1980)
23. Kornfield, I., Smith, P.F.: African Cichild Fishes: Model Systems for Evolutionary Biology. Annual Review of Ecology and Systematics 31, 163–196 (2000)
24. Sturmbauer, C., Meyer, A.: Genetic divergence, speciation and morphological stasis in a lineage of African cichlid fishes. Nature 358, 578–581 (1992)
25. Hewitt, G.M.: Genetic consequences of climatic oscillations in the Quaternary. Philosophical Transactions: Biological Sciences 359, 183–195 (2004)
26. Butlin, R.K., Walton, C., Monk, K.A., Bridle, J.R.: Biogeography of Sulawesi grasshoppers, genus Chitaura, using DNA sequence data. In: Biogeography and geological evolution of Southeast Asia, pp. 355–359. Backhuys Publishers, Leiden, The Netherlands (1998)
27. Cowling, S.A., Maslin, M.A., Sykes, M.T.: Paleovegetation Simulations of Lowland Amazonia and Implications for Neotropical Allopatry and Speciation. Quaternary Research 55, 140–149 (2001)

28. Bennett, K.D.: Continuing the debate on the role of Quaternary environmental change for macroevolution. Philosophical Transactions: Biological Sciences 359, 295–303 (2004)
29. Coope, G.R.: Several million years of stability among insect species because of, or in spite of, Ice Age climatic instability? Philosophical Transactions: Biological Sciences 359, 209–214 (2004)
30. Gavrilets, S., Vose, A.: Dynamic patterns of adaptive radiation. Proceedings of the National Academy of Sciences USA 102, 18040–18045 (2005)
31. Paperin, G., Green, D.G., Dorin, A.: Fitness Landscapes in Individual-Based Simulation Models of Adaptive Radiation. In: CMLS 2007. 2007 International Symposium on Computational Models for Life Science, Gold Coast, Australia (2007)
32. Russell, P.J.: Fundamentals of genetics. HarperCollinsCollege Publishers, New York (1994)
33. Green, D.G.: Simulated effects of fire, dispersal and spatial pattern on competition within forest mosaics. Plant Ecology 82, 139–153 (1989)
34. Barton, N.H., Hewitt, G.M.: Analysis of hybrid zones. Annual Review of Ecology and Systematics 16, 113–148 (1985)
35. Grimm, V., Revilla, E., Berger, U., Jeltsch, F., Mooij, W.M., Railsback, S.F., Thulke, H.-H., Weiner, J., Wiegand, T., DeAngelis, D.L.: Pattern-Oriented Modeling of Agent-Based Complex Systems: Lessons from Ecology. Science 310, 987–991 (2005)
36. Purvis, A., Hector, A.: Getting the measure of biodiversity. Nature 405, 212–219 (2000)
37. Rojas, M.G.S.: Measures of diversity: a comparison of spatial patterns in a marine fouling community. Marine Ecology, vol. Graduate Thesis. Göteborg University, Göteborg (2004)

Directed Evolution of an Artificial Cell Lineage

Nicholas Geard[1,2] and Janet Wiles[2]

[1] School of Engineering and Computer Science, University of Southampton
Southampton SO17 1BJ, UK
nlg@ecs.soton.ac.uk
[2] ARC Centre for Complex Systems, The University of Queensland
Brisbane, Queensland 4072, Australia
j.wiles@itee.uq.edu.au

Abstract. Biological development is a complex process that mediates between genotypes, to which mutations occur, and phenotypes, on which selection acts. Properties of development can therefore have considerable impact on evolution. However, in many existing simulation models of development, the developmental process itself is difficult to recover and/or analyse. We have previously introduced a model of development in which the developmental process is represented as a cell lineage. Here we use this model to further explore the control of development, and the influence that development has on shaping an adaptive landscape.

1   Introduction

Novel phenotypic forms arise from gene mutations that reprogram developmental trajectories [1]. Evolution by natural selection occurs because certain individuals, by virtue of some heritable phenotypic trait, stand a better chance of surviving to pass on their genes to offspring than others. The specific phenotypic traits that increase an organism's chance of reproduction will depend on the nature of the ecological niche it inhabits. In a relatively stable environment, it is therefore possible to imagine an adaptive gradient mapped to phenotypic space.

The idea of an adaptive phenotypic space was introduced by Simpson [2], who described a two-dimensional landscape representing the possible combinations of two phenotypic characters in which elevation corresponded to fitness. The highest point in the landscape represents the phenotype that is most adapted to the current environment. Because environments are dynamic, the location of this optimum point will move over time. Simpson's adaptive phenotypic landscape is a descendant of the fitness landscape described by Wright [3] but differs in two respects. First, the axes of Wright's fitness landscape represent gene frequencies rather than phenotypic characters. Second, the structure of fitness landscapes is typically more complex due to epistatic interactions between genes.

There is an important relationship between genotypic and phenotypic landscapes. The adaptive phenotypic landscape specifies the direction of evolution favoured by selection. However, any movement from phenotype A to phenotype B in phenotypic space is contingent upon genotype B being mutationally accessible from genotype A in genotypic space (Figure 1). The mapping from a genotype

M. Randall, H.A. Abbass, and J. Wiles (Eds.): ACAL 2007, LNAI 4828, pp. 144–155, 2007.
© Springer-Verlag Berlin Heidelberg 2007

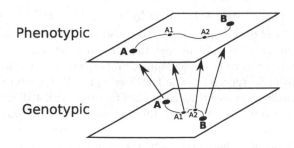

Fig. 1. Phenotypic adaptation depends on mutational accessibility. In order for phenotype adaptation to proceed from phenotype **A** to phenotype **B**, there must be a mutationally accessible path of genotypes between genotypes **A** and **B**. The mapping from genotypic to phenotypic space will be affected by the nature of development.

to a phenotype is defined by the developmental process; therefore, properties of the developmental process will affect adaptation. Determining the impact that development has on adaptive landscapes requires a better understanding of the mapping between genotypic and phenotypic space.

This study explores the effect on evolution of a developmental mapping based on the dynamics of a gene regulatory network. The following section describes the artificial cell lineage model. Two series of simulations are then used to explore the effects of different phenotypic constraints, and different target complexities, on adaptive search difficulty. Finally, the results of these simulations are analysed to provide insight into the characteristics of the adaptive landscape.

2   The Artificial Cell Lineage Model

The artificial cell lineage model consists of two components: a network component that generates the gene expression dynamics controlling development and a cell lineage component that defines how these dynamics are interpreted to define an ontogeny. The model is described briefly here; a more thorough description and justification can be found elsewhere ([4,5]).

The genetic component of the model is defined by a network of interacting nodes, based on a standard recurrent network architecture. Three layers of nodes represent N_I input, N_R regulatory and N_O output genes respectively. All input nodes are connected to all regulatory nodes, all regulatory nodes are connected to all output nodes, and each regulatory node is connected to, on average, K other regulatory nodes (including self connections). The interactions between two network layers are represented by a weight matrix, in which the entry at row i, column j specifies the influence that gene j has on gene i. For the simulations, random networks were created by setting each weight to a value drawn at random from a Gaussian distribution with mean zero and standard deviation W. The state of the network was updated synchronously in discrete time steps, with the activation of node i at time $t + 1$, $a_i(t + 1)$, given by $a_i(t + 1) = \sigma\left(\sum_{j=1}^{N} w_{ij} a_j(t) - \theta_i\right)$ where w_{ij} is the level of the interaction from node j to node i, θ_i is the activation threshold of node i, and $\sigma(x)$ is the logistic sigmoid function.

A cell lineage is a record of a developmental trajectory in the form of a binary tree [6]: the root node represents the fertilised egg cell; the non-terminal nodes represent the transient states that cells pass through whilst differentiating; and the terminal nodes represent the final differentiated cells that exist at the end of the developmental process. Therefore, the terminal nodes of the cell lineage that represent an organism's phenotype, and the topology of the tree describes the relationship between all cells that existed at some point during development.

The network model described above is a general purpose computing device. In a developmental system, the computation performed is the transformation of a temporal sequence of contextual inputs into an ordered pattern of cell division and differentiation events. Two input nodes specified the relative position of a cell with respect to its sibling. After division, the activation of these nodes was set to $\{0, 1\}$ in the left daughter and $\{1, 0\}$ in the right daughter. The output nodes were used to determine cell division and differentiation. If the activation of the first output node was above a certain division threshold θ_d, that cell would divide, otherwise it would differentiate. In development, the likelihood of a cell continuing to divide decreases over time. To simulate this, the division threshold was scaled dynamically, according to $\theta_d = 1 - 0.01e^{\lambda d}$ where d was the depth of the current cell and λ was a scaling parameter. Once a cell stopped dividing, the remaining $N_O - 1$ output nodes were used to determine its differentiation type via a 'one-hot' or exclusive encoding scheme.

3   Evolving Complex Cell Lineages

We have previously demonstrated that the artificial cell lineage model is capable of generating a diverse range of ontogenies of varying levels of complexity [7,8,4]. The aims of these simulations were to investigate the effect of different phenotypic distance metrics on the type of lineages located by adaptive search, and to investigate how the difficulty of adaptive search increased as phenotypic targets became more complex. The specific targets for the adaptive tasks used in this study are derived from the lineages of the organisms *C. elegans* and *H. roretzi*. The use of targets derived from real lineages is important because we know that they have been evolved once, and hence are of a biologically plausible level of complexity. Defining and measuring biological complexity are difficult issues: a full description of the metric employed here can be found in [5], and further exploration of the complexity of cell lineages can be found in [9].

We make a simplifying assumption that adaptation is occurring in a fixed environment, and the target phenotype is the most highly adapted to that environment. Fitness was calculated in terms the distance between the current and target phenotypes. In a real environment, ecological niches are highly dynamic, changing as environments change or according to fluctuations in co-evolutionary

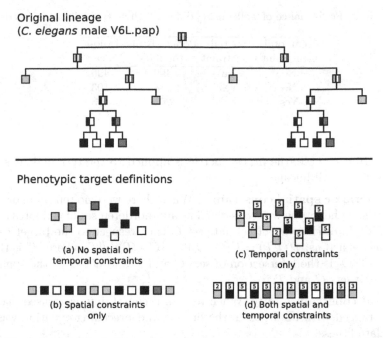

Fig. 2. The four phenotypic distance metrics as applied to the *C. elegans* male V6L.pap lineage [10]. See text for a full description of each metric.

relationships. However, when environmental change is slower than adaptation, the assumption of a static fitness landscape is not implausible.

3.1   Measuring Fitness

Cell lineages are an organisational, rather than morphological, description of a phenotype and can be quantified and compared in an automated fashion. We defined four metrics based on the phenotypic component of a cell lineage (*i.e.*, the terminal cells) in terms of the intersection between three types of constraint: on the set of cell identities, the relative spatial location of each cell, and the point in developmental time at which they appear. The first and most basic constraint is on the cell fate distribution: the requirement that a certain number of cells of each specific type are present at the end of development. The second and third constraints require that each terminal cell be correctly positioned in relation to the other cells in the phenotype, and appear at the correct time during development. We do not suggest that natural selection acts to explicitly satisfy these constraints, but rather that they may serve as surrogates for a broad range of selective criteria operating on development.

For each of the fitness metrics used in this study, the identity constraint was considered fundamental and always used, in addition to which temporal and spatial constraints could be applied either separately or together. The practical implication of each of these constraints and their intersection is illustrated in

Table 1. Performance of walks using different phenotypic distance metrics

Temporal Constraint	Spatial Constraint	Perfect Runs (of 500)	Unique Lineages
No	No	499	496
No	Yes	288	103
Yes	No	201	113
Yes	Yes	27	1

Figure 2. In each case, the fitness metric is applied to the terminal cells of the fully developed cell lineage.

No temporal or spatial constraints. When there were no temporal or spatial constraints, a phenotype was considered as an unordered set of cell fates and the fitness $f(C,T)$ of the current cell fate set C with respect to the target cell fate set T was defined as: $f(C,T) = \big(|(C \cap T)| - |(C \ominus T)|\big)/|T|$ where $|T|$ is the size of set T, $C \cap T$ is the intersection of sets C and T and $C \ominus T$ is the symmetric difference of sets C and T.

Temporal constraints only. When temporal constraints were used, each cell fate was tagged with its depth in the lineage and preceding equation was used to calculate fitness.

Spatial constraints only. When spatial constraints were used, a phenotype was considered as an ordered sequence of cell fates and the fitness $f(C,T)$ of the current cell fate sequence C with respect to the target cell fate sequences T was defined as: $f(C,T) = \big(\mathrm{LEV}(C,T)\big)/|T|$ where $\mathrm{LEV}(C,T)$ was the Levenshtein distance between sequences C and T (see [5] for the algorithm used to calculate this metric) and $|T|$ was the length of sequence T.

Both temporal and spatial constraints. When both temporal and spatial constraints were used, each cell fate was tagged with its depth in the lineage and the preceding equation was used to calculate fitness.

3.2   Comparison of Different Phenotypic Distance Metrics

An initial set of adaptive walks compared the effect of using the four different phenotypic distance metrics described above as fitness measures. For each metric, an ensemble of 500 networks with eight fully connected regulatory nodes ($N = 8, K = 8, W = 2.0$) were created and adaptive walks were performed. Each adaptive walk consisted of 20,000 steps; at each step, a new network was created by replacing one weight at random with a new value drawn from a Gaussian distribution with mean zero and standard deviation W. The newly created network replaced the current network if its fitness was equal to or greater than that of the current lineage.

As anticipated, as phenotypic definition became more constrained, the difficulty of the search process increased (Table 1). With no spatial or temporal constraints, only one of 500 walks failed to find a perfect solution (*i.e.*, a cell lineage whose terminal nodes consisted of the correct quantity of each cell type). In

Table 2. Target Lineage Details

Lineage	Number of Cells	Number of Cell Types	Maximum Depth	Weighted Complexity
C. elegans maleV6Lpap	12	4	5	6.55
C. elegans C	48	4	6	11.23
C. elegans MSp	46	5	7	22.49
C. elegans MSa	48	5	7	26.55
H. roretzi (half)	55	7	6	31.57

contrast, with both spatial and temporal constraints, only 27 of 500 walks were able to find lineages that produced the target phenotype. When the phenotypic definition incorporated either spatial or temporal constraints, around half of the runs found lineages that produced the target phenotype. Spatial constraints were moderately easier to satisfy than temporal constraints (288 compared to 201 perfect solutions).

The phenotypic definition had a significant effect on the variety of lineages that were found. Of the 499 solutions found with no spatial and temporal constraints, 496 of the lineages generating these phenotypes were unique. In contrast, the intersection of spatial and temporal constraints restricted the space of possible solutions to a single lineage, that of the original data set. One explanation for the lower rate of success under this phenotypic definition appears to be the structure of the adaptive landscape. Using the least constrained phenotypic definition means that a greater number of lineages map to the target phenotype, and hence a larger proportion of genotypic space maps, via ontogeny, to a perfect fitness value. When the most constrained phenotypic definition is used, only a single lineage maps to the target phenotype, and hence a much smaller proportion of genotypic space maps to a perfect fitness value.

3.3   Comparison of Different Phenotypic Targets

The second series of adaptive walks compared the performance of adaptive walks on five target lineages derived from real data sets (Table 2). The first target lineage was the *C. elegans* male V6L.pap used above (shown in Figure 2). Three further target lineages from *C. elegans* were also used: the sublineage of the C founder cell, which produces the muscle and epidermis cells in the posterior region of the worm's body; and two sublineages, MSa and MSp, of the MS founder cell, which primarily produces the pharynx (a digestive organ), but also some muscle cells and the somatic gonad precursors [11]. The final target lineage was taken from the ascidian *H. roretzi* [12].

For each phenotypic target, an ensemble of 50 random networks ($N = 16, K = 16, W = 2.0$) was generated and adaptive walks were performed as above. The second phenotypic definition (spatial constraints only) was used to evaluate the fitness of each phenotype. Each adaptive walk consisted of 60,000 steps.

The results of these simulations demonstrate that adaptive search becomes more difficult as the complexity of the target lineage increases (Table 3). While

Table 3. Performance of walks using targets of varying complexity

Target Lineage	Best Fitness	Remaining Errors	Avg. Fitness (Std. Dev.)	Perfect Runs (of 50)
C. elegans maleV6Lpap	1.0	-	0.938 (0.071)	24
C. elegans C	1.0	-	0.950 (0.038)	6
C. elegans MSp	0.956	3	0.852 (0.068)	-
C. elegans MSa	0.958	3	0.834 (0.076)	-
H. roretzi (half)	0.982	1	0.745 (0.074)	-

almost half of the walks were able to locate the simplest lineage (*C. elegans* maleV6Lpap), the best performing walk on the most complex lineage (*H. roretzi*) contained a single incorrect cell after 60,000 steps. In order to demonstrate that the MSp, MSa and *H. roretzi* tasks were in fact achievable, the best performing networks on each of these targets were re-run with no limitations on the maximum length of the walk. At least one walk was able to locate each of the target lineages; however the search times required were on the order of 300,000 steps.

3.4   Analysis of an Adaptive Walk

The progress of an adaptive walk towards a target may be measured in several ways (Figure 3 shows the first 60,000 steps of a successful adaptive walk using the *H. roretzi* target, after which only one incorrect terminal cell remained).

Fitness followed a hyperbolic trajectory over the duration of the walk; a pattern commonly observed in evolution under both computational and *in vitro* conditions [13].

Complexity (as defined in [5]) tended to increase over the course of the adaptive walk, achieving the complexity of the target lineage after approximately 7,000 steps and thereafter fluctuating about that value. Comparing the fitness and complexity plots, it is evident that there is a degree of neutrality in the mapping from ontogenetic space (measured by complexity) to the fitness landscape. Clearly it is possible for multiple lineages to share an equal fitness value, and for an adaptive walk to move between these equivalent lineages via mutation.

Genotypic substitution rate measures the acceptance of newly created networks. Initially, around 60% of mutations are accepted (*i.e.*, are either beneficial or neutral). This probability decreases at a constant rate until around step 7,000. After this point, approximately 20% of mutations are accepted with a moderate decrease over the remainder of the walk. Should this statistic ever reach zero, it is possible that no further adaptation could occur as the network weights would be so finely tuned that any mutation would be detrimental. In practice, this phenomenon was never observed in any of the simulations reported here: there was sufficient neutrality in the gene network to lineage mapping to ensure that some change was possible.

Phenotypic substitution rate measures the acceptance of networks that generated a different phenotype to the previous network. Initially, around 10%

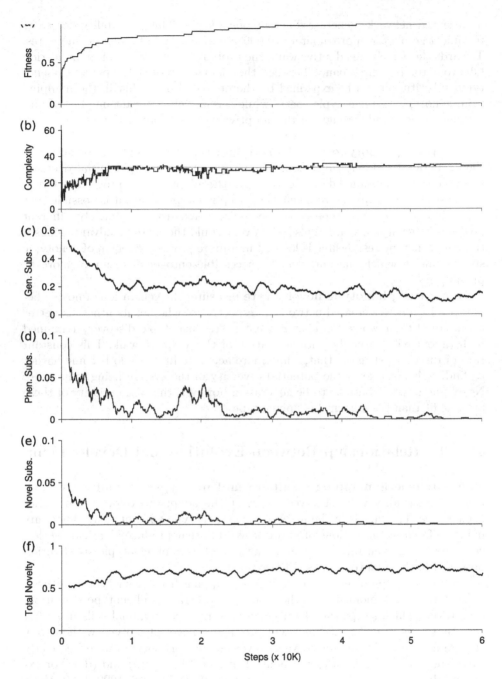

Fig. 3. Analysis of a single adaptive walk using the *H. roretzi* target lineage. From top to bottom, the plots show: (a) fitness; (b) complexity; (c) genotypic substitution rate; (d) phenotypic substitution rate; (e) accepted phenotype novelty rate; (f) generated phenotype novelty rate. See text for further details.

of accepted networks generate different phenotypes. This probability decreases to almost zero after approximately 10,000 generations and thereafter fluctuates. Towards the end of the adaptive walk, the probability of phenotypic substitution falls to zero. The discrepancy between the probability of genotypic and phenotypic substitution can be explained by the degree of neutrality in the mapping from genotypic to phenotypic space: while a relatively constant number of mutations are accepted throughout the adaptive walk, the proportion of these that result in phenotypic change decreases.

Accepted phenotype novelty rate measures the acceptance of networks that generated a previously unseen phenotype. Again, a rapid initial decrease was followed by a gradual decrease to zero as the adaptive walk proceeded. Given the many-to-one mapping from genotypic to phenotypic space, it is possible that a previously seen phenotype could be rediscovered from an entirely different position in genotypic space. This rediscovery could therefore be advantageous if the new genotype responsible is located in a more promising region of genotypic space—one in which the mutationally accessible ontogenies result in more fit phenotypes.

Generated phenotype novelty rate measures the generation of novel phenotypes by a newly created network, irrespective of whether its fitness is better than, equal to or worse than the current best. Phenotypic discovery remained high (above 50%) over the entire duration of the adaptive walk. This constant rate of discovery suggests that, while more accurate lineages do become harder to find, it is not due to the potential diversity of the system being exhausted. Novel phenotypes continue to be generated; however, the vast majority of these are less fit than the current best phenotype.

4   The Relationship Between Evolution and Development

The ontogenetic mapping results in multiple types of neutrality. Two types of neutrality were observed to affect the adaptive exploration of genotypic space. The first is in the mapping from genotype and ontogeny. There are many different combinations of network weights that produce identical cell lineage trees. This neutrality accounts for the high rate of genotypic substitution observed in the adaptive walks (Figure 3(c)).

The second type of neutrality is in the mapping from phenotype to fitness. Considering for a moment just the spatially constrained phenotype definition: a mutation which swaps the identities of two incorrect terminal cells in such a way that they are still incorrect will produce a novel phenotype without any change in fitness. The adaptive walks revealed that phenotypes were frequently substituted while on a plateau of neutral fitness (Figure 3(a) and (d)). For example, during one long period of stasis (approximately steps 1200–1800) there was considerable neutral substitution until, around step 1800, a burst of novel phenotypic substitution resulted in further fitness increases. Two interpretations of this dynamic are possible. First, the neutrality may have been beneficial, as it allowed search to continue where it would otherwise have become trapped at a

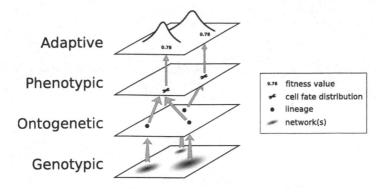

Fig. 4. Summary of different types of neutrality affecting adaptive search. Many different genotypes map to a single ontogeny. More than one ontogeny may map to a given phenotype. Finally, multiple phenotypes have equivalent fitness values.

local optima. Second, the neutrality may have been a hindrance, introducing a long period of drift where a more rapid transition to a more fit phenotype could otherwise have been achieved. Distinguishing between these two possibilities is difficult, as it implies a comparison with a search landscape that lacks neutrality, but is otherwise identical.

A third type of neutrality—in the mapping from ontogeny to phenotype— is known to be possible: In the first series of adaptive walks, under all but the strictest set of phenotypic constraints, multiple lineages were located that mapped to the target phenotype. In practice, none of the adaptive walks in the second series were observed to exploit this form of neutrality. One possible explanation for this is that these neutral lineages are located at some distance from one another with respect to genotypic space, such that they are not mutationally accessible to one another. Figure 4 summarises the different types of neutrality that were observed or inferred from the adaptive walks.

Phenotypic improvements occur across a range of scales. Analysis of the accepted mutations over the adaptive walk shown in Figure 3 revealed that mutations can cause phenotypic improvement across a wide range of scales. At the lower end of the spectrum were those frequently accepted mutations that modified the identity of a single terminal cell, and those that added or removed a single cell. At the upper end of the spectrum were those more rarely accepted mutations that introduced or removed a new cell type, and those that added or removed an entire branch of the cell lineage. The size of a phenotypic improvement was estimated by applying the fitness function using the pre-mutation lineage as the current solution and the post-mutation lineage as the target solution. The sizes of such changes follow a power law distribution (Figure 5).

The scale of evolutionary change is a subject of ongoing debate in evolutionary biology [14]. The essence of the debate concerns how to explain the evolution of species as inferred from the fossil record: is the selection of individual mutations a sufficient mechanism, or are higher-level evolutionary forces necessary?

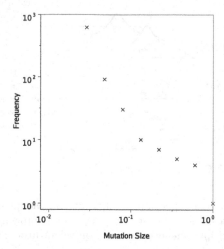

Fig. 5. The distribution of phenotypic improvements indicates that beneficial mutations occur across a range of scales. Mutation size was measured as the distance between the initial and mutant lineages for each accepted mutations and sorted into exponentially scaled bins.

Fisher [15] argued that mutations of large effect would be far less likely to be beneficial, and hence only mutations of small effect were likely to be significant. Kimura [16] challenged this claim, pointing out that if very rare large beneficial mutations *did* occur, they would be more likely to be fixed, and hence the distribution of mutation sizes would be skewed. The distribution observed in Figure 5 supports the claim that mutations causing both large and small phenotypic changes can be accepted during an adaptive walk.

Figure 5 also highlights one of the benefits of the gene network approach to modelling ontogeny. If the cell lineage representation had been modified directly by the adaptive process, we would have needed to specify the sizes and types of mutations that were possible (*e.g.,* swapping sublineages, or adding and deleting terminals) As it is, we did not need to impose a preconceived step size on the adaptive process—it emerged naturally as a consequence of the dynamic mapping.

5   Conclusions

Our investigations demonstrate that adaptive search is capable of locating networks whose dynamics generate specific complex developmental patterns derived from real cell lineages. Search difficulty is affected both by the types of constraint (spatial and temporal) applied to the phenotypic targets, as well as their complexity. The dynamics of search suggest that the adaptive landscapes resulting from the proposed developmental mapping are dominated by the presence of several different levels of neutrality.

Acknowledgements. The authors would like to thank R. B. R. Azevedo, R. Lohaus and K. Willadsen for useful discussions. This research was supported by an APA and an ACCS scholarship to NG and an ARC grant to JW.

References

1. Arthur, W.: The concept of developmental re-programming and the quest for an inclusive theory of evolutionary mechanism. Evolution & Development 2, 49–57 (2000)
2. Simpson, G.G.: Tempo and Mode in Evolution. Columbia University Press, New York, NY (1944)
3. Wright, S.: The roles of mutation, inbreeding, crossbreeding and selection in evolution. Proceedings of the 6th International Congress on Genetics 1, 356–366 (1932)
4. Geard, N.: Artificial Ontogenies: A Computational Model of the Control and Evolution of Development. PhD thesis, School of Information Technology and Electrical Engineering, The University of Queensland (2006)
5. Geard, N., Wiles, J.: LinMap: Visualising complexity gradients in evolutionary landscapes. special issue of Artificial Life (to appear, 2007)
6. Stent, G.: Developmental cell lineage. International Journal of Developmental Biology 42, 237–241 (1998)
7. Geard, N., Wiles, J.: A gene network model for developing cell lineages. Artificial Life 11(3), 249–268 (2005)
8. Geard, N., Wiles, J.: Investigating ontogenetic space with developmental cell lineages. In: Rocha, L.M., et al. (eds.) Artificial Life X, Cambridge, MA, pp. 56–62 (2006)
9. Lohaus, R., Geard, N., Wiles, J., Azevedo, R.B.R.: A generative bias towards average complexity in artificial cell lineages. Proceedings of the Royal Society of London, Series B 274(1619), 1741–1750 (2007)
10. Braun, V., Azevedo, R.B.R., Gumbel, M., Agapow, P.M., Leroi, A.M., Meinzer, H.P.: ALES: cell lineage analysis and mapping of developmental events. Bioinformatics 19, 851–858 (2003)
11. Sulston, J.E., Schierenberg, E., White, J.G., Thompson, J.N.: The embryonic cell lineage of the nematode *Caenorhabditis elegans*. Developmental Biology 100, 64–119 (1983)
12. Nishida, H.: Cell lineage analysis in ascidian embryos by intracellular injection of a tracer enzyme. III. Up to the tissue restricted stage. Developmental Biology 121, 526–541 (1987)
13. Lenski, R.E., Travisano, M.: Dynamics of adaptation and diversification: a 10,000 generation experiment with bacterial populations. Proceedings of the National Academy of Science, USA 91, 6808–6814 (1994)
14. Leroi, A.M.: The scale independence of evolution. Evolution & Development 2(2), 67–77 (2000)
15. Fisher, R.A.: The Genetical Theory of Natural Selection. Clarendon Press, Oxford (1930)
16. Kimura, M.: The Neutral Theory of Molecular Evolution. Cambridge University Press, Cambridge (1983)

An Integrated QAP-Based Approach to Visualize Patterns of Gene Expression Similarity

Mario Inostroza-Ponta, Alexandre Mendes,
Regina Berretta, and Pablo Moscato

Centre for Bioinformatics,
Biomarker Discovery and Information-based Medicine
The University of Newcastle
Callaghan, NSW, 2308, Australia
Pablo.Moscato@newcastle.edu.au

Abstract. This paper illustrates how the Quadratic Assignment Problem (QAP) is used as a mathematical model that helps to produce a visualization of microarray data, based on the relationships between the objects (genes or samples). The visualization method can also incorporate the result of a clustering algorithm to facilitate the process of data analysis. Specifically, we show the integration with a graph-based clustering algorithm that outperforms the results against other benchmarks, namely k−means and self-organizing maps. Even though the application uses gene expression data, the method is general and only requires a similarity function being defined between pairs of objects. The microarray dataset is based on the budding yeast (*S. cerevisiae*). It is composed of 79 samples taken from different experiments and 2,467 genes. The proposed method delivers an automatically generated visualization of the microarray dataset based on the integration of the relationships coming from similarity measures, a clustering result and a graph structure.

1   Introduction

The analysis of gene expression data coming from microarray technologies has become an important challenge for computer scientists working in bioinformatics. Among the techniques available, the visualization of microarray data is crucial to assist the analysis. Currently, it is mainly carried out by the use of *heat maps* of the gene expression, which give the user a clear appreciation of how a set of genes are expressed along a set of samples, experiments or conditions. Another approach is the use of graph visualization algorithms. After modeling the microarray data as a graph (which is generally obtained with different *ad hoc* procedures), different graph layout algorithms are applied, like the *force-direct* [1] or the *circular layout* [2,1]. An important step in these visualization methods is the definition of the graph, since the result of the graph layout algorithms will depend on the structure of the graph, which in general disregard any other information from the dataset. For example, a generally employed method to build such a graph is to define a distance or similarity between genes and then

M. Randall, H.A. Abbass, and J. Wiles (Eds.): ACAL 2007, LNAI 4828, pp. 156–167, 2007.

a threshold value such that a graph will be constructed with a one-to-one map between vertices and genes; an edge connects a pair of genes if, and only if, they are at a distance smaller than the threshold value. A force-direct layout algorithm can subsequently be applied with forces that attract vertices connected by these edges, and in some cases the pairs of vertices not connected tend to be separated via *ad hoc* repulsion forces. As a result, it is expected that the layout of the graph will have the elements with similar expression pattern linked by an edge, and consequently closer in the layout.

In this work we propose a different visualization method for microarray data based on the modeling of the layout problem as an instance of the Quadratic Assignment Problem (QAP). Briefly, the QAP considers a set of objects to be assigned on a set of available locations, considering the flow between all the objects and the distance between all the locations, aiming to minimize the overall flow cost. The layout solution will provide a visualization of the objects (representing genes or samples) where the position of each will depend on the relationships between them in the dataset. Moreover, we combine our visualization technique with a graph-based clustering algorithm that uses a combination of k Nearest Neighbors and Minimum Spanning Tree, showing the versatility of our visualization method to display the components with the integration of several levels of information. In order to illustrate the method's performance, we considered a dataset used by Eisen *et al.* [3] composed of the expression of 2,467 genes over 79 samples from time courses during different experiments on the budding yeast *S. cerevisiae*. We use the visualization method proposed on both genes and samples. The output shows the ability of the method to produce a layout based on the integration of different levels of information: **a) similarity between objects** (genes or samples), **b) a representative graph structure** and **c) a clustering result**. Also, because of the formulation of the problem, we can mathematically guarantee that all the relationships will be considered in the generation of the layout.

The paper is organized as follows: first we present the mathematical formulation of the layout problem as an instance of the QAP. Next, the graph-based clustering method is described. Furthermore, we show the computational results on the microarray dataset and the comparison with two other clustering algorithms: k-Means [4] and self organizing maps (SOM) [5]. Finally, the analysis and conclusions are drawn.

2   QAP-Based Layout Method

The Quadratic Assignment Problem (QAP) belongs to the *NP-hard* class and it is a well-studied combinatorial optimization problem [6,7,8]. In this problem, the task is to assign a set of n objects to m locations ($m \geq n$). A matrix $L = \{l_{ij}\}$ of *distances* between the m locations is given as input, as it is also given a matrix $F = \{f_{ij}\}$ of *flows* between the n objects. The objective is to minimize the overall transportation cost between all the objects considering both the flow between each pair of them and the distances between the locations in which objects are being assigned. Formally, the objective function to be minimized is the following:

$$Cost(S) = \sum_{i=1}^{n} \sum_{j=1}^{n} f_{ij} l_{S(i)S(j)}, \qquad (1)$$

where $S(i)$ represents the assigned location of the object i in solution S. In this paper, we are using a mathematical formalization of the visualization problem in which we have QAP instances with n objects and $m \gg n$ available locations. In order to create one of these instances we use the following procedure:

1. each of the n objects represents either a sample or a gene;
2. a matrix D of distances between each pair of objects is computed using a distance metric, for example based on Pearson Correlation, Euclidean or Cosine distances, among others;
3. the flow f_{ij} (matrix F) between any pair of objects i and j is defined as $f_{ij} = \frac{1}{d_{ij}}$ ($\forall\ i \neq j$, 0 otherwise);
4. a grid of m locations ($m \gg n$) is defined;
5. the distance between each pair of locations in the grid (matrix L) is calculated using Euclidean distance.

Clearly, higher flow values will be assigned to objects that are very similar and lower flow values to samples that are dissimilar. A good solution for the QAP will put the objects with a high flow closer in the layout, which is exactly our goal.

2.1 Proximity Graph Clustering Method

As we mentioned in the introduction, there are several alternatives to define a graph that represents the most important proximity relationships in the dataset. Our *ad-hoc* proximity graph is built using information from a *minimum spanning tree* (G_{MST}) and a *k-nearest neighbors* (G_{kNN}) graphs. A *minimum spanning tree* is a connected, acyclic subgraph $G_{MST}(V, E_{MST})$ containing all the nodes of G and whose edges total sum has minimum weight. On the other hand, a *k-nearest neighbors* correspond to the graph $G_{kNN}(V, E_{kNN})$, where $e_{ij} \in E_{kNN}$ iff j is one of the k nearest neighbors of i.

Initially, we create a complete undirected weighted graph $G(V, E, w)$ using the distance matrix D, such that the weight $w_{ij} = d_{ij}$. We define our proximity graph, namely $G_{cluster}(V, E_{cluster})$, such that $E_{cluster} = E_{MST} \cap E_{kNN}$. This type of proximity graph was also used in González-Barrios and Quiroz (2003) [9]. In their paper, the authors fixed $k = \lceil ln(n) \rceil$ as the parameter for the kNN graph, where n represents the number of vertices in the graph G. After many tests, we adopted a variant of that expression for the value of k, which is shown in expression 2.

$$k = \min \{\lceil ln(n) \rceil ; \min\ k\ /\ G_{kNN}\ \text{is connected}\} \qquad (2)$$

The graph $G_{cluster}$ has $c \geq 1$ disconnected subgraphs $(G^1_{cluster}, ..., G^c_{cluster})$. The process is applied recursively on each subgraph of $G_{cluster}$ until no further partition is found. In Figure 1 we show the algorithm of this process.

MST-kNN (D: distance matrix)
   compute G.
   compute G_{MST}.
   compute G_{kNN} according to expression 2.
   $G_{cluster} = \{V_{cluster} = V, E_{cluster} = E_{MST} \cap E_{kNN}\}$
   if (# of subgraphs of $G_{cluster}$ > 1) **then**
     for each subgraph $G^i_{cluster} \in G_{cluster}$
      $G_{cluster} = \bigcup_i MST\text{-}kNN(submatrix(D, G^i_{cluster}))$
     end for
   end if
   return $G_{cluster}$

Fig. 1. Pseudo-code of the proximity graph based clustering method

The final output is a partition of the set of elements based only on the data provided (the distance matrix D between the elements) and with no user-determined parameter. However, it is also possible to provide a maximum value for k and thus have more control over the algorithm. In addition to the clusters themselves, the edges of the graph $G_{cluster}$ represent elements that are close to each other.

2.2 Integration of Proximity Graph Clustering with the Visualization Method

In order to integrate the result of the graph-based clustering presented in 2.1 with the visualization method proposed, we consider each subgraph $G^i_{cluster} \in G_{cluster}$ as a QAP instance that it will be independently optimised. In addition, for each of them, we also redefine the flow for each pair of samples (or genes) in the cluster as show in 3, since the edges that belong to the $G_{cluster}$ are responsible for keeping the cluster together. After several tests, we define 1000 as the multiplicative factor for those edges, because it allows the method to better represent the graph structure on the layout.

$$f_{ij} = \begin{cases} \frac{1000}{d_{ij}} & \text{, if } e_{ij} \in E_{cluster}; \\ \frac{1}{d_{ij}} & \text{, otherwise.} \end{cases} \tag{3}$$

Following the creation of several QAP instances, one for each cluster, we have to place all clusters in a single layout. To do so, another QAP instance is created. In this case, each object represents one of the clusters $G^i_{cluster}$. This instance is created by building a fully connected graph $G_C(V_C, E_C, w_C)$ where $|V_C| = c$

(number of subgraphs in $G_{cluster}$) and the weight $w_{C_{ij}}$ corresponds to the average flow between the subgraphs $G^i_{cluster}$ and $G^j_{cluster}$, calculated as:

$$w_{C_{ij}} = \frac{\sum\limits_{p \in G^i_{cluster}} \sum\limits_{q \in G^j_{cluster}} d_{pq}}{|V^i_{cluster}| * |V^j_{cluster}|} \tag{4}$$

From G_C, the new QAP instance is created as was describe in section 2. To tackle the QAP problem, we use a memetic algorithm described in [10]. The memetic algorithm will produce a solution for the QAP, which will correspond to the layout of the data. The method produces a layout in two stages: first, the samples (or genes) are located according to the relationships between all the components of each cluster and finally, the position of each cluster is determined by the relationships between different clusters, considering the average distances among all components of each pair of clusters.

3   Experiments

3.1   Data Description

The dataset used in this work originates from the yeast (Saccharomyces cerevisiae) gene expression microarray used by Eisen et al. [3]. The original yeast whole-genome data contains the cDNA sequences associated to 9,800 ORFs (open reading frames) and 79 samples. The samples are divided into eight reference groups associated to experiments on the budding yeast S. cerevisiae, namely *alpha factor* - 18 samples (alpha factor arrest and release); *CDC15* - 15 samples (cdc15 arrest and release); *cold shock* - 4 samples; *diauxic shift* - 7 samples; *DTT shock* - 4 samples; *elutriation* - 14 samples; *heat shock* - 6 samples; and *sporulation* - 11 samples. We use the freely available data (from http://www.pnas.org/cgi/content/full/95/25/14863/DC1) corresponding to the Fig. 2 of the work presented by Eisen *et al.* [3]. It contains a subset of 2,467 genes with functional annotations.

3.2   Computational Experiments

The experiments were performed on samples and genes separately. Firstly, we calculate the distance matrix D, using as distance metric the Pearson Correlation as shown in 5, where ρ_{ij} represents the Pearson correlation either between a pair of genes or a pair of samples. This produces a distance matrix with values between 0 (if the elements are perfectly correlated) and 2 (if the elements are perfectly anti-correlated).

$$d_{ij} = 1 - \rho_{ij}. \tag{5}$$

To find a solution to the QAP instances, we used the Memetic Algorithm described in [10]. The methods were coded in Java, and run in a Pentium IV (2,3 GHz) workstation. The program generates a GML file with the location of each sample in a grid. To visualize the file, we use the yEd Graph Editor, freely available at *http://www.yworks.com/*.

In the case of samples, since they are from eight different experiments on yeast, it is supposed that samples belonging to the same experiment, should have similar expression patterns, consequently they should be put closer in the final layout. We performed two tests: first, we applied the visualization method on the complete dataset without the use of the graph-based clustering method, to illustrate its ability to produce a layout with the samples with similar expression patterns together, and second, we combined the output of the graph-based clustering algorithm presented in 2.1 with the visualization method.

In the case of genes, we expect that genes with similar functional annotation have a similar expression profile, so, they should be assigned closer in the visualization. We present the result for the 2,467 genes in the dataset using the complete method proposed (visualization + clustering) to show how it performs in a larger dataset. We show how the method is capable of place closer genes with similar functional annotation even if they are in different clusters.

Results: In the first experiment, we applied the visualization method on the 79 samples, without the use of any extra parameter. In Figure 2 we show the layout produced by the memetic algorithm. The samples have different shapes to indicate the experiment to which they belong. It is clear in the figure that

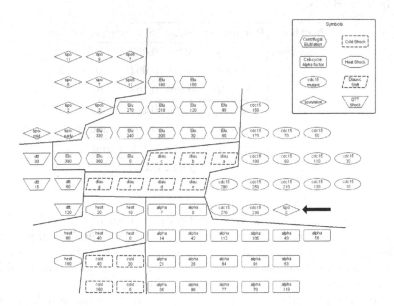

Fig. 2. QAP-based visualization of the 79 samples of yeast, considering the expression similarity between the samples

the organization of the samples shows elements with a similar gene expression pattern (with higher correlation) located closer, giving to the user an informative display of the data.

In the second experiment with samples, we integrate the visualization method with the graph-based clustering algorithm. It is important to clarify that we could integrate the result of any clustering algorithm with the visualization method. We first compared the solution of the graph-based clustering algorithm against k-means [4] and SOM [5], using the *homogeneity* (**H**) and *separation* (**S**) indexes [11], which give us an idea of how similar and dissimilar are the elements into a cluster and among the clusters respectively. To run the two algorithms, we used the software Expander [11], available online at *http://www.cs.tau.ac.il/~rshamir/expander/* and also to evaluate the indexes **H** and **S** for each solution. Since the algorithm implementations are non deterministic, we ran each of them 10 times for different parameters. For the k-means algorithm, we tested for $k = 8, 9$ and 10, and in the case of SOM, we used grids of sizes 2x4, 3x3, and 2x5. In Table 1 we show the average **H** and **S** indexes for each of the algorithms with different input parameters.

Table 1. Results for k-means, SOM and the MST-kNN approach in terms of *homogeneity* and *separation*. Several parameters configurations are shown for k-means and SOM, along with the average results for 10 runs. The MST-kNN approach had the best results in both criteria.

Method	Parameter	H_{avg}	S_{avg}	#clusters
k-means	$k = 8$	0.487	0.099	8
	$k = 9$	0.511	0.094	9
	$k = 10$	0.526	0.108	10
SOM	2 x 4	0.589	0.111	7 and 8
	3 x 3	0.579	0.107	8 and 9
	2 x 5	0.619	0.116	9 and 10
MST-kNN	—	**0.642**	**0.057**	9

We also ran the fuzzy k-means algorithm presented in [12] and the MOCK algorithm presented by Handl and Knowles in [13], for the 79 samples. In the first case, the program provided by the author gave an "out of memory" error after more than one hour processing and the latter could not recommend a specific solution from a set of solutions found by the algorithm.

We chose the clustering algorithm presented in 2.1 because it has a better performance in comparison with the two other classical algorithms, and it also provides a graph structure that allows us to show the capacity of integrate this information in the visualization. Once the clusters are obtained, we proceed to the visualization of the clusters, applying the layout process described in section 2. The resulting graph layout using the clustering generated by the MST-kNN is shown in Figure 3. We emphasize that the layout algorithm arranges first the objects within each cluster according to the similarity between them, and then the position of the clusters relative to each other taking into account the similarity between their components.

From a computational cost point of view, the method took less than 30 sec to produce the layout, including the clustering algorithm (which for this dataset takes less than one second).

The final experiment corresponds to the visualization of the 2,467 genes in the dataset, as a proof of the scalability of the method. The result is showed in Figure 4. In this case, the clustering algorithm took 27 seconds and the layout less than 10 minutes, running 30 generations of the MA for each QAP instances created (52 for clusters, plus one for the clusters layout).

4   Discussion

Figures 2 and 3 show that the layout correlates well with the type of experiment to which the individual samples belong to. However, there is one particular sample (*spo-0*), that seems to be close to samples of other type. In both figures sample *spo-0* is closer to *alpha* experiments. At first sight this may indicate an error, as in general the correlation with sample type is strong, however the expression pattern of sample *spo-0* is closer to *alpha* than *sporulation* experiments with an average gene expression correlation of 0.112 to *alpha* samples and -0.010 to *sporulation* samples. In Fig. 5 we show the heatmap of the gene expression of *alpha* and *sporulation* experiments. The hierarchical tree on the left also has *spo-0* in the same subtree that contains all *alpha* samples as leaves. The most similar sporulation sample to *spo-0* corresponds to *spo5_2* and it is the 23rd nearest neighbor of *spo-0* , indicating that indeed *spo-0* should not be assigned close to the other *sporulation* samples. On the other hand, one of the issues of clustering algorithms is that they show different degrees of sensitivity, separating in more classes than necessary (showing probable subclasses) or keeping elements from different classes together. Both situations are covered by this visualization method. In the first case, if a cluster is split the layout method manages to put them together as can be seen in Fig. 3. The clustering algorithm split the experiment cdc-15, but the layout method put them together in the result. In the second case, the layout method manages to put together samples from different experiments closer (see middle-bottom cluster in Fig. 3).

For genes, we use the functional annotation of genes to show the main characteristics of the method: when a group is split the visualization manages to put it together in the grid (Protein Degradation and Protein Synthesis) and, into a cluster, it manages to organize similar genes together (Glycolysis). This result confirms the ability of the visualization method to layout together clusters with similar components, and arrange elements within a cluster according to their similarity. The groups mentioned are the main groups found by Eisen et al. [3], but with our method, we are able to find larger groups of genes with the same functional annotation than in the original work.

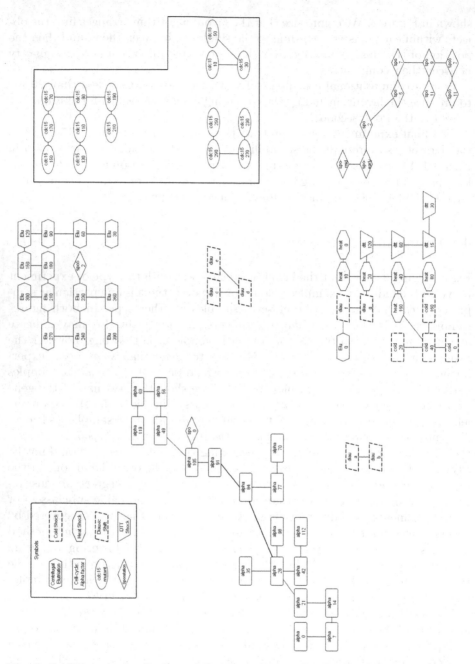

Fig. 3. QAP-based visualization of the clusters identified by the MST-kNN method. One can see that the result of the clustering method is mainly according with the experiments from which the samples come. Important, is that even when the MST-kNN approach assigns samples from the same reference group to two or more clusters, the QAP-based layout puts them in adjacent positions, e.g. cdc15 (enclosed clusters), evidencing their similarity.

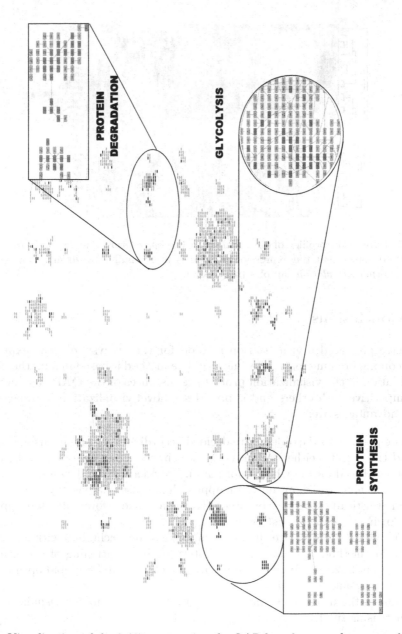

Fig. 4. Visualization of the 2,467 genes using the QAP-based approach presented. The same groups highlighted by Eisen et al in [3] are found here, but our method also put together clusters composed with a majority of genes with the same functional annotation.

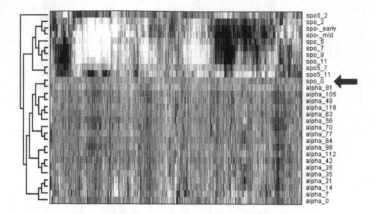

Fig. 5. Expression profiles of the samples in the *sporulation* and *alpha factor* experiments. It shows that the expression of *spo0* is more similar to *alpha factor* samples than to other *sporulation* samples (see arrow).

5   Conclusions

We have presented a visualization method for the analysis of gene expression data coming from microarray technology. The method is based in a mathematical formulation of the visualization problem as instances of the QAP. The proposal is competitive with others and it provides a novel visualization approach. The main advantages are:

1. The position of the objects are decided according to the relationships among *all* the objects (either samples or genes) in the dataset, giving to the user a layout based on the whole information provided by the dataset.
2. The combinatorial optimization approach proposed here is novel, and can be synergistically "seeded" by other algorithms, like a force-directed approach allowing it to have a better starting point.
3. The ability to integrate the result of a clustering technique with the visualization method provides an easily interpretable positioning of the elements, with optimization for the layout within a cluster, and a global optimization of the layouts of the clusters.
4. Finally, the method can integrate a graph structure in the visualization by increasing the flows of some relationships.

Most of the software packages that are available for gene expression analysis lack of visualization systems that provide a reliable and information-based layout that can assist the interpretation of the datasets. Most of them consider only some graph-layout algorithms and generally they only use heatmaps. The method proposed arises as a good alternative to be integrated in the current software packages for microarray data analysis, because it has a good performance, it is scalable and its characteristics make it unique.

Acknowledgment. This work has been supported in part by an ARC Discovery Project (DP0559755, Evolutionary algorithms for problems in functional genomics data analysis) and the ARC Centre in Bioinformatics. Mario Inostroza-Ponta also acknowledges the support of the Department of Informatics Engineering, University of Santiago of Chile and the Newcastle Bioinformatics Inititative.

References

1. Shannon, P., Markiel, A., Ozier, O., Baliga, N., Wang, J., Ramage, D., Amin, N., Schwikowski, B., Ideker, T.: Cytoscape: A software environment for integrated models of biomolecular interaction networks. Genome Research 13, 2498–2504 (2003)
2. Kohler, J., Baumbach, J., Taubert, J., Specht, M., Skusa, A., Ruegg, A., Rawlings, C., Verrier, P., Philippi, S.: Graph-based analysis and visualization of experimental results with ondex. Bioinformatics 22(11), 1383–1390 (2006)
3. Eisen, M., Spellman, P., Brown, P., Botstein, D.: Cluster analysis and display of genome-wide expression patterns. Proc. Natl. Acad. Sci. USA 95, 14863–14868 (1998)
4. Tavazoie, S., Hughes, J., Campbell, M., Cho, R., Church, G.: Systematic determination of genetic network architecture. Nat. Genet. (22), 281–285 (1999)
5. Tamayo, P., Slonim, D., Mesirov, J., Zhu, Q., Kitareewan, S., Dmitrovsky, E., Lander, E., Golub, T.: Interpreting patterns of gene expression with self-organizing maps: methods and application to hematopoietic differentiation. Proc. Natl. Acad. Sci. (96), 2907–2912 (1999)
6. Burkard, R., Çela, E., Pardalos, P., Pitsoulis, L.: The quadratic assignment problem. In: Pardalos, P., Du, D. (eds.) Handbook of Combinatorial Optimization, pp. 241–338. Kluwer Academic Publishers, Dordrecht (1998)
7. Taillard, E.: Robust taboo search for the quadratic assignment problem. Parallel Computing 17(4-5), 443–455 (1991)
8. Oliveira, C., Pardalos, P., Resende, M.: Grasp with path-relinking for the quadratic assignment problem. In: Ribeiro, C.C., Martins, S.L. (eds.) WEA 2004. LNCS, vol. 3059, pp. 356–368. Springer, Heidelberg (2004)
9. González-Barrios, J., Quiroz, A.: A clustering procedure based on the comparison between the k nearest neighbors graph and the minimal spanning tree. Statistics & Probability Letters 62(1), 23–34 (2003)
10. Inostroza-Ponta, M., Berretta, R., Mendes, A., Moscato, P.: An automatic graph layout procedure to visualize correlated data. In: Bramer, M. (ed.) Artificial Intelligence in Theory and Practice: Ifip 19th World Computer Congress. IFIP International Federation for Information Processing, vol. 217, pp. 179–188. Springer, Heidelberg (2006)
11. Shamir, R., Maron-Katz, A., Tanay, A., Linhart, C., Steinfeld, I., Sharan, R., Shiloh, Y., Elkon, R.: Expander-an integrative program suite for microarray data analysis. BMC Bioinformatics 6(232) (2005)
12. Gasch, A., Eisen, M.: Exploring the conditional coregulation of yeast gene expression through fuzzy k-means clustering. Genome Biology 3(11) (2002)
13. Handl, J., Knowles, J.: Multiobjective clustering with automatic determination of the number of clusters. Technical Report TR-COMPSYSBIO-2004-02, UMIST, Manchester, UK (2004)

Complement-Based Self-Replicated, Self-Assembled Systems (CBSRSAS)

Mostafa M.H. Ellabaan

Cairo University
Doki, Giz, Egypt
Tel.: (+2) (018) 256-3816
mostafa.mhashim@gmail.com

Abstract. Self-assembly and self-replication are the main common thematic features of living organism. The life of most living organisms bases on biomolecular systems such as DNA and RNA. Self-replication and assembly are emergent properties of such systems. These biomolecular systems have common thematic features that enable them from self-replicating and self-assembling themselves. In this paper, I try to generalize these common thematic features to generate a generalized model of the complement-based self-replicated, self-assembled systems. In this model, I explained the main requirements for achieving these systems in terms of basic system components and how these components interact with each other through self-assembling rule set and complement-based replication rule set generating systems with potential to be self-replicated machines. After generating this model, I have applied this model to DNA and simulated its basics.

Keywords: Self-assembly, Self-replication, DNA, Wang Tiles, self-organization, artificial life, simulation.

1 Introduction

Building artificial systems that exhibit the features of living organisms is one of most active research areas in today's world [5]. Scientists around the world try with a high-level of cooperation to generate machines with features of biological systems.

Self-assembly, self-replication, self-healing and adaptation to the surrounding environment are the main common distinguished features of living cells. First, self-assembly is a process by which simple objects autonomously assemble into complex systems [6]. It is an advantageous manufacturing approach where with an appropriate set of components, the target system will be generated [6]. Scientists, especially Astrobiologists such as Deamer, consider it as the main reason behind the existence of living systems. Deamer described the development of life as a sequence of self assembly processes. This sequence of self-assembly processes transformed a soap of molecules into a cell-like membranous structure capable of capturing energy and nutrients from surrounding environments that began to grow and reproduce [7].

Second, self-replication which is considered as one of the main features of living systems came under intensive research investigations since 1940 when J. von Neumann

M. Randall, H.A. Abbass, and J. Wiles (Eds.): ACAL 2007, LNAI 4828, pp. 168–178, 2007.

introduced his theory of self-reproducing automata [2, 3 and 4]. Since then, it has become one of the most active challenges of research in the field of artificial life. A lot of research has been done to investigate the process of self-replication such as the work done by Hod Lipson at Cornell university [8, 9 and 10], and Hiroki Sayama [1].

The most powerful feature of the living cells or organisms is their ability to self-heal themselves as well as their ability to adapt to new conditions. These features will be handled in future investigation.

In this paper, I gave a simple overview about different self-replication approaches in categories of the main important themes of self-replication such as universal constructor suggest by von Neumann, self-replication loops suggested by Langton and Artificial Chemistry approaches introduced by Hutton in section two, and followed by detailed description of CBSRSAS in section three. In section four, I described DNA as an example of CBSRSAS, followed by a simple simulation of CBSRSAS basics in section five, and finally in the last section by a detailed discussion of CBSRSAS and its characteristics with a summary of suggested future investigation.

2 Approaches to Self-Replication

It is around six decades since Von Neumann; the father of self-replication as consider-ed by most of researchers, suggested his first model about self-replication. Since then, Self-replication attracted a lot of scientists and researchers who have done a lot of work to investigate this area which classified into Universal constructor, Self-replication loops, and artificial chemistry.

2.1 Universal Constructor

Von Neumann self-replication approaches based on designing or building a universal constructor that can build any thing even if this thing is the universal constructor in itself [4]. Von Neumann self-replication suggested five approaches for building a universal constructor. The most the successful one was the cellular automata model in which the universal constructor was composed of a group of thousand cells that began in a specific configuration of initial states. In addition, there is a line of cells or a tape for any given finite set of configuration or cells. The tape can instruct the universal constructor to build the given configuration. Therefore, the tape of universal constructor configuration can cause the universal constructor to build a copy of itself, thereby self-replicating.

2.2 Self-Replicating Loops

Self-replicating loops is a simple cellular automat model suggested by Langton. This model is more efficient and simpler than the Von Neumann cellular automata model of self-replication. It has the advantages of generating replicates of the systems without the trail of generating other systems. The cellular automata model of Langton approach consists of eight states other than twenty nine states as in Von Neumann's approach. The constructor consisted of a group about hundred cells in a specific configuration. Langton's constructor cells are arranged in a loop. The states of the cells in the loop go through a cycle, periodically creating a copy of the original system loop starting from the initial loop increasing numbers of copies spread across the grid. This approach has been developed by a lot of studies such as [15], [16], [17] and [18]. Self-replicating loops approaches have not only proven superiority offer the universal constructor approaches, but they have also proven success in generating an interesting behavior such as evolution self-repair, in performing computation, and at last but not the least, in generating a structures.

2.3 Artificial Chemistry

Self-replication in artificial chemistry simulation using a template based approach [19] was introduced by Hutton. Simulation of self-replication in artificial chemistry begins with a seed chain in a soup of molecules which generate another parallel chain of molecules near by. When the parallel chain is completed, it separates from the seed chain and the process repeats. Hutton tried to use a cellular automata model but the discrete space constrained the mobility of the simulated molecules. So he suggested another model designed based on the continuous space. In this model, molecules move in a continuous two-dimensional space, following linear trajectories until an obstacle such as the container wall or another molecules is encountered. When molecules make contact with each other, they undergo a chemical reaction that bonds them together, according to the rules of artificial chemistry.

3 Complement-Based Self-Replicated, Self-Assembled System (CBSRSAS)

In this section, I will explain the main components and rules suggested for generating complement-based self-replicated and self-assembled systems. I will explain what the basic system components and what the main characteristics of these main components are, and how these characteristics may lead to autonomous replication and assembling of the systems.

3.1 System Component

System component is an item composed of two types of sections (see fig 1.): the first type of sections is the complement interaction section. It is the place where complements can interact. This section type determines relationships between different system components and plays an important role in generating replicates; second type of sections is the self-assembly section type.

In this type, objects take specific colors which determine the interaction between system components in the same system, and help generating specific segments of a system or the system itself. System component may have one or more replication section and one or more self-assembly section.

3.2 System Component Set

Basic system components represented as (f) are all the components of which system will generate. The system can be described as β where β is a spatial order of $I \in (f)$.

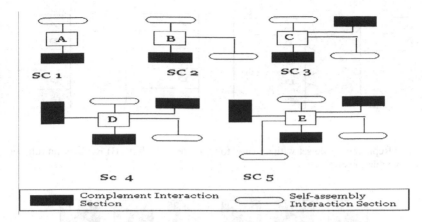

Fig. 1. Shows examples of system component where SC1 is a system component consist a complament interaction and a self-assembly interaction system, SC2 is composed of two self-assembly sections and one complement interaction section, and SC5 is composed of of three self-assembly interaction sections and three complement interaction sections.

3.3 Complement-Based Replication Rule Set

Each item in (f) may have one or more complements. Each pair of items (item and its complement) is called a complement-based replication rule. Complement-based replication rule set which is donated by $\Re$ includes all system complement-based replication rules. For example, the complement-based rule set of in fig 2 section B where items or system components have a one to one relationship is $\Re b$ = { {A,B} , {C,C} , {D,F}}, while complement-based replication rule set of fig 2 section A is $\Re a$ = {{A, A}, {A, B}, {A, C}, {B, D}, {C, C}, {D, F}}. This rule set may generate the same system but there is a probability of mutation. This probability depends on the average number of the complements pair of a single system component. The more the object has complements, the less probable the target systems are generated and the higher probability for mutations to occur at the side of this object. Consequently, complement-replication rule set $\Re a$ is not a good complement-based replication rule set, while complement replication rule set $\Re b$ is a one to one relationship. Therefore no mutations are expected in case of replication because each system component has its complement.

3.4 Assembling Rule Set

The Assembling rule set determines how system components interact with each other in case of collision on self-assembling sections of system components. This idea is driven from Wang tile or Wang dominoes (see fig 3) proposed by Hao Wang in 1961 [14]. His main purpose was to use a given finite set of geometrical tiles to determine whether they could be arranged using each tile as many times as necessary to cover the entire plan without gap [13].

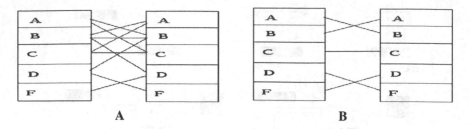

Fig. 2. (A) Replication rule set with a many to many relationship; (B) Replication rule set with a one to one relationship

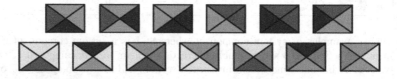

Fig. 3. Wang Tiles or Wang dominoes [13]

In the assembling rule set, the interactions between colors are stored in a symmetric matrix called a stickiness matrix (see fig 4).

To illustrate, assume we have two system components. Each one has one self-assembling section with color A for the first system component and color B for the second. If the interaction between color A and B in the case of collision satisfies a specific rule such as equation 1, they either stay together if they satisfy stability condition defined in the rule or continue moving according their previous states.

3.5 Kinematics Model

Each system component should have capabilities to move freely and autonomously. This movement should not burden the capabilities of objects to interact with each other from the side of either self-assembly or self-replication section.

The kinematics model is not only dedicated to the kinematics model of the system components, but it may also include a kinematics model of the complement interaction sections and self-assembly interaction sections.

Colors \ Colors	Color 1	Color 2	....	Color N
Color 1	C11	C12	....	C1N
Color 2	C12	C22	....	C2N
....	....	....	....	....
Color N	C1N	C2N	....	CNN

Fig. 4. Interaction Matrix or stickiness matrix [13]

3.6 Replication-Initiation Rule Set

Replication-initiation rule sets are the main signals that may be required to initiate the replication process. By these signals, the complementary relations of the system are broken leading to breaking the system into its complementary parts. The number of complementary parts depends on the number of the complement interaction sections in the system components. If the system component has N complement interaction sections, then the process of breaking system component will lead to N complementary parts. Each complementary part can generate the system again. Although the advantages of having more than one complement interaction sections, generating many replicates, it may require complex procedures for setting replication-initiation rule set.

3.7 Replication Machinery

To replicate systems of this type, it is required to break the system into two complementary parts and bring the system components for each of the two parts. There are two machinery types for handling this type of replication. The first type of machinery depends totally on system components and its kinematics model which I called the autonomous replication machinery. The second machinery depends on the interaction between the system and other systems or agents. This machinery is called the agent-based replication machinery. The agent-based replication machinery is the most famous in biological or natural systems.

4 DNA as an Example of Complement-Based Self-Replicated, Self-Assembled Systems

DNA is behind the most of living organisms. DNA is not only a storage area of biological information, but it is also behind the most biological processes in living organisms; its characteristics enable living organism from self-replicating and self-assembling themselves. [12]

4.1 System Components of DNA

System components of DNA are cytosine, thymine, guanine and adenine which are called by chemists as nucleotides. These components are mainly composed of phosphate group, five-carbon sugar group and nitrogen base as shown in table 1.

4.2 DNA Self-Assembling Rule Set

Each nucleotide or system component has two sides where self-assembly occurs. The first one is the OH group at the third carbon atom in sugar group which I call τ, and $HOPO3$ or phosphate group which I call ε.

In the interaction matrix above (fig 5), the interaction between τ and ε (i.e. F) represent the minimum energy required to break the bond between τ and ε. So DNA stability or assembling requires the free energy around DNA to be less than F and

Table 1. Basic system components of DNA (nucleotides) [12]

Cytosine 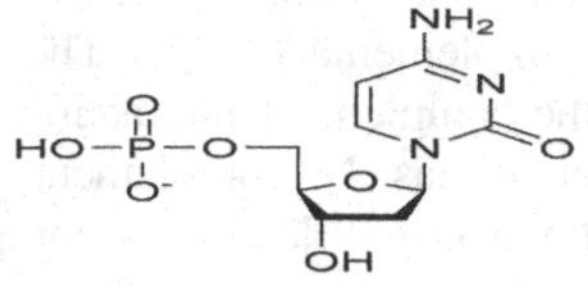	Thymine 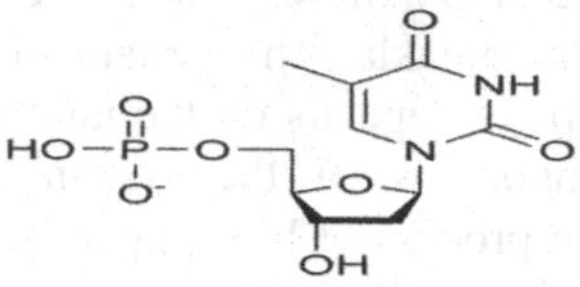
Guanine 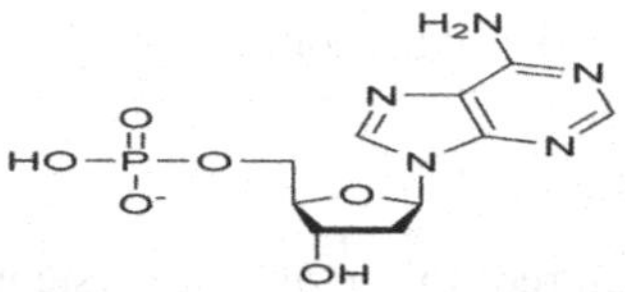	Adenine 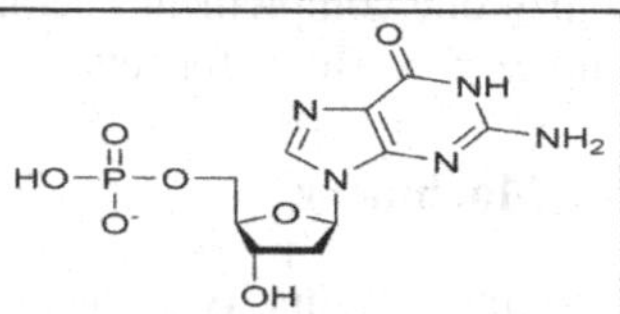

Item \ Item	τ	ε
τ	K	F
ε	F	L

Fig. 5. Interaction Matrix

greater for disassembling. K represents the interaction between two groups of τ and L represents the interaction between two groups of ε. Naturally, No interaction usually occurring between two groups of the same type in case of DNA assembling then L and K, So I suggest that L and K values to be zero.

4.3 DNA Complement-Based Replication Rule Set

Nucleotides or system components of DNA have two complement-based replication rules which are considered as the main reason behind self-replication of most living organisms. These rules can be summarized as $\mathfrak{R} = \{\{$guanine, cytosine$\}, \{$adenine, thymine$\}\}$ as shown in table 2. The rules emerge from the number of hydrogen bonds between different pairs. For the first pair {Guanine and cytosine}, the number of hydrogen bonds is three, while the number of hydrogen bonds for the second pair is two [12].

4.4 Replication-Initiation Rule Set

Replication-initiation rule set differs according to the biological organism in which DNA exists. For example, in E. coli DNA replication is mediated by DnaA, while in archaea DNA replication is mediated by Cdc6/Orc1 [11].

Table 2. DNA Complement-Based Replication Rule Set [12]

Guanine H Cytosine	Adenine Thymine
First complement-based Replication Rule in DNA is {Guanine, Cytosine}	Second Complement-based Replication Rule in DNA is {Adenine, Thymine}

4.5 Replication Machinery

Most of the DNA in biological cells depends on a protein or a set of proteins to be replicated (i.e. Agent-based replication machinery). Check [11] for more information.

5 Simulation of CBSRSAS

I have implemented a software tool that enables its users from mimicking the process of creating self-assembled and self-replicated systems. It enables them form creating a set of system components. Each system component has two side where self-assembling occurs. These sides are on the right and left sides of the system components. Each side of self-assembling section has a color ξ which is selected randomly from a set of colors called assembling controlling set ψ. It also enables them from generating the assembling controlling set (i.e. colors) and the interactions between each pair of ξ. These interactions are stored in two dimensions array called the assembling rule set. System component also has another section that enables system component to be integrated with its complement. Finally, this software tool also enables users from determining complement-based replication rule set (see fig 6).

5.1 Kinematics of System Components

After generating system components, assembling rule set and self-replication rule set, and determining the number of objects to be modeled in the simulator, the objects will move randomly in Brownian motion up, down, left, and right. If these objects collide together from the side of self-assembly section, they either stay together forming a new system or continue moving according to their previous states as shown in equation (1) [13].

$$State\,(O_I, O_J) = \begin{cases} 1 & St(C(O_I, F), C(O_J, K)) > T \\ 0 & Otherwise \end{cases} \quad Equation\ [1]$$

Equation 1 states the result of interaction between Objects (O_I, O_J), if interaction between colors on their self-assembly section are greater than temperature–environmental variable affecting self-assembly of objects- objects will stay together forming a new system. Otherwise, they will continue moving according to their previous state.

5.2 Replication Initiation and Machinery

In this simulation, the replication is initiated by replication breaking items. They are hard coded items that move randomly with the same container in which system components move. Once it is attached to any system, the system breaks into its complementary parts. After this division occurs every complementary part is considered as a blue print for the former system. The autonomous movement of system component takes the responsibility of generating replicates by attached to either of complementary parts. This section require a lot of future investigations to handle possibility of creating specific replication breaking items for specific systems and to add agents to support process of replication.

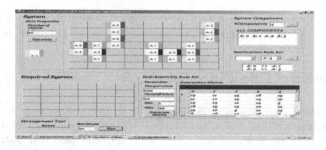

Fig. 6. Snapshot of the simulator

6 Discussion and Future Work

In this paper, I suggested a model of self-assembled and self-replicated systems which may be classified as an artificial chemistry self-replication approach, and applied it to one of the most famous bimolecular systems, DNA. This model may be considered as a blue print for systems that can exhibit the living organisms' features of self-assembling and self-replication.

I defined the main requirements of the CBSRSAS such as the main system compon ents and their parts which are classified as a self-assembly interaction section and a complement interaction section. To build a complex system, it may be possible to define more than one interaction section for either self-assembly section or complementary section. CBSRSAS with many complement interaction sections has capability to produce a more replicas but require more complex procedures to replicate such as checking whether all complementary parts are complete and handling the breaking of complementary relations or interactions. While CBSRSAS with many self-assembly sections has ability to build a chain with complex structure.

Relative to the simulation suggested in fifth section, it was very simple. It was to just visualize a very simple model of CBSRSAS. Unfortunately, I failed to get a good result of this simulator relative to the limited space. But I suggested improvements to this simulation to be handled as future investigations.

Relative to the future research investigation, I think that CBSRSAS have a lot of potentials for future investigations. I classified it according to area of investigation into: Robotics, Simulation, and Chemistry.

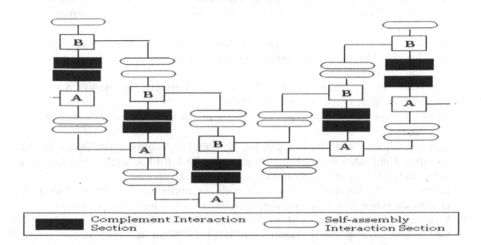

Fig. 7. Represent an example for CBSRSAS

Robotics
Applying CBSRSAS to robotics may be an interesting future investigation. CBSRSAS can help building self-assembled and self-replicated Robotics. If robotics are built with a CBSRSAS characteristics such as one presented in the fig 7, defining self-assembly interaction sections and complement interaction sections and defining self-assembling rule set as well as self-replication rule set, It may generate a powerful robotics with interesting behavior and capabilities of generating themselves either through self-assembly or self-replication.

Simulation
One of the main future investigations regarding this paper is to build a complex graphical representation of CBSRSAS with more complex dynamics assigned to self-assembly interaction sections and complement interaction sections, defining agents to bring system components to complementary parts and handling the basics of physics such as repulsive field, momentum and viscosity.

Chemistry
Investigating the possibility of generating CBSRSAS at atomic or molecular details in chemistry may lead to discovering other systems with higher assembling and replication rate than the existed ones such as DNA and RNA and, consequently, creating a new type of living systems. Currently, I investigate in the possibility of generating CBSRSAS system dedicated to generate specific protein and taking the advantages of DNA as powerful encoding system and RNA as protein-manufacturing factory.

References

1. Sayama, H.: Von Neumann's Machine in the Shell: Enhancing the Robustness of Self-Replication Processes. In: Artificial Life VIII: Proceedings of the Eighth International Conference on Artificial Life (in press)
2. Bedau, M.A., McCaskill, J.S., Packard, N.H., Rasmussen, S., Adami, C., Green, D.G., Ikegami, T., Kaneko, K., Ray, T.S.: Open Problems in Artificial Life. Artificial Life 6, 363–376 (2000)
3. von Neumann, J.: The General and Logical Theory of Automata. In: Cerebral Mechanisms in Behavior—The Hixon Symposium, pp. 1–41. John Wiley, New York, NY (1951) (originally presented in 1948)
4. von Neumann, J.: Theory of Self-Reproducing Automata. In: Burks, A.W. (eds.) University of Illinois Press, Urbana, IL (1966)
5. Brooks, R.: The relationship between matter and life by Rodney Brooks. Nature 409, 409–411 (2001)
6. Terrazas, G., Krasnogor, N., Georghe, M., Kendall, G.: Automated Tile Design for Self-Assembly Conformations. In: Proceedings of the 2005 IEEE Congress on Evolutionary Computation, Edinburgh, Scotland (September 2-5, 2005)
7. Deamer, D.: The origin of life, Evolutionary theory conference (November 14-19, 1999), http://www.esalenctr.org/display/confpage.cfm?confid=5&pageid=50&pgtype=1
8. Mytilinaios, E., Desnoyer, M., Marcus, D., Lipson, H.: Designed and Evolved Blueprints for Physical Self-Replicating Machines. In: ALIFE IX. Ninth Int. Conference on Artificial Life, pp. 15–20 (2004)
9. Adams, B., Lipson, H.: A universal framework for self-replication. In: Banzhaf, W., Ziegler, J., Christaller, T., Dittrich, P., Kim, J.T. (eds.) ECAL 2003. LNCS (LNAI), vol. 2801, Springer, Heidelberg (2003), http://ccsl. mae.cornell. edu/papers/ ECAL03_Adams.pdf
10. Studer, G., Lipson, H.: Spontaneous emergence of self-replicating, competing cube species in physical cube automata. In: GECCO Late Breaking Paper, Proceedings, Washington DC, USA, June 25-29, 2005, ACM, New York (2005), http://ccsl. mae. cornell. edu/papers/GECCO05_Studer.pdf
11. Wikipedia: DNA Replication (accessed 20/4/2007), URL: http://en.wikipedia.org/wiki/DNA_replication
12. Wikipedia: DNA (accessed 1/5/2007), URL: http://en.wikipedia.org/wiki/DNA
13. Ellabaan, M.: 3D Simulation Of Self-Assembly. M.Sc, Dissertation, the University of Nottingham, Nottingham, UK (2006)
14. Wang, H.: Bell System Tech. Journal 40, 1–42 (1961)
15. Sayama, H.: A New Structurally Dissolvable Self-Reproducing Loop. Artificial Life 5, 343–365 (1999)
16. Reggia, J.A., Lohn, J.D., Chou, H.-H.: Self-replicating structures: Evolution, emergence and computation. Artificial Life 4, 283–302 (1998)
17. Sayama, H.: A new structurally dissolvable self-reproducing loop evolving in a simple cellular automata space. Artificial Life 5, 343–365 (1999)
18. Smith, A., Turney, P., Ewaschuk, R.: Self-replicating machines in continuous space with virtual physics. Artificial Life 9, 21–40 (2003)
19. Hutton, T.J.: Evolvable self-replicating molecules in an artificial chemistry. Artificial Life 8, 341–356 (2002)

Self-maintained Movements of Droplets with Convection Flow

Hiroki Matsuno[1], Martin M. Hanczyc[2], and Takashi Ikegami[1]

[1] Department of General Systems Sciences, The Graduate School of Arts and Sciences, University of Tokyo, 3-8-1 Komaba, Tokyo 153-8902, Japan
hiroki@sacral.c.u-tokyo.ac.jp
[2] ProtoLife Srl, Via della Liberta 12, Marghera, Venezia 30175, Italy

Abstract. Running droplets have been studied recent years as dissipative macroscopic structures with locomotive capability, a characteristic of which is shared with biological systems. We constructed a numerical model of a droplet that integrates fluid dynamics and chemical reaction. Our results show that the chemical gradient generates droplet's motion, accompanied with convection flow. This convection flow contributes sustaining the chemical gradient, making a positive feedback loop. The simulated droplet self-maintains a chemical gradient, a prerequisite for locomotion, which constitutes a prototype of autonomous movement.

1 Introduction

Since the pioneering work of von Neumann [1] and succeeding studies by stimulated researchers, self-reproduction has been intensively studied in the field of artificial life [2,3]. While self-reproduction is, beyond question, essential to life for its relevance to evolution, motility of biological systems is no less essential than self-reproduction.

Biological systems are non-equilibrium macroscopic structures, many of which show locomotive capability. As a dissipative structure, the natural cell receives resources from its environment and coverts them into waste through metabolism. The accumulated waste unless removed will saturate the local environment effectively slowing or stopping the metabolism that created it. Moving through the environment to obtain new resource allows biological systems to avoid the equilibrium. A self-movement becomes particularly important when a system acquires sensors and the adequate coupling between sensors and motors. A self-movement with sensors will differentiate context of the environment, which is food and where is enemy, to increase the value of survivability. We thus think the locomotion or exploratory behaviors is a basis of further evolution to take off.

In the field of artificial life, movement has drawn much attention through the recent enthusiasm for embodied cognition and situatedness. On the other hand there has not been much work to date on movement in simple chemically embodied systems such as protocells. Suzuki and Ikegami [4] have constructed an abstract model on a running cell in which metabolism and motility is coupled,

M. Randall, H.A. Abbass, and J. Wiles (Eds.): ACAL 2007, LNAI 4828, pp. 179–188, 2007.

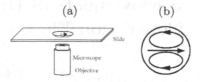

Fig. 1. (a) Experimental setup. The concave glass slide is filled with aqueous solution. An oil droplet is then added and observed by a microscope. (b) Schematic representation of the droplet with convection flow inside. This droplet moves to the right.

employing a stochastic automata. Numerical result shows that the cell demonstrates the motility driven by inhomogeneous configurations of chemicals in the cell, which can be distinguished from Brownian motion.

From laboratory experiments it has been shown that oil droplets placed in aqueous media sometimes exhibit spontaneous movements. Difference in interfacial tension at the droplet boundary is responsible for the movement, known as Marangoni effect. For example, an oil droplet [5] is driven by receiving surfactants from the environment to create Marangoni effect, constituting a non-equilibrium structure with locomotive capability, the property of which is shared with living things, thus providing a prototype of biological movements. Recently we conducted a series of experiments on spontaneous motions of oil droplet with its boundary covered with fatty acid [Fig. 1 (a)] [6]. The oil droplet contains fatty acid anhydride that reacts with water at the boundary to produce fatty acid. Since the interfacial tension depends largely on the fatty acid concentration, inhomogeneity in the distribution of the chemical causes a Marangoni effect resulting in droplet's locomotion. Along with the movement, convection flow is observed as shown in Fig. 1 (b), whose axis coincides with the direction of movement.

Convection flow has been also observed in some of other studies but it remains unclear whether the convection flow is merely a byproduct of droplet motion or it contributes to the motion in a positive manner. Since in our experiment convection flow carries fatty acid anhydride toward the boundary or reaction region, it is implied that convection flow plays a role in enhancing locomotive behavior. It is then interesting in a sense that chemical gradient generates motion together with convection flow, which, in turn, sustains chemical gradient by providing reaction substrate, forming a circular relation and self-maintained movement. This paper aims to evaluate this conjecture. We have constructed a numerical model of a droplet that integrates fluid dynamics and chemical reaction. In the subsequent sections we explain our model in detail, report the result of droplet's motions and conclude with summary and discussion.

2   Numerical Simulation

We introduce a model that qualitatively reproduces the behaviors observed in the experiment. As we saw, chemical reaction and convection flow are coupled and play a key role to sustain droplet motion. For decades, spatial structures formed through chemical reaction have been typically studied as reaction diffusion. To model running droplet we have to introduce the flow of the reaction field itself. Therefore, the model must include both fluid dynamics and a chemical reaction.

On the similar line Kitahata et al. [7] conducted numerical simulations which incorporate both convection flow and reaction diffusion system. In their model chemical gradient produced by BZ reaction generated the differentiated interfacial tension on the liquid-liquid interface. Assuming that the interface is a fixed straight line, it was shown that convection flow is generated near the interface due to Marangoni effect and moves along with chemical traveling wave. Our model differs in that it deals with the movement of droplet itself, which have the curved geometry and can change its shape. The governing equations are as follows.

Governing Equations

$$\nabla \cdot \mathbf{u}(\mathbf{x}, t) = 0 \tag{1}$$

$$(\frac{\partial}{\partial t} + \mathbf{u}(\mathbf{x}, t) \cdot \nabla)\mathbf{u}(\mathbf{x}, t) = -\frac{1}{\rho}\nabla P(\mathbf{x}, t) + \nu\nabla^2\mathbf{u}(\mathbf{x}, t) + aF_s\delta \tag{2}$$

$$F_s(\mathbf{x}, t) = \gamma(v(\mathbf{x}, t))\kappa\mathbf{n} + \nabla\gamma(v(\mathbf{x}, t)) \tag{3}$$

$$\gamma(v(\mathbf{x}, t)) = v(\mathbf{x}, t) + b \tag{4}$$

$$(\frac{\partial}{\partial t} + \mathbf{u}(\mathbf{x}, t) \cdot \nabla)v(\mathbf{x}, t) = G(v(\mathbf{x}, t))\delta + D_v\nabla^2 v(\mathbf{x}, t) \tag{5}$$

$$G(v(\mathbf{x}, t)) = \begin{cases} c, & \text{if } 0 \le x < 0.8 \\ 0.1c, & \text{else if } 0.8 \le x < 1 \\ 0, & \text{otherwise} \end{cases} \tag{6}$$

Equations (1)-(4) describe the dynamics of an incompressible fluid and (5) and (6) are for chemical reaction. Eq.(1) represents the conservation of mass. Eq.(2) represents the conservation of momentum or Navier-Stokes equation. The third term in the right hand side is a force at the interface which is defined in eq.(3). The first term in RHS of (3) constitutes interfacial tension. $\kappa, \mathbf{n}, \delta$ are curvature, normal vector of interface, delta function, respectively. Delta function takes a positive value at an interface , otherwise 0. γ is an intensity of interfacial tension which depends on chemical concentration. Here we simply assume it is linear to chemical concentration(4). The second term in eq.(3) is a force generated due to the difference of γ, that is, Marangoni effect. Eq.(5) is the same as reaction diffusion equation except for the advection term which enable the flow of chemical field. For simplicity, we deal with a single chemical species whose amount is represented with v. Chemical reaction is defined as in (6).

Numerical Procedure

Space is two dimensional and described by a square mesh 64 by 64. We solve partial differential equations by finite difference method. Density function ϕ is defined to discriminate the droplet from the surrounding medium. It is set to 1 inside the droplet, otherwise 0 as an initial condition. ϕ is advected as (7). Interface between the two fluids can be detected by the difference in ϕ. δ function in eq.(2) and (5) can be substituted by $\nabla \phi$ (8). To obtain $\nabla \phi$, the stepwise-value ϕ needs to be smoothed out beforehand as numerical treatment. Here we apply eq.(10) to ϕ eight times, which gives about four mesh wide boundary. Two fluids can have different physical properties. When the densities of droplet fluid and surrounding media are ρ_1 and ρ_2 respectively, the density at the position $\mathbf{x}$ at time t is expressed as in (10). The kinematic viscocity is written similarly (11). Using these values, the governing equations can be solved as if we dealt with a single type of fluid. As a simulation scheme we used cubic-polynomial interpolation method. It interpolates the values between two neighboring mesh points using cubic polynomial to suppress numerical errors and is known for its simple algorithm [8]. Numerical procedure is summarized as follows.

$$\left(\frac{\partial}{\partial t} + \mathbf{u}(\mathbf{x}, t) \cdot \nabla\right)\phi(\mathbf{x}, t) = 0 \tag{7}$$

$$\delta = \nabla \phi \tag{8}$$

$$\phi_{i,j}^{new} = \frac{1}{2}\phi_{i,j} + \frac{1}{2} \cdot \frac{1}{1 + 4C_1 + 4C_2}\{\phi_{i,j} + \tag{9}$$
$$C_1(\phi_{i-1,j} + \phi_{i+1,j} + \phi_{i,j-1} + \phi_{i,j+1}) +$$
$$C_2(\phi_{i-1,j-1} + \phi_{i-1,j+1} + \phi_{i+1,j-1} + \phi_{i+1,j+1})\}$$
$$\left(C_1 = 1/(1 + 1/\sqrt{2}), C_2 = C_1/\sqrt{2}\right)$$

$$\rho(\mathbf{x}, t) = \phi(\mathbf{x}, t)\rho_1 + (1 - \phi(\mathbf{x}, t))\rho_2 \tag{10}$$

$$\nu(\mathbf{x}, t) = \phi(\mathbf{x}, t)\nu_1 + (1 - \phi(\mathbf{x}, t))\nu_2 \tag{11}$$

1. set an initial condition as will be described in (12).
2. calculate the interfacial force F_s, using eq. (3).
3.   – solve the Navier-Stokes equation (2) and continuity equation (1) to get u and P updated.
     – ϕ and v are also updated through advection and chemical reaction by eq. (7) and (5).
4. Iterate 2 and 3.

3   Results

In this section we report the simulation results of the above-mentioned model. We impose the chemical distribution with gradient as an initial condition. Chemical concentration is highest at left end of the droplet and is decreasing to the right end. Besides, Chemical is set dense near droplet's boundary (Fig. 2 (a)). It is written as follows.

$$v = \begin{cases} |\theta|/\pi \cdot (|x|/R)^3, & \text{if } |x| \leq R \\ 0, & \text{otherwise} \end{cases} \tag{12}$$

(a)   (b)   (c)

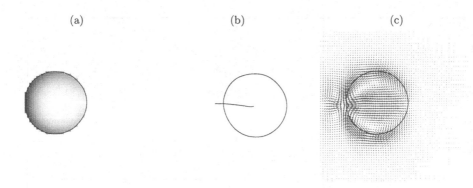

Fig. 2. (a)an initial condition: the chemical is asymmetrically distributed inside the droplet. The line showing the boundary of the droplet is drawn as the contour where $\phi = 0.5$. (b)The droplets marches from left to right. The line shows the trajectory of the center of mass. (c)The velocity field is shown. Each of short lines at mesh points represents a velocity vector at its location. Convection flow is observed whose axis is directed to the droplet's movement.

Here, x, R, θ, are a positional vector, diameter of the droplet and radian measured from the center of droplet as the original. The parameters used in our simulations are listed at the end of this paper [table 1].

As the simulation proceeds, we can observe that the droplet moves to the right (Fig. 2 (b)). This rightward motion is produced because interfacial tension is stronger at the left boundary than at right one due to chemical gradient. During its locomotion velocity field is formed as shown in Fig. 2 (c). Inside the droplet we can see convection generated whose axis coincides with the direction in which droplet moves. The chemical flows within the droplet and then reaches the right side and then flows along the upper or lower boundary, gradually approaching the left side. This structure of convection flow can be equated with that observed in the experiments and reproduces it well.

As mentioned before, it is implied that chemical gradient is being sustained possibly because chemical reaction at an interface is balanced with convection flow which transports fresh chemical for the reaction. To verify this conjecture, we run simulations with the following two scenarios and compared them one another.

no reaction condition. If there is no reaction at the interface, chemical gradient is expected to decline, flowing away from the boundary due to convection flow. This condition is realized simply by ignoring the reaction term in Eq. 5.

no convection condition. We conducted a simulation in which the initial chemical distribution is not altered by convection flow. The droplet's center of mass velocity $\int \phi \mathbf{u} dV / \int \phi dV$ (V: volume) is calculated from velocity field. Then we drift the droplet with this velocity uniformly. Chemical

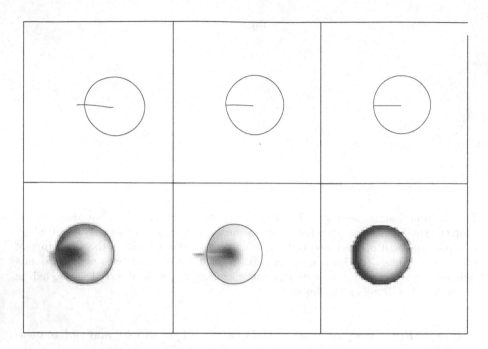

Fig. 3. (upper) Trajectory of droplets' movement. Each figure corresponds to a regular case (left), no chemical reaction case (middle) and no convection case (right). (lower) the snapshot of chemical gradient for each of three cases at the same time elapsed since the start.

distribution moves in space without being deformed as if it were rigid body because its velocity gets independent from the position. While convection cannot affect the chemical gradient, chemical reaction can increase chemical product. Though this situation is physically impossible, it is conducted as virtual setting for comparison.

Results are shown in Fig. 3 and 4 along with the result of the regular case in which both reaction and convection are intact. Fig. 3 (upper) shows the distance of locomotion. In all cases droplet begins to travel and after a while cease to move. While the difference in the distance of displacement is not significant, the droplet travels longer distance in the regular case than in the other two scenarios. Fig. 3 (lower) shows the snapshot of chemical gradient for each of three cases at the same time elapsed since the start. We can see that chemical gradient remains similar along the interface in the regular case. In no reaction case chemical is diluted and shows diminished gradient at the interface. On the contrary, chemicals are accumulated in no convection case, showing isotropic distribution. The reason of higher motility in the regular case is that chemical production is balanced with convection flow which takes away the chemical from the interface. Fig. 4 shows the speed of locomotion. The regular case demonstrates higher

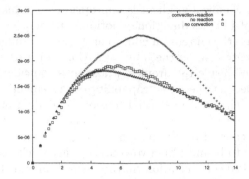

Fig. 4. The horizontal and vertical axes represent simulation time and the speed of droplet's center of mass, respectively. Regular case shows the highest speed among three scenarios.

(a)   (b)

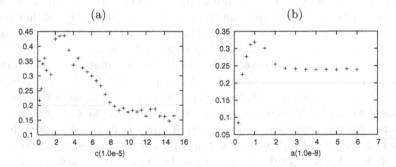

Fig. 5. The parameter dependencies on reaction rate c (a) and a, the coefficient of the force F_s (b). The horizontal and vertical axes stand for each of parameter values and the displacement distance, respectively. The horizontal dotted line represents the length of droplet's radius for comparison.

speed than the other two. These results indicate that droplet can have higher locomotive capability when convection and reaction works cooperatively.

To understand further the relation between the chemical reaction and convection flow, we tested how the droplet's behavior depends on the reaction rate c and the parameter a which determines the convection flow. Fig. 5 (a) shows dependencies on c with a fixed. We can see that around a specific parameter region the droplet accomplishes the maximum displacement distance. Fig. 5 (b) illustrates dependencies on c. Again, specific c region is favorable for locomotion. Both reaction rate and Marangoni effect have an influence on motility.

4   Conclusion and Future Work

We qualitatively reproduced the movement of a droplet, using the model which couples the fluid dynamics and chemical reaction. Staring from the initial condition in which chemical gradient is appropriately set inside the droplet, we observed that droplet shows directional motion. This movement is generated by the imbalance in interfacial tension which can be locally different, depending on the chemical concentration on the interface. Along with its motion, convection flow is also observed inside the droplet, whose axis coincides with the direction in which the droplet marches. The capability of movement is influenced by the two parameters, intensity of interfacial tension depending on the chemical and reaction rate of chemical at the interface. We checked that the maximum traveling distance is achieved at the parameter region where the two parameters take specific values to balance the two contrary effects: chemical reaction which increases the chemical concentration till the saturation and convection flow that carries away chemicals from the interface. Lack of either one would cause faster decline of chemical gradient, resulting in poor motility. Besides, to see whether the convection flow contributes to the droplet's motion in a positive manner, we compare the observed running behavior with that in a virtual scenario in which convection flow doesn't affect the initial distribution of the chemical. While the droplet exhibits directional movement in both cases, it displays higher motility when convection flow is incorporated. Therefore we can say that there exists a case where convection flow enhances the running behavior.

Chemical gradient is a prerequisite for generating motion. In addition, the chemical gradient is being sustained and generated by the droplet itself in the course of time. Decades ago, autopoiesis was suggested as biological model which sustains itself by reproducing its components and boundary conditions. In our model the chemical gradient and convection flow constitute a positive feedback loop which sustain the droplet's motion. This is a self-sustenance of motion, as autopoiesis is for reproduction, and can be regarded as extension to a more general conception where the motion is also self-maintained as reproduction is.

A special emphasis has been put on self-reproduction in the field of artificial life both from theoretical and empirical points of views [2,9,10,11,12,13], while movements have not drawn much attention until recent years. However, when it comes to autonomy, the essential characteristics in which life manifests itself, we can hardly imagine life without movement.

In biological experiments chemotaxis of bacteria and other organisms has been studied, which is considered to be the prototype of cognitive behaviors [14]. From very different points of view in psychology, Gibson[15] and more recently O'Regan and Noë [16] in particular contributed much to the shift of paradigms in cognition from a passive view that sensor input is a signal fed from environments to agents to an alternative view that input is generated by agents through their exploratory movements.

Adaptive behavior and cognitive capabilities are now considered in terms of correlation between movements and sensor input. Various computational models and robotic experiments have been suggested to illuminate sensor-motor loops [17]. Our research on the droplet serves as a model which exemplifies the self-maintenance of movement and bridges the gap between protocell models and

Table 1. Parameters: Values above are used in our simulation, unless specified in the text. To determine Reynolds number by $Re = uL/\nu$, the diameter of the droplet $2R$ and the average velocity at the center of droplet's mass are used as representative length L and velocity u, respectively.

parameter	description	value
ρ_1	density (droplet)	1.2
ρ_2	density (surrounding media)	1.0
ν_1	kinematic viscosity (droplet)	1.7×10^{-6}
ν_2	kinematic viscosity (surrounding media)	1.0×10^{-6}
R	the radius of a droplet	0.2
δh	mesh size	1.56×10^{-2}
Re	Reynolds number	15
D_v	chemical diffusion coefficient	0.2
a	the coefficient of force F_s	1.0×10^{-8}
b	the constant in the interfacial tension term	0.2
c	chemical reation rate	1.0×10^{-5}

locomotive agent models, which have been sometimes considered unrelated to one another.

Some issues remain to be addressed as future work. Spatio-temporal structures formed by reaction diffusion system can be introduced in our model. An interplay between diffusion and convection might have some significance. In laboratory experiment Kitahata [7], for example, showed that droplets loaded with BZ reaction diffusion system demonstrate oscillatory movements driven by oscillatory chemical patterns. Besides, it is widely reported that convection and diffusion take place inside the cell. Our model can be used to simulate and analyze such situations. Secondly, we imposed a chemical gradient as an initial condition to initiate movement. On the other hand, droplets in the experiment spontaneously give rise to symmetry break without special treatment of chemical gradient as an initial condition. Symmetry breaking and initiation of movement are left to future work. Thirdly, The shape of the object, in general, can affect the mode of its movement or be altered along with the movement. For example, camphor, which is also known to show running behavior, generates directional or circular motion, depending on its shape. Relation between the shape and its mode of motion can also be of importance.

Acknowledgements

This work is partially supported by the 21st Century COE (Center of Excellence) program(Research Center for Integrated Science) of the Ministry of Education, Culture, Sports, Science, and Technology, Japan, and the ECAgent project, sponsored by the Future and Emerging Technologies program of the European Community (IST-1940).

References

1. Neumann, J.v.: Theory of Self-Reproducing Automata: Illinois, University of Illinois Press Urbana (1966)
2. Ono, N., Ikegami, T.: Self-maintenance and Self-reproduction in an Abstract Cell Model. J. Theor. Biol. 206, 243–253 (2000)
3. Ono, N., Madina, D., Ikegamni, T.: Origin of Life and Lattice Artificial Chemistry. In: Rasumussen, S., Chen, L., Packard, N., Bedau, M., Deamer, D., Stadler, P., Krakauer, D. (eds.) Protocells: Bridging Nonliving and Living Matter, MIT Press, Cambridge (in press, 2007)
4. Suzuki, K., Ikegami, T.: Self-repairing and Mobility of a Simple Cell. In: Artificial Life IX. Proceedings of the Ninth International Conference on the Simulation and Synthesis of Living Systems, pp. 421–426. MIT Press, Cambridge (2004)
5. Sumino, Y., Magome, N., Hamada, T., Yoshikawa, K.: Self-Running Droplet: Emergence of Regular Motion from Nonequilibrium Noise. Physical Review Letters 94, 68301 (2005)
6. Hanczyc, M.M., Toyota, T., Ikegami, T., Packard, N., Sugawara, T.: Chemistry at the oil-water interface: Self-propelled oil droplets. JACS (in press, 2007)
7. Kitahata, H., Aihara, R., Magome, N., Yoshikawa, K.: Convective and periodic motion driven by a chemical wave. J. Chem. Phys. 116, 5666–5672 (2002)
8. Yabe, T., Ishikawa, T., Wang, P.Y., Aoki, T., Kadota, Y., Ikeda, F.: A universal solver for hyperbolic equations by cubic-polynomial interpolation II. Two- and three-dimensional solvers. Computer Physics Communications 66, 233–242 (1991)
9. Varela, F.R.: Principles of Biological Autonomy, New York, North Hollandk (1979)
10. Langton, C.G.: Self-reproduction in cellular automta. Physica D 10, 135–144 (1984)
11. Rasmussen, S., Chen, L., Deamer, D., Krakauer, D., Packard, N., Stadler, P., Bedau, M.: Transitions from nonliving to living matter. Science 303, 963–965 (2004)
12. Luisi, P.L., Varela, F.J.: Self-Replicating Micelles - A Chemical Version of Minimal Autopoietic Systems. Origins of Life and Evolution of the Biosphere 19, 633–643 (1990)
13. Takakura, K., Toyota, T., Sugawara, T.: A Novel System of Self-Reproducing Giant Vesicles. J. Am. Chem. Soc. 125, 8134–8140 (2003)
14. Adler, J., Tso, W.: Decision-Making in Bacteria: Chemotactic Response of Escherichia Coli to Conflicting Stimuli. Science 184, 1292–1294 (1974)
15. Gibson, J.J.: Observations on active touch. Psychol. Rev. 69, 477–491 (1962)
16. O'Regan, J.K., Noë, A.: A sensorimotor account of vision and visual consciousness. Behavioral and Brain Sciences 24, 939–973 (2001)
17. Pfeifer, R., Scheider, C.: Understanding Intelligence. MIT Press, Cambridge, MA, USA (2001)

Structural Circuits and Attractors in Kauffman Networks

Ken Hawick, Heath James, and Chris Scogings

Computer Science, Massey University Albany, North Shore 102-904, Auckland, New Zealand
Tel.: +64 9 414 0800; Fax: +64 9 441 8181
{k.a.hawick,h.a.james,c.scogings}@massey.ac.nz

Abstract. There has been some ambiguity about the growth of attractors in Kauffman networks with network size. Some recent work has linked this to the role and growth of circuits or loops of boolean variables. Using numerical methods we have investigated the growth of structural circuits in Kauffman networks and suggest that the exponential growth in the number of structural circuits places a lower bound on the complexity of the growth of boolean dependency loops and hence of the number of attractors. We use a fast and exact circuit enumeration method that does not rely on sampling trajectories. We also explore the role of structural self-edges, or self-inputs in the NK-model, and how they affect the number of structural circuits and hence of attractors.

Keywords: Kauffman networks; Random boolean functions; Circuit enumeration; Loops; Attractors.

1 Introduction

Random Boolean Network (RBN) models are effectively a generalisation of the 1-dimensional Cellular Automata model [1]. Kauffman's NK-Model [2] of an N-node network with K-inputs to a boolean function residing on each node has found an important role in the study of complex network properties. RBNs have found important applications in biological gene regulatory networks [3] but also in more diverse areas such as quantum gravity through their relationship with ϕ^3-networks [4,5]. RBNs have many interesting properties [6] and have been amenable to various analyses [7] including mean-field theory. They continue to be an important and interesting tool in studying biological and artificial life problems.

A key property of RBNs is the now well established existence of a frozen phase and a chaotic phase [8,9] and the critical phase transition lies at the integer value of connectivity $K_c = \frac{1}{2p(1-p)} = 2$ for unbiased networks with a mean boolean function output value of $p = 0.5$. It is therefore of most interest to study RBNs at or around this critical value.

The Random Boolean Network or graph G is expressed as a four-tuple $\mathbf{G} = (V, E, F, x)$ and has $N = |V| = |F| = |x|$ nodes or vertices, and $N_e = |E|$ directed edges or arcs, which express the K_i inputs for node i. The Kauffman NK-Network is constructed with fixed $K = 1, 2, 3, ..$ and a boolean function f_i of K_i inputs is assigned to each

M. Randall, H.A. Abbass, and J. Wiles (Eds.): ACAL 2007, LNAI 4828, pp. 189–200, 2007.

node. All the nodes of the network carry boolean variables x_i which may be initialised randomly and which are updated (usually, but not necessarily) synchronously so that:

$$x_i(t) \leftarrow f_i\left(x_j(t-1)\right), j = 1, 2, ..., K_i \qquad (1)$$

The NK-network model assigns the K_i inputs for node i randomly and with uniform probability across all nodes. Even for a large network there is still a non-zero probability of assigning a node as one of its own inputs. In the case of $K_i > 1$ there is also a possibility of assigning a node j as an input of i more than once. These self-edges or multiple edges can have a subtle effect on the behaviour of the NK-network model.

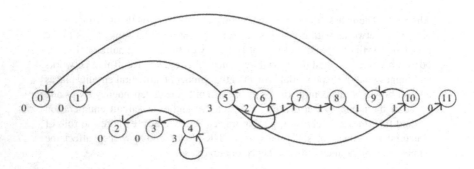

Fig. 1. 12 Node Network with $K = 1$ inputs, showing the output degree of each node

Figure 1 shows a small example Random Boolean Network of 12 nodes, each with $K = 1$ inputs. The out-degree varies depending upon the number of other nodes that depend upon each node, and for this network self-arcs are allowed. This example is fragmented into two independent components and there are interesting structural changes in the component composition of RBNs between $K = 1$ and $K = 2$.

Work has been carried out on a number of different update mechanisms for boolean networks including asynchronous algorithms [10]. In this paper we consider only synchronous updating where all nodes execute their boolean function once, together, and at every time step. A significant body of work has now been carried out on the roles of different sub-classes of boolean functions including the so called canalizing functions [11] and in particular the effect of the frozen or fixed-value boolean functions on particular elements of the network.

A consequence of the boolean functions in RBNs is the formation of attractor basins [12]. These are observed in RBN models whereby diverse initial starting conditions will still lead to statistically similar behaviour. The state of the network falls into attractor cycles whereby a chain of interdependence of nodes (via their boolean functions) leads to the network periodically repeating its state. The number and length of these periods or attractors is of great importance in understanding the behaviour of the NK-model and associated application problems. This can be seen quantitatively by tracking a metric such as changes in the normalised Hamming distance between the network's successive boolean states.

Of particular recent interest in the literature has been the uncertainty concerning the number of attractors [13, 14] and how their number and lengths varied with the size of the network. Scaling was initially believed to be $O(\sqrt{N})$ [15]. It was later reported as linear [16], and then as "faster than linear" [17] and subsequently as "stretched exponential" in [18, 19] but is now known to be faster than any power law [20].

A recent review of the RBN model [7] discusses the attractor behaviour in terms of the loops of boolean variable states that form and several exact results concerning these loops have been obtained [21]. Important observations concern the distribution of components with particular sub-classes of possible boolean functions. These "relevant elements" are defined as those nodes that are not frozen and that control at least one other relevant element in the system [19]. A number of important results have been obtained using particular sub-classes of the possible boolean functions. Drossel et al. have considered networks with only non-fixed boolean functions thus making all elements relevant and have therefore shown the equivalence of $K = 1$ and $K = 2$ networks under appropriate restrictions on the boolean functions used.

In this paper we use numerical methods to investigate the role that structural circuits play in the complex structure of the network and the resulting attractor behaviour of RBNs. Recent work in the literature has used trajectory sampling. The combinatorics of RBN models means that the number of boolean functions grows as 2^{2^K} and a rapid growth in the number of possible network states with network size. Taking limited numbers of sample trajectories through this state space can lead to very misleading results. Numerical sub-sampling of attractor trajectories seems to be the main difficulty behind obtaining a good understanding of attractor scaling. We focus on the structural properties of RBNs including the number and length distribution of elementary circuits and of components. We compute these properties exactly using brute force enumeration techniques for a range of network sizes and connectivities. Our statistical sampling is only over different randomly configured networks, not over attractor trajectories.

In [22] we described the D Code we developed to simulate very large-scale Random Boolean Networks. In this paper we exploit this capability to study the cluster and monomer populations in large systems for which even an $O(n^2)$ cluster labelling algorithm is feasible. However, in the study of circuits we are severely limited by the time complexity of even the best circuit enumeration algorithm we have been able to find (see section 3).

As we discuss in section 2, there is a close relationship between the number of elementary circuits in the underpinning structural network and the number of attractors that can be supported in an associated RBN. We also analyse the role that self-edges play in the structural networks and the consequences for both the number of circuits and disconnected components. We present some results for various network sizes and connectivities, with and without self-edges in section 4 and draw some conclusions on scaling with N and K in section 5.

2   Attractor Numbers

Figure 2 shows a network with $N = 16, K = 2$. The construction algorithm has allowed self-arcs - in other words the inputs for each node have been chosen according to a flat

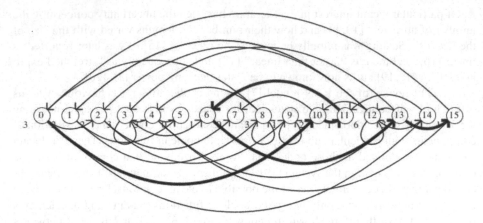

Fig. 2. 16 Node Network with $K = 2$ inputs, showing the output degree of each node and one of the circuits in the graph, connecting node 0 to node 15. This network allows self-arcs.

uniform distribution so they can connect to themselves. The consequent self-edges allow self-inputs in the corresponding RBN. These are known to play a vital role in supporting the number of attractors. A self-input or "self-ancestor" in the input dependence chain of boolean variables anchors the periodic or attractor behaviour [7] of RBNs.

We felt intuitively that the presence of structural circuits would also be vital to the periodic attractor behaviour. Figure 2 shows one such circuit or loop in the network structure. In fact, exact enumeration (as shown in figure 3) indicates that there are 22 circuits composed as follows: 4 of length 1; 2 of length 3; 4 of length 5; 2 of length 6; 6 of length 8; and 4 of length 10. If self-edges are disallowed we would obtain a higher number of circuits present in the network.

As Drossel et al. have shown there are definite relationships between the number of attractors and the number of loops. Qualitatively summarizing, the number of struc-

0	10	11	6	13	3	4	15	0			1	8	4	13	12	1
0	10	11	6	13	12	1	8	0			1	8	4	13	12	1
0	10	11	6	13	12	1	8	4	15	0	3	3				
0	10	11	6	13	12	1	8	0			3	4	13	3		
0	10	11	6	13	12	1	8	4	15	0	3	6	13	3		
0	10	11	6	13	15	0					6	13	12	10	11	6
0	14	11	6	13	3	4	15	0			6	13	12	14	11	6
0	14	11	6	13	12	1	8	0			8	8				
0	14	11	6	13	12	1	8	4	15	0	9	9				
0	14	11	6	13	12	1	8	0			12	12				
0	14	11	6	13	12	1	8	4	15	0						
0	14	11	6	13	15	0										

Fig. 3. 22 Circuits found in the network shown in figure 2 which has 16 nodes and 32 arcs and allows self-arcs. Note there are repeated circuits due to the presence of a multiple-arc connecting nodes 12 and 1.

tural circuits provides a lower bound on the number of possible attractors. It therefore gives insight into the controversy over the number of attractors in RBNs to consider the exactly enumerated number and distribution of circuits in the underlying networks.

An elementary circuit is a closed path along a subset of the edges of the graph such that no node, apart from the first and last, appears twice. The number of elementary circuits for a fully connected graph is bounded by $\sum_{i=1}^{N-1} C^N_{N-i+1}(N-i)!$, [23]. This expression represents the limit for the number of structural circuits in an NK-network when $K \to N$.

3 Graph Algorithms

RBNs can be represented in a number of different data structures in computer simulation programs. We used a neighbour list approach [24] and have experimented with various structures for boolean variables and boolean functions using the D programming language [22]. D is essentially a systems-oriented programming language derived from C and C++. It is a good platform for custom simulation codes that must be efficient in both time and space to tackle problems with high computational complexity.

The problem of component labelling or clustering is well known and we used a simple colour propagation algorithm [25] which was readily re-implemented as part of our custom RBN code.

Various algorithms have been formulated to count the circuits in a graph but these either use infeasible amounts of memory or are time exponential [26, 27] with a time bound of $O(N.e(c+1))$. We count circuits using a variation of Johnson's algorithm [28] implemented in D. For graphs of N vertices, e edges, c circuits and 1 fully connected component, Johnson's algorithm is time bounded in time by $O((N + e)(c + 1))$ and space bounded by $O(N + e)$. Unlike Johnson's algorithm our code copes with partially connected graphs without resorting to the need to treat each of the possible $N_c > 1$ components separately [29]. This is still a highly expensive process since the number of circuits c itself grows very rapidly with (N, e).

In the graph literature the term loop is unfortunately sometimes used to describe a self-edge or a circuit of length 1. In the NK-networks we study the number of self-edges is much less than N, even for low K. However we do count them and observe the effect of allowing them in the number of possible circuits and their length histograms.

On a modern compute server with 4GBytes memory and a speed of 2.66GHz, we found it was entirely feasible to enumerate circuits exactly in networks of up to $N \approx$ 100 for $K = 1, 2$. Smaller networks were required for higher K. We were able to count components quite easily up to networks of around $N \approx 20,000$. We were able to exploit the near-linear speed-up of parallel job-farming to average our exact enumeration/counting results over many independently generated networks.

4 Numerical Results

In this section we present results for the number of component clusters; the number of monomers and the number and length distribution of circuits. The sample numbers are

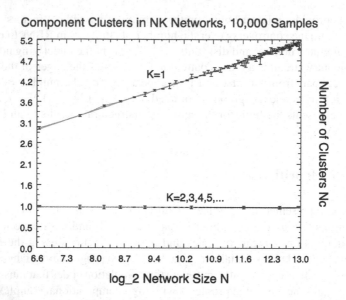

Fig. 4. Number of Cluster Components for Kauffman Networks of $K = 1, 2, 3, 4, 5, 6$

shown but typically these are over 100,000 independently generated NK-networks for K=1,2 and 100 for K=3,4 and 10 for K=5,6 and higher.

Figure 4 shows the variation of the mean number of clusters in NK-networks. For $K \geq 2$ the system is dominated by a single giant cluster, although even in large networks it is still possible for a single isolated monomer to exist. The figure shows that for $K = 1$ that the number of component clusters $N_c \approx 0.345 \log_2 N + 0.65$, showing that a range of different cluster sizes are co-existing even in arbitrarily large networks.

It is instructive to consider what size distribution makes up the number of components. Figure 5 shows the cluster size distribution averaged over one million sample networks of size $N = 256$. Interestingly this shows a relationship between the mean population $\langle P \rangle$ for cluster component size s such that $\log \langle P \rangle \approx As^{-0.89}$. This relationship appears to hold true for arbitrarily large network sizes, but only for sizes up to one half of the total network size. For $K > 1$ the size distribution is completely dull - being just one single cluster of size N.

One might expect that in those regimes where there are multiple components, single isolated monomers or other very small-sized components dominate the component distribution. However the number of monomers is almost entirely flat, independent of the network size. In the case of $K = 1$ it is small but definitely non-zero. In the case of $K > 1$ the number of monomers is almost completely zero on average. Figure 6 shows the variation in the number of monomers over the same networks sampled for components in figures 4 and 5. This suggests an interesting cluster size multi-scale behaviour for $K = 1$ and in particular that the cluster composition is influenced by two competing effects. As the network grows in size there are more nodes and therefore there is a growing possibility of some being disconnected, but conversely there are also more arcs available in the network and therefore a higher probability of each node being

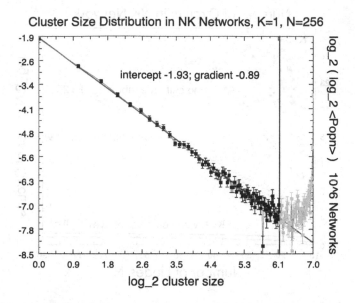

Fig. 5. The cluster size distribution for a $K = 1$ network, generated using the algorithm described. Using $N = 256$ and 1,000,000 sample networks. Note the plot does not show the single giant component that occurs for $K > 1$.

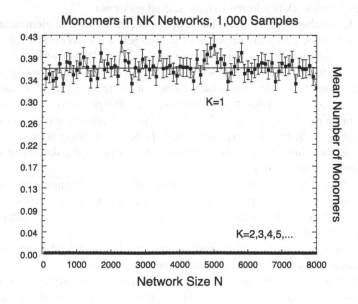

Fig. 6. Number of Monomers for Kauffman Networks of $K = 1, 2, 3, 4, 5, 6$

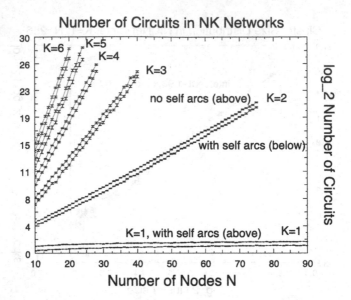

Fig. 7. Growth of the number of circuits with network size for Kauffman Networks of $K = 1, 2, 3, 4, 5, 5$

connected to some other node. The number of monomers is affected surprisingly little by whether the network is allowed to have self-edges or not.

Unlike the number of monomers and components, the number of elementary circuits is significantly affected by the presence of self-edges in the network. Figure 7 shows the growth of the number of circuits with network size in NK-networks. It is clear that for $K > 1$ the number of circuits grows very rapidly - much faster than any power law. This would appear to confirm the present view that the related number of attractors in an RBN will grow at least this fast. A least-squares fit again reveals that the number of circuits or loops varies as $N_L \approx A_L e^{bN}$. It is not entirely clear from our data what the exact relationship between exponent b and K is. While there is clearly a positive monotonic relationship, our data are not good enough to distinguish b linear with K or $\log K$ or some power law in K.

It is interesting that for $K > 1$ the presence of self-edges in the network considerably lowers the number of circuits present. It does not appear to influence the value of b or its relationship with K. However for $K = 1$ networks, the presence of self-edges actually raises the number of circuits.

This is intriguing since the self-arcs are vital for loops in RBNs but this entirely structural behaviour crossover occurs exactly at the RBN critical K_c value.

Figure 8 shows the growth of the number of circuits with network size in Kauffman networks for the special case of $K = 1$. This definitely does not exhibit the same relationship as for $K > 1$. Even over the relatively small network sizes of $N = 10, ... 100$ the data are consistent with a power law $N_L \approx N^x$ where $x \approx 0.225 \pm 0.003$ for a network with no self-edges, and $x \approx 0.187 \pm 0.003$ when self-edges are present.

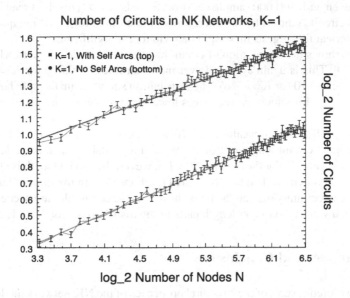

Fig. 8. Growth of the number of circuits with network size for Kauffman Networks of $K = 1$

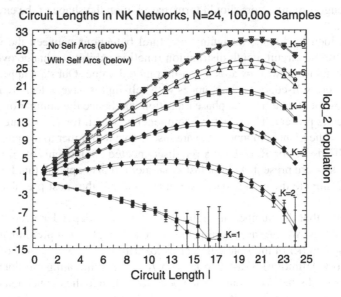

Fig. 9. Circuit length distribution in a 24 Node Kauffman Network for $K = 1, 2, 3, 4, 5, 6$ with and without self-edges allowed in the network

Although even with 100,000 sample networks there is still a sizeable spread in the mean number of circuits (shown by the uncertainty bars on the plot), the high quality nature of the numerical fit suggests these split behaviours are significant.

Figure 9 shows the distribution of circuit lengths in 24 node NK-networks for $K = 1, 2, 3, 4, 5, 6$. This is again averaged over many samples (100,000 for $K = 1, 2$; 100 for $K = 2, 3$; and 10 for $K = 5, 6$). The distribution shows again the split behaviour for the cases $K > 1$ for which the self-edges lower the number of circuits of each length, and for $K = 1$ for which the self-edges raise the number of circuits. The shape of the distribution itself if quite revealing. For $K > 1$ there will be circuits of lengths up to the Hamiltonian circuit length of $L_H \equiv N$, with a modal value at some lesser length that increases with K. For the case of $K = 1$ however, the modal length is always unity and the maximum circuit length is truncated (perhaps only in the limit of large N ?) to $N/2$. As the uncertainty bars on the plots show, very large samples are needed to extract smooth mean values the longer length parts of the distributions for $K = 1, 2$.

5   Discussion and Conclusions

We have explored several of the structural properties of the NK-network model and have found that the number of circuits grows faster than any power law with network size. This confirms that the number of attractor loops in Random Boolean Networks should also grow faster than any power law. This behaviour also gives some insights into why particular lengths of attractors should form, based upon the shape of the circuits length distribution.

We have identified some intriguing structural behaviour between the values $1 < K < 2$ where the circuit length distribution function exhibits a change over from exponential decay to growth towards a non-unit modal value. Our data appears to show that the network self-edges or RBN nodes with self-inputs have a decisive role to play in influencing the location of the phase transition and hence the number of circuits and hence attractors present. The structural components, which for $K > 1$ are completely dominated by the giant component and are insensitive in number to the presence or absence of self-edges. For $K = 1$ however, disconnected components of sizes up to half the network size are present, and not just monomers. It also appears that for $K = 1$ the self-edges dominate the circuit size distribution and are the most prevalent loop type present.

We observe that the number of circuits displays some dependence on the presence of multiple-edges. We are investigating this more thoroughly, but preliminary data suggests that the number of multiple-edges grow logarithmically with N and of course can only affect (by definition) NK-networks with $K > 1$. Eliminating multiple-edges like self-edges from the network generation model does change the connection distribution probability away from a flat uniform one.

We have determined some of the growth behaviours for monomers, components and circuits. We might expect the results we have found to hold well, on average for finite practicably sized networks as well as for the large N limit. We are investing more computational effort into studies of higher K values to investigate the exponent dependence on K for the number of circuits.

References

1. Wolfram, S.: Theory and Applications of Cellular Automata. World Scientific, Singapore (1986)
2. Kauffman, S.A.: The Origins of Order. Oxford University Press, Oxford (1993)
3. Kauffman, S., Peterson, C., Samuelsson, B., Troein, C.: Random boolean network models and the yeast transcriptional network. Proc. Natl. Acad. Sci. USA 100, 14796 (2003)
4. Baillie, C.F., Johnston, D.: Damaging 2d quantum gravity. Physics Letters B 326, 51–56 (1994)
5. Baillie, C., Hawick, K., Johnston, D.: Quenching 2d quantum gravity. Physics Letters B, 284–290 (1994)
6. Gershenson, C.: Introduction to random boolean networks. Available at arXiv:nlin/0408006v3 (2004)
7. Kadanoff, L., Coppersmith, S., Aldana, M.: Boolean dynamics with random couplings. In: Kaplan, E., Marsden, J., Sreenivasan, K. (eds.) Perspectives and Problems in Nonlinear Science, Springer, Heidelberg (2003)
8. Derrida, B., Pomeau, Y.: Random networks of automata: A simple annealed approximation. Europhys. Lett. 1, 45–49 (1986)
9. Derrida, B., Stauffer, D.: Phase transitions in two-dimensional Kauffman cell automata. Europhys. Lett. 2, 739 (1986)
10. Harvey, I., Bossomaier, T.: Time out of joint: Attractors in asynchronous random boolean networks. In: Husbands, P., Harvey, I. (eds.) ECAL 1997. Proc Fourth European Conference on Artificial Life, pp. 67–75. MIT Press, Cambridge (1997)
11. Szejka, A., Drossel, B.: Evolution of canalizing boolean networks. The European Physical Journal B 56, 373–380 (2007)
12. Wuensche, A.: Discrete dynamical networks and their attractor basins. In: Standish, R., et al. (eds.) Proc. Complex Systems 1998, UNSW, Sydney, Australia, pp. 3–21 (1998)
13. Drossel, B.: On the number of attractors in random boolean networks. Technical Report arXiv.org:cond-mat/0503526, Institute fur Festkorperphysik, TU Darmstadt (2005)
14. Drossel, B., Mihaljev, T., Greil, F.: Number and length of attractors in a critical Kauffman model with connectivity one. Phys. Rev. Lett. 94, 88701 (2005)
15. Kauffman, S.A.: Metabolic stability and epigenesis in randomly constructed genetic nets. Journal of Theoretical Biology 22, 437–467 (1969)
16. Bilke, S., Sjunnesson, F.: Stability of the Kauffman model. Phys. Rev. E 65, 16129 (2001)
17. Socolar, J.E.S., Kauffman, S.A.: Scaling in ordered and critical random boolean metworks. Phys. Rev. Lett. 90, 068702–1 (2003)
18. Bastolla, U., Parisi, G.: Relevant elements, magnetization and dynamical properties in Kauffman networks: A numerical study. Physica D 115, 203–218 (1998)
19. Bastolla, U., Parisi, G.: The modular structure of Kauffman networks. Physica D 115, 219–233 (1998)
20. Samuelsson, B., Troein, C.: Superpolynomial growth in the number of attractors in Kauffman networks. Phys. Rev. Lett. 90, 98701–1 (2003)
21. Flyvbjerg, H., Kjaer, N.J.: Exact solution of Kauffman's model with connectivity one. J. Phys. A: Math. Theor. 21, 1695–1718 (1988)
22. Hawick, K.A., James, H.A., Scogings, C.J.: Simulating large random boolean networks. Technical Report CSTN-039, Information and Mathematical Sciences, Massey University, Albany, North Shore 102-904, Auckland, New Zealand (2007)
23. Harary, F., Palmer, E.M.: Graphical Enumeration. Academic Press, New York (1973)
24. James, H.A., Hawick, K.A.: Computational data structures for lattice-based small-world simulations. Technical report, Institute for Information and Mathematical Sciences, Massey University (2005)

25. Dewar, R., Harris, C.K.: Parallel computation of cluster properties: application to 2d perco-lation. J. Phys. A Math. Gen. 20, 985–993 (1987)
26. Tiernan, J.C.: An efficient search algorithm to find the elementary circuits of a graph. Com-munications of the ACM 13, 722–726 (1970)
27. Tarjan, R.: Enumeration of the elementary circuits of a directed graph. SIAM Journal on Computing 2, 211–216 (1973)
28. Johnson, D.B.: Finding all the elementary circuits of a directed graph. SIAM Journal on Computing 4, 77–84 (1975)
29. Hawick, K.A., James, H.A.: A fast code for enumerating circuits and loops in graphs. Massey University Technical Note (2005)

The Effects of Learning on the Roles of Chance, History and Adaptation in Evolving Neural Networks

Grant Braught and Ashley Dean

Department of Mathematics and Computer Science
Dickinson College
Carlisle, PA 17013 USA
braught@dickinson.edu

Abstract. The course of the evolution of a population is affected by chance events, the population's genetic history and adaptation via selection. The presence of individual lifetime learning is also known to influence the course of a population's evolution. The experiments reported here, examine the effects that lifetime learning has on the roles played by chance, history and adaptation in the evolution of populations of simple neural networks. The effects of chance, history and adaptation on both learned fitness (fitness after learning) and innate fitness were considered both when learning incurred no cost and when a fitness cost was incurred for learning. When learning was cost-free it was found to decrease the influence of adaptation, history and chance on learned fitness, while having the opposite or possibly no effect on innate fitness. When a fitness cost was incurred for learning, the role of adaptation in determining innate fitness increased, while the roles of chance and history decreased for both learned and innate fitness. These observed effects are interpreted in light prior results on the effects of learning on evolution.

Keywords: Evolution, Chance, History, Adaptation, Learning, Baldwin Effect.

1 Introduction

The state of an evolving population can be attributed to the effects of three primary forces: chance, history and adaptation. Adaptation, defined as the cumulative effects of ongoing selection, tends to increase a population's fitness as it is tailored to its ecological niche. Both chance and history have the effect of opening and/or closing potential avenues along which a population might adapt. History constrains the evolutionary paths that are accessible to a population based upon ancestry and prior selection. Similarly, chance events such as random mutations, genetic drift or other stochastic events alter the evolutionary paths accessible to a population.

While the majority of studies of evolution, in both biology and artificial life, have focused on adaptation as the primary agent of evolutionary change, a number of studies have begun to reveal that both history and chance can play non-trivial roles. Most notably, Travisano, Mongold, Bennett and Lenski [1] studied populations of evolving *E. coli* bacteria in several controlled experiments that allowed them to directly assess the roles of chance, history and adaptation. They found, not

M. Randall, H.A. Abbass, and J. Wiles (Eds.): ACAL 2007, LNAI 4828, pp. 201–211, 2007.
© Springer-Verlag Berlin Heidelberg 2007

unexpectedly, that adaptation is the largest influence on the reproductive success of a population (i.e. fitness). However, they also report that in the evolution of traits less directly associated with fitness, *E. coli* cell size in their experiments, chance and history played a much more significant role. In addition, there is a growing number of reports of biological, ecological and genetic studies that investigate the roles of chance and/or history in addition to adaptation [2,3,4,5,6].

Investigations and applications of chance and history have also appeared in the artificial life literature. In an experiment paralleling that of Travisano et. al., Wagenaar and Adami evolved populations of "digital organisms" in the AVIDA system, obtaining comparable results [7]. In the area of evolutionary robotics, researchers have made practical application of the effects of history. They have used carefully designed fitness functions that change over time so as to steadily shape the population in such a way as to make accessible behaviors that evolution would otherwise be highly unlikely find [8,9,10].

The experiment presented here parallels that of Travisano et. al. and Wagenaar and Adami. It differs however, in that it begins to investigate how lifetime learning influences the effects of chance, history and adaptation. There is ample evidence that the course of evolution can be altered by lifetime learning [11,12,13]. Thus, examining the roles of chance, history and adaptation in the presence of lifetime learning provides an opportunity to examine the effects of learning on evolution from a new angle. The essence of what is reported here is an experiment in which the effects of chance, history and adaptation are observed in three populations of evolving neural networks. One population uses feed-forward networks to model non-learning individuals. The other two populations use back-propagation to model individuals capable of lifetime learning. In one learning population there is no cost for learning, while in the other there is a cost associated with learning. The observed effects of chance, history and adaptation are compared among the three populations and interpreted in light of known effect of learning on evolution.

2 Methods

The experimental design used here is based on that developed by Travisano et. al [1] and later applied by Wagenaar and Adami [7]. Figure 1(a) provides a schematic representation of the experiment. A number of parent populations are evolved in one environment (the *pre-transfer environment*) until they are well adapted. Each parent population is then cloned, producing a set of *child populations* that will be referred to as *sibling populations*. All of the child populations are then evolved in a significantly different environment (the *post-transfer environment*). Figure 1(b) illustrates this process, showing the mean fitness for a parent population it evolves in the pre-transfer environment, followed by the mean fitness of the sibling populations derived from it as they evolve in the post-transfer environment. As will be discussed below, this design makes it possible to isolate the effects due to chance, history and adaptation.

2.1 Measuring Chance, History and Adaptation

The effects of each of the components of evolutionary change, chance, history and adaptation, can be extracted from the results of this experiment. To measure

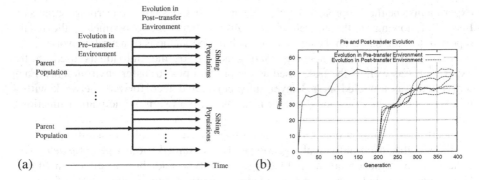

Fig. 1. (a) A schematic representation of the experiment. (b) The fitness of a single parent population during 200 generations of evolution in the pre-transfer environment and the fitness of 5 of its child populations during 200 generations of evolution in the post-transfer environment.

adaptation, the mean fitness of each child population in the post-transfer environment, prior to any evolution in that environment, is noted as its ancestral fitness. At any subsequent time, the contribution of adaptation is given by the difference between the current mean fitness of a child population and its ancestral fitness averaged across all child populations.

Chance and history are measured by the degree to which they affect the variability of the mean fitness of the child populations. Chance is given by the standard deviation in mean fitness among sibling populations averaged across the parent populations. To see why this measure reflects chance, consider that when each population begins its evolution in the post-transfer environment it is genetically identical to all of its sibling populations. Thus, any differences that arise in the mean fitness between two sibling populations can only be due to the occurrence of different sequences of random mutations, i.e. chance. The effect of history is computed by first averaging the mean fitness values within each set of sibling populations. The standard deviation of these values then represents the effect of history. By averaging the mean fitness values within sets of sibling populations, the effects of chance can be canceled out. Thus, any differences in average fitness between sets of sibling populations derived from different parents can be attributed to differences in their parents, i.e. history.

The experiment design described above is a two-level nested ANOVA [14] with the replicated measurements being the fitness values of the individuals in the populations. The effect of chance is given by the square root of the within groups variance component. The effect of history is given by the square root of the between groups variance component. In addition, 95% confidence intervals have been computed for each of the variance components. The square roots of those confidence intervals then represent 95% confidence intervals on the contributions of chance and history.

2.2 Implementation Details

Travisano et. al performed their experiment using *E. coli* bacteria. Wagenaar and Adami performed their experiment using virtual organisms in the Avida system. While it would have been preferable to make direct comparisons to the results from these earlier experiments, neither approach readily lent itself to the study of learning organisms. Instead, following in the model of so many artificial life experiments, the results reported here are based on the analysis of evolving populations of neural networks.

Figure 2 shows the phenotype and genotype of the individuals used in the experiments as well as the tasks used as the pre and post-transfer environments. Each individual was modeled as a 3 layer fully connected feed forward network with 4 inputs, 4 hidden neurons and 2 output neurons, all with sigmoid activation functions. The genotype for each individual contained genes that encoded the weight of each connection, including a bias for each neuron (30 weights in all). In addition to genes for the weights, each individual's genotype also contained another gene (*gene0*). This gene had no influence on the individual's phenotype and thus no effect on fitness. Gene0 was included as a control; allowing the examination of the effects of chance, history and adaptation on a selectively neutral trait (see section 3.1). It plays the same role in these experiments as cell size did for Travisano et. al and program length did for Wagenaar and Adami.

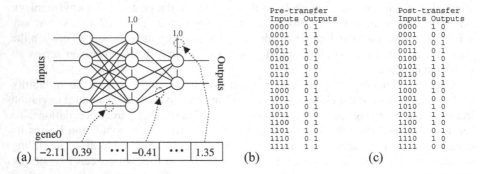

Pre-transfer		Post-transfer	
Inputs	Outputs	Inputs	Outputs
0000	0 1	0000	1 0
0001	1 1	0001	0 0
0010	1 0	0010	0 1
0011	1 0	0011	0 1
0100	0 1	0100	1 0
0101	0 0	0101	1 1
0110	1 0	0110	0 1
0111	1 0	0111	0 1
1000	0 1	1000	1 0
1001	1 1	1001	0 0
1010	0 1	1010	1 0
1011	0 0	1011	1 1
1100	0 1	1100	1 0
1101	1 0	1101	0 1
1110	0 1	1110	1 0
1111	1 1	1111	0 0

(a) gene0 | -2.11 | 0.39 | ••• | -0.41 | ••• | 1.35 | (b)   (c)

Fig. 2. (a) The structure of an individual's phenotype and genotype. (b) The pre-transfer environment and (c) post-transfer environment used for fitness evaluation.

In the experiments reported here 20 parent populations of 100 individuals each were used. The individuals in the parent populations were generated at random and were evolved in the pre-transfer environment until they were well adapted (< 10% error in the pre-transfer environment). Once well adapted, the best individual from each parent population was cloned 100 times to produce a child population. Each child population was then repeatedly cloned to produce 20 sibling populations from each parent. Each set of sibling populations were then cloned 3 times producing populations for non-learning, no-cost-learning and costly-learning. In all, 1200 child populations were evolved in the post-transfer environment.

Individual fitness during evolution was given by the percentage of input patterns from the environment (pre or post-transfer as appropriate) for which the individual's neural network produced outputs within 0.2 of the correct responses. In the non-learning populations, an individual's genes directly coded the neural network used for its fitness evaluation. For individuals in the learning populations an innate fitness and a learned fitness were computed. An individual's innate fitness was found in the same way as for the non-learning individuals. For no-cost-learning, an individual's learned fitness was given by the percentage of correct outputs generated by the neural network after 25 epochs of back-propagation. For costly-learning, an individual's learned fitness was the average of its innate fitness and its fitness after 25 epochs of back-propagation.

Additional details of the evolutionary and learning algorithms were as follows. Roulette wheel selection was used. Selection for learning populations was performed using the learned fitness, however reproduction was performed using the gene values from before back propagation. Individuals were reproduced asexually with a mutation rate of 15%. Genes were mutated by addition of a pseudo-random value drawn from a Gaussian distribution with zero mean and unit standard deviation. During back propagation, a learning rate of 0.05 and a momentum term of 0.95 were used.

3 Results

3.1 Results for Gene0

The absence of selective pressure on gene0 would imply that its value should undergo a random walk and consequently adaptation should have no consistent effect on the value of gene 0. Figure 3 confirms this intuition, showing that the effect of adaptation wanders randomly about its initial value of 0. The traces for chance and history in figure 3 also confirm intuition. Initially all of the populations within a set of sibling populations are identical, making the influence of chance initially 0. Each set of sibling populations, however derives from a different parent population. Thus, because the parent populations do differ from one another, the effect of history is initially non-zero. However, over time, the accumulation of unique sequences of random mutations causes the populations within a set of siblings to differ from one another, increasing the influence of chance. This accumulation of mutations also decreases the differences in mean value between sets of sibling populations, causing the effect of history to diminish over time, eventually becoming negligible. Though not shown, the results for no-cost-learning and costly-learning gave similar results. These results are qualitatively similar to those reported for *E. coli* cells size by Travisano et. al. [1] and for program length by Wagenaar and Adami [7].

3.2 Adaptation

In all of the cases where fitness was analyzed, adaptation was the dominant influence. However, there were differences in its effect between the non-learning, no-cost-learning and costly-learning populations. There were also differences in the effect of adaptation on innate and learned fitness. These differences can be understood on the basis of prior results on the Baldwin effect [15, 16] and on the evolution of robustness with respect to mutations [17, 18, 19, 20].

Figure 4(a) compares the effects of adaptation on fitness in the non-learning populations to learned fitness in both types of learning populations. In the trial shown, the effect of adaptation is largest on the non-learning population. In other trials, the role of adaptation on learned fitness was occasionally greater for costly-learning populations than for non-learning populations. However, the role of adaptation on the non-learning and costly-learning populations was always greater than for the no-cost-learning populations.

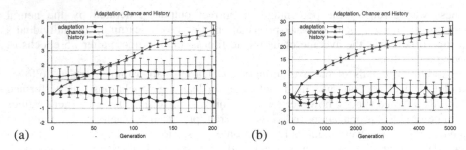

Fig. 3. The effects of adaptation, chance and history on a selectively neutral trait (gene0) (a) over the first 200 generations of evolution in the post-transfer environment (b) over 5000 generations. Error bars show 95% confidence intervals. Note: The traces on this and subsequent graphs have been offset slightly along the generation axis to allow the effective display of overlapping error bars.

Initially, the results in figure 4(a) were surprising. Intuition suggests that the learned fitness of the learning populations would be greater than the fitness of the non-learning populations, which also suggested that adaptation would be playing a larger role in their evolution. However, because learning plays a role in determining learned fitness this is not necessarily true. Learning populations have not only higher learned fitness values but they also have higher ancestral fitness values (also due to learning). Thus, because the learning populations begin with higher ancestral fitness values and because there is a ceiling on the fitness that can be achieved in this experiment the effect of adaptation can actually be diminished by learning.

Figure 4(b) considers the role of adaptation in determining innate fitness. It shows that the effect of adaptation on the innate fitness of the costly-learning populations is greater than for the non-learning populations, which in turn is greater than for the no-cost-learning populations. The significant increase in the role of adaptation in determining innate fitness of the costly-learning population seen in figure 4(b) can be attributed to the Baldwin effect. Specifically, this increase in the role of adaptation is driven by the cost of learning [15]. Because all individuals are of approximately equal learning ability, reducing the cost of learning produces a selective benefit. This benefit results in innate fitness values being pulled closer and closer to the level of the learned fitness values. Thus, because the ancestral value of innate fitness is lower than the ancestral value of learned fitness, the effect of adaptation on innate fitness is greater than for learned fitness. As seen in figure 5(a), the difference in the effect of adaptation on learned and innate fitness for costly-learning populations approaches a constant value. This is a result of the learned fitness reaching a plateau and the innate fitness rising to the same plateau.

For the no-cost-learning populations the case is somewhat different. As shown in figure 5(b) the effect of adaptation on innate fitness is slightly less than its effect on learned fitness early in the evolution, and this difference diminishes over time. In this case, because learning has no cost it serves to hide differences in innate fitness from selection. This hiding effect greatly reduces the evolutionary pressure that would otherwise narrow the gap between the innate and learned fitness values [16].

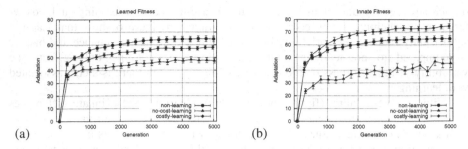

Fig. 4. The influence of adaptation on learned and innate fitness. (a) A comparison of the effect of adaptation on fitness in the non-learning populations to the learned fitness of the learning populations. (b) The same comparison using the innate fitness of the learning populations.

However, the effect of adaptation on innate fitness is still increasing both over time and relative to its effect on learned fitness. This increase is driven by the evolution of robustness with respect to mutation, which leads to more conservative transmission of fitness from parents to offspring [17, 18, 19]. When the innate and learned fitness values are close, less of the learning capability is necessary to achieve the learned fitness. Thus, there is learning capability "in reserve" that can be used to compensate for mutations inherited during reproduction. Therefore, parents with small differences in innate and learned fitness are more likely to have offspring that are able to maintain the same level of learned fitness as their ancestors despite the accumulation of mutations. If the simulation were run for a longer period of time it is expected that the effect of adaptation on innate fitness would, via this effect, become greater than its effect on learned fitness. It is also worth mention that this selective pressure is likely at work in the costly-learning case as well but is significantly weaker than, and thus masked by, the effect due to the cost of learning.

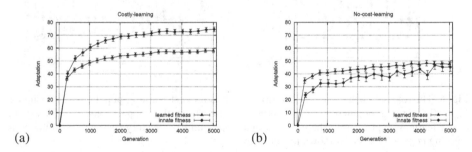

Fig. 5. The effects of adaptation on learned and innate fitness (a) in costly-learning populations and (b) in no-cost-learning populations

3.3 Chance and History

As with adaptation, the roles of chance and history were significantly affected by lifetime learning. Also, as with adaptation, there were dramatic differences in the roles of chance and history between the no-cost-learning populations and the costly-learning populations. The graphs in figure 6 illustrate the effects of chance and history on learned and innate fitness and for non-learning, no-cost-learning and costly-learning populations. Taken collectively, the graphs in figure 6 show that learning significantly reduces the variability in learned fitness due to both chance and history as compared to non-learning populations. However, learning only affects the variability of innate fitness due to chance and history when there is a cost associated with learning.

With no-cost-learning (figures 6(a) and 6(c)), learning hides innate differences between individuals, allowing individuals with differences in innate fitness to achieve similar levels of learned fitness. Thus, with no-cost-learning there is more variability in innate fitness than in learned fitness both within a set of siblings (chance), and between sets of siblings (history). However, when costly-learning is used (figures 6(b) and 6(d)), the selective pressure to bring the innate fitness closer to the learned fitness reduces the variability of the innate fitness.

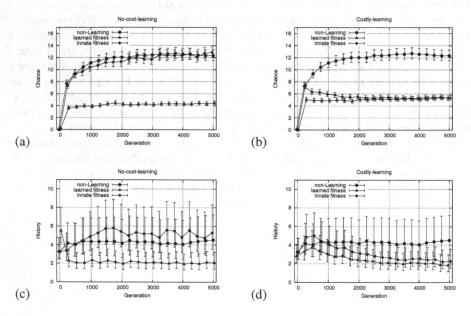

Fig. 6. The roles of chance and history. (a) Chance in no-cost-learning-populations. (b) Chance in costly-learning populations. (c) History in no-cost-learning populations. (d) History in costly-learning populations. The role of chance (in a and b) or history (in c and d) in the non-learning populations is also shown for reference.

Another effect of costly-learning that can be seen in figures 6(b) and 6(d) is that chance and history affect innate and learned fitness differently at different times. Early in the post-transfer evolution, both chance and history have a larger effect on innate fitness than on learned fitness. Later in the evolution, this difference is largely erased. The early differences appear because immediately post-transfer accumulating mutations increase the genetic diversity of the populations. This increase in genetic diversity is directly reflected by greater variability in the innate fitness of individuals and consequently in the mean innate fitness of their populations (i.e. chance). Further, the propensity for increased genetic variability to translate into increased variability in mean population fitness will vary depending upon the parent population (i.e. history). This same increase in genetic diversity affects the learned fitness as well, though less dramatically, resulting in the effects of chance and history being smaller for learned fitness than innate fitness. These early differences in the effects of chance and history diminish over time due to the increasing similarity of the innate and learned fitness values caused by the Baldwin effect, as described earlier.

Another feature of the results in figure 6 is the similarity of the effects of chance on innate fitness in no-cost-learning and non-learning populations. If no-cost-learning is hiding innate differences from selection, it would be reasonable to expect more variability in the innate fitness of the no-cost-learning populations than in the fitness of the non-learning populations. It seems likely that this observed similarity in the effects of chance on innate fitness in the no-cost-learning and non-learning populations is simply an artifact of the strength of the learning mechanism. It is expected that if more learning epochs were used, learning would mask genetic differences to a greater degree, thus lessening the selective pressure to eliminate them. The result being that effect of chance on the innate fitness of no-cost-learning populations would increase to a level greater than that for non-learning populations. Further investigation would be required to test this hypothesis.

4 Discussion

The results and analysis presented above have described how known effects of lifetime learning on evolution are expressed in the roles played by chance, history and adaptation in evolution. Costly-learning significantly increased the role played by adaptation in a population's innate fitness via the second phase of the Baldwin effect, where genetic assimilation reduces learning costs. Conversely, no-cost learning significantly reduced the role of adaptation on innate fitness, via the ability of learning to hide individual genetic differences. However, over time, the role of adaptation on innate fitness of the populations using no-cost-learning did increase due to the improved genetic robustness provided by learning. Learning, both costly and no-cost, significantly reduced the effects of both history and chance on learned fitness. Thus, two learning individuals, with distinct genetic histories and experiencing unique chance events, will be more likely to have similar learned-fitness than two similar non-learning individuals. With costly-learning, the roles of history and chance in determining innate fitness were also decreased.

Gould has famously asked what would happen if we were able to "replay life's tape" from the arrival of unicellular organisms, would the result look anything like it does now [21]? Of course the neural networks evolved here are far from unicellular organisms and the static environments are equally distant from the complex ecosystems of natural evolution. However, in their small way, these results suggest that by reducing the roles played by chance and history, the arrival of lifetime learning may increase the odds that evolution would follow a similar course if it were rerun from that point forward. Other researchers have suggested that the effects of history will differ between sexual and asexual reproduction [22]. Thus, further experiments involving more complex virtual organisms, a variety of reproductive mechanisms, and a dynamic environment incorporating coevolution, perhaps through adaptation of Kauffman's NKC landscape [23], are necessary before drawing any conclusions about the generality of these results.

References

1. Travisano, M., Mongold, J., Bennett, A., Lenski, R.: Experimental Tests of the Roles of Adaptation, Chance and History in Evolution. Science 267, 87–90 (1995)
2. Teotonio, H., Rose, M.: Variation in the reversibility of evolution. Nature 408, 463–466 (2000)
3. Joshi, A., Castilo, R., Mueller, L.: The contribution of ancestry, chance, and past and ongoing selection to adaptive evolution. Journal of Genetics 82, 147–162 (2003)
4. Losos, J., Jackman, T., Larson, A., de Quieroz, K., Rodriquez-Schettion, L.: Contingency and Determinism in Replicated Adaptive Radiations of Island Lizards. Science 279, 2115–2118 (1998)
5. Emerson, S.: A macroevolutionary study of historical contingency in the fanged frogs of Southeast Asia. Biological Journal of the Linnean Society 73, 139–151 (2001)
6. Vanooydonck, B., Irschick, D.: Is Evolution Predictable? Evolutionary Relationships of Divergence in Ecology, Performance and Morphology in Old and New World Lizard Radiations. In: Aerts, P., D'Aout, K., Herrel, A., Van Damme, R. (eds.) Topics in Functional and Ecological Vertebrate Morphology, pp. 191–204. Shaker Publishing (2002)
7. Wagenaar, D., Adami, C.: Influence of Chance, History, and Adaptation on Digital Evolution. Artificial Life 10, 181–190 (2004)
8. Harvey, I., Husbands, P., Cliff, D.: Seeing the Light: Artificial Evolution; Real Vision. In: Cliff, D., Husbands, P., Meyer, J.-A., Wilson, S. (eds.) From Animals to Animats 3: Proceedings of the Third International Conference on Simulation of Adaptive Behavior, pp. 392–401. MIT Press, Cambridge (1994)
9. Mondada, F., Floreano, D.: Evolution of Neural Control Structures: Some Experiments in Mobile Robots. Robotics and Autonomous Systems 16, 183–195 (1995)
10. Lee, P., Hallam, J., Lund, H.: Learning Complex Robot Behaviours by Evolutionary Computing with Task Decomposition. In: Birk, A., Demiris, J. (eds.) Learning Robots. LNCS (LNAI), vol. 1545, pp. 155–172. Springer, Heidelberg (1998)
11. Belew, R., Mitchell, M. (eds.): Adaptive individuals in evolving populations: models and algorithms. Addison-Wesley Longman, Boston (1996)
12. Turney, P., Whitley, D., Anderson, R. (eds.): Evolution, Learning, and Instinct:100 Years of the Baldwin Effect. Special Issue of: Evolutionary Computation 4 (1996)
13. Hinton, G., Nowlan, S.: How Learning can Guide Evolution. Complex Systems 1, 496–502 (1987)
14. Sokal, R., Rohlf, F.: Biometry, 3rd edn. W.H. Freeman (1994)
15. Mayley, G.: The evolutionary cost of learning. In: Mayes, P., et al. (eds.) From Animals to Animats 4: From Animals to Animats 4: Proceedings of the Fourth International Conference on Simulation of Adaptive Behavior, pp. 458–467. MIT Press, Cambridge (1996)
16. Mayley, G.: Guiding or Hiding: Explorations into the Effects of Learning on the Rate of Evolution. In: Husbands, P., Harvey, I. (eds.) Proceedings of the 4th European Conference on Artificial Life, pp. 135–144. MIT Press, Cambridge (1997)
17. Suzuki, R., Arita, T.: The Dynamic Changes in Roles of Learning Through the Baldwin Effect. Artificial Life 13, 31–43 (2007)

18. Glickman, M., Sycara, K.: Evolutionary Algorithms: Exploring the Dynamics of Self-Adaptation. In: Koza, J., et al. (eds.) Proceedings of the 3rd Annual Conference on Genetic Programming, pp. 762–769. Morgan Kauffmann, San Francisco (1998)
19. Altenberg, L.: The Evolution of Evolvability in Genetic Programming. In: Kinnear, K. (ed.) Advances in Genetic Programming, pp. 48–74. MIT Press, Cambridge (1994)
20. Bäck, T.: Self-Adaptation in Genetic Algorithms. In: Varela, F., Bourgine, P. (eds.) Proceedings of the 1st European Conference on Artificial Life, pp. 263–271. MIT Press, Cambridge (1992)
21. Gould, S.: Wonderful Life: The Burgess Shale and the Nature of History. Norton, New York (1998)
22. Teotonio, H., Rose, M.: Perspective: reverse evolution. Evolution 55, 653–660 (2001)
23. Kauffman, S., Johnsen, S.: Coevolution to the edge of chaos: Coupled fitness landscapes, poised states, and coevolutionary avalanches. Journal of Theoretical Biology 149, 467–506 (1991)

Unsupervised Acoustic Classification of Bird Species Using Hierarchical Self-organizing Maps

Edgar E. Vallejo[1], Martin L. Cody[2], and Charles E. Taylor[2]

[1] ITESM-CEM, Computer Science Dept.
Atizapan de Zaragoza, Edo. de México, 52926, México
vallejo@itesm.mx
[2] UCLA, Dept. of Ecology and Evolutionary Biology
Los Angeles, CA, 90095, USA
mlcody@ucla.edu
taylor@biology.ucla.edu

Abstract. In this paper, we propose the application of hierarchical self-organizing maps to the unsupervised acoustic classification of bird species. We describe a series of experiments on the automated categorization of tropical antbirds from their songs. Experimental results showed that accurate classification can be achieved using the proposed model. In addition, we discuss how categorization capabilities could be deployed in sensor arrays.

1 Introduction

We are engaged in a research program that aims to explore the capabilities of sensor arrays for the acoustic monitoring of bird behavior and diversity. Our long term goal is to create sensor arrays that behave as a single ensemble (Taylor, 2002). In this idealization, sensor nodes can recognize concepts and discourse intelligently about them (Lee et al, 2003). Constructing autonomous sensor arrays possessing problem solving capabilities in a variety of environments remains a challenge for artificial life research.

We believe that creating adaptive sensor arrays would be a major step towards realizing the full potential of this emerging technology (Estrin et al, 2001). Similarly, we think sensor arrays are excellent platforms for studying fundamental aspects of living systems such as emergence, self-organization and the evolution of communication systems (Collier and Taylor, 2004).

Pervasive in living entities is the remarkable ability to distinguish among different elements of the environment. This involves the identification of meaningful categories describing different aspects of the environment and is often critical for the viability of an organism (Pfeifer and Bongard, 2007). Moreover, we believe that the emergence of learnable languages in sensor arrays would be contingent to the ability of associating symbolic descriptions to cognitive salient categories (Stabler et al, 2003).

Several computational models have proven to be highly effective for the accurate classification of acoustic signals, such as hidden Markov models, among

M. Randall, H.A. Abbass, and J. Wiles (Eds.): ACAL 2007, LNAI 4828, pp. 212–221, 2007.

others (Rabiner, 1993). However, the later methods possess the limitation that they often require the explicit intervention of a teacher (i. e. supervised learning). Much of the categorization is developed in living systems without the explicit intervention of a teacher (i.e. unsupervised learning). If we are to develop adaptive sensor arrays, our computational methods should adhere to unsupervised learning, whenever possible.

In this work, we explore the capabilities of self-organizing maps for categorizing different species of birds from their songs. Moreover, we would like to pose here that a hierarchy of self-organizing maps provides an effective method for the unsupervised acoustic classification of bird species.

We conducted a series of computational experiments in which bird songs are transformed into strings of symbols and then classified from this representation using hierarchical self-organizing maps. Experimental results show that the proposed method is capable of categorizing four species of antbirds accurately.

2   Methods

2.1   Hierarchical Competitive Learning

The simplest form of a self-organizing map is the competitive learning network (Kohonen, 1997). This network consists of a single layer of output units c_i, each fully connected to a set of inputs o_j via excitatory connections w_{ij} (Hertz et al, 1991). Figure 1 shows an example of such a network.

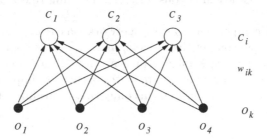

Fig. 1. Simple competitive learning network

Given an input vector $\mathbf{o}$, the winner is the unit c_{i*} with the weight vector $\mathbf{w}_{i*}$ that satifies the condition:

$$|\mathbf{w}_{i*} - \mathbf{o}| \leq |\mathbf{w}_i - \mathbf{o}| \text{ (for all } i)$$

The learning process consists of updating weights w_{i*j} for the winning unit c_{i*} only, using the standard competitive learning rule:

$$\Delta w_{i*j} = \eta(o_j - w_{i*j})$$

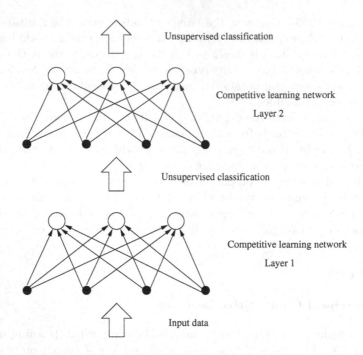

Fig. 2. Hierarchical competitive learning network

where $\eta \in [0,1]$ is the learning constant. This learning rule moves w_{i*j} directly towards o_j.

A hierarchical competitive learning network has two or more layers, each consisting of a simple competitive learning network (Kohonen, 1997). In such a network, each layer produces a new representation of the input data. Each layer in the network is expected to elucidate features that are implicit in the original representation. A two-layer hierarchical competitive learning network is depicted in Figure 2.

2.2   The Model

We propose a model for the unsupervised acoustic classification of bird species. The overall approach consists of transforming the acoustic signal of bird songs into strings of symbols. This transformation is achieved by the unsupervised classification of syllables of the original acoustic signal using a competitive learning network. Unsupervised species classification is achieved using a second competitive learning network that classifies strings of symbols from their syllable structure features.

A block diagram describing the proposed model is shown in Figure 3. Each module composing the diagram will be described in the experiments section of this paper.

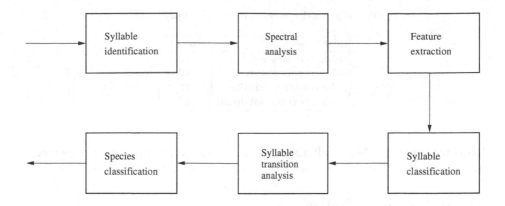

Fig. 3. Block diagram

3   Experiments and Results

3.1   Dataset

The samples used in the experiments reported here came from two sources: the Macaulay Library of Natural Sounds of the Cornell Laboratory of Ornithology and from recordings obtained in the field by one of the co-authors of this paper. The dataset consists of songs from four different antbird species that are abundant at the Montes Azules Biosphere Reserve in Chiapas, Mexico. They are listed in Table 1.

The spectrograms describing the songs of each species are shown in Figure 4. It can be appreciated that the songs from different species posses a similar structure. In effect, they consist of repetitive segments of sounds that spawns over similar

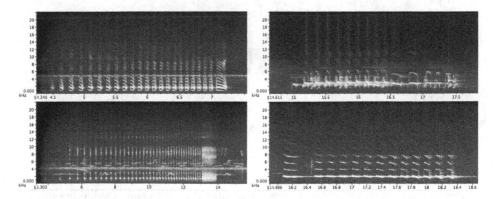

Fig. 4. Spectrograms for BAS (upper left), DAB (upper right), GAS (lower left) and MAT (lower right)

Table 1. Bird species used in the experiments

label	species	samples
BAS	Barred antshrike	12
DAB	Dusky antbird	12
GAS	Great antshrike	12
MAT	Mexican antthrush	12

frequency spectra. These similarities pose challenges for automated species recognition; especially for those methods that rely on unsupervised classification.

3.2   Syllable Identification

Bird song is thought to possess a hierarchical organization similar to that used for describing human language. In effect, bird song is typically described as consisting of phrases, syllables and elements (Catchpole and Slater, 1995). Compared to that of other singing birds, the structure of antbird songs is relatively simple. As a consequence, we believe that a two-level description consisting of songs and syllables would provide sufficient information elements for accurate automated recognition.

For experiments reported here, songs were segmented using the Raven bird song analysis program (Charif, 2004). Syllables were identified by small discontinuities in the corresponding spectrogram as shown in Figure 5.

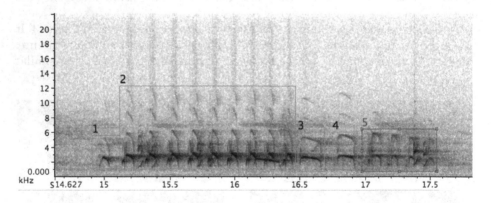

Fig. 5. Syllable identification

3.3   Spectral Analysis

Using the procedure described above, we obtained a collection a syllable samples as listed in Table 2. For each sample, we obtained a series of temporal and spectral measurements using Raven. These parameters are extracted from the sound signal using the short-time Fourier transform (STFT) (Charif, 2004).

Table 2. Number of syllable samples per species

label	samples
BAS	216
DAB	129
GAS	339
MAT	117

3.4   Feature Extraction

Previous work on species recognition using discriminant feature analysis have demonstrated the existence of minimal subsets of features for accurate discrimination of different bird species (Nelson, 1989). These subsets of features have proven to depend heavily on the species to be discriminated. However, some parameters such as high frequency and duration are commonly present in these results. From these observations, we select a collection of measurements for each syllable. These measurements are described in Table 3.

Table 3. Acoustic features

parameter	description
Low frequency	The lower frequency bound of the syllable
High frequency	The upper frequency bound of the syllable
Delta time	The duration of the syllable
Max amplitude	The upper amplitude bound of the syllable
Max power	The upper power bound of the syllable

A normalization process was applied to this data as the selected measurements spawn different orders of magnitude. Using the mean and the standard deviation of each measurement, we obtained a collection of feature vectors described as z-scores.

3.5   Syllable Classification

The collection of feature vectors describing syllables were classified using a simple competitive learning network. Once the syllables have been categorized we proceeded to represent the original songs as strings of symbols using the label from each syllable category. Table 4 shows the string representation of a subset of the songs obtained using a two-unit competitive learning network, each representing an hypothetical syllable, with $\eta = 0.1$ and epochs $= 1000$.

Similarly, Table 5 shows the string representation of a subset of the songs obtained using a four-unit competitive learning network, each representing an hypothetical syllable, with $\eta = 0.1$ and epochs $= 1000$.

Table 4. String representation of songs with 2 syllables

label	string
BAS$_1$	BBBBBBBBBBAAABABBBBBBA
BAS$_2$	BBBBBBAAAABBBAAABBBBAAAAA
BAS$_3$	BBBBBBBBBAABAABBAABBBBBBA
DAB$_1$	BBBBBBBBBBBBBBBB
DAB$_2$	BBBBBBBBBBBBBBB
DAB$_3$	BBBBBBBBBBBBBBB
GAS$_1$	BBAAAAAAAAAAAAAAAAAAAAAAAAAAAAAAAAAAAAAB
GAS$_2$	BBAAAAAAAAAAAAAAAAAAAAAAAAAAAAAAAAAAAB
GAS$_3$	BBBAAAAAAAAAAAAAAAAAAAAAAAAAAAAAAAAAAAAB
MAT$_1$	BBBBBBBBBBBBBB
MAT$_2$	BBBBBBBBBBBBBB
MAT$_3$	BBBBBBBBBBBBBB

Table 5. String representation of songs with 4 syllables

label	string
BAS$_1$	DDDAAAAAABAAAAAAAAABAC
BAS$_2$	DDDAAAAAAAABBBABBBBBBBAAB
BAS$_3$	DDDAAAAAAAAABAAAAAAAAAAC
DAB$_1$	DBBBBBBBBBBDDDDD
DAB$_2$	DBBBBBBDDDDDDD
DAB$_3$	DDBBBBBBBBBDDDD
GAS$_1$	DDDDCCCDCCCCCCCCCCCBCCACBAACCAAAABACCD
GAS$_2$	DDACCCCCBCCCCBCCCCCACAACABBABBBBAAD
GAS$_3$	DDDCDCCCCCCBCCCCBBBCCBCCAABAABABBBBBCD
MAT$_1$	DDDDDDDDDDDDDD
MAT$_2$	DBBBBBBDDDBBB
MAT$_3$	BBBBBBBBBBBBBB

3.6   Syllable Transition Analysis

Once represented as strings of syllables, bird songs are more amenable to their syntax analysis. This abstract representation of songs hides much detail of the acoustic signal and emphasizes others. For example, the calculation of syllable composition of songs is straightforward from this representation.

In this work, we propose to describe the structure of each song using the length (l), the number of different syllables (Σ) and the frequency of each pair of syllable combinations in the song. In this way, we obtained an additional collection of feature vectors as shown in Table 6 for the two-syllable experiment.

Similarly, Table 7 shows the feature vectors obtained for the four-syllable experiment. For this experiment, there are 16 (4×4) two-syllable combinations. These feature vectors were again normalized as z-scores.

Table 6. Syllable transition analysis with two syllables

label	l	Σ	AA	AB	BA	BB
BAS$_1$	22	2	2	2	3	14
BAS$_2$	25	2	9	2	3	10
BAS$_3$	25	2	3	3	4	14
DAB$_1$	15	1	0	0	0	14
DAB$_2$	14	1	0	0	0	13
DAB$_3$	14	1	0	0	0	13
GAS$_1$	39	2	35	1	1	1
GAS$_2$	36	2	32	1	1	1
GAS$_3$	38	2	33	1	1	2
MAT$_1$	13	1	0	0	0	12
MAT$_2$	13	1	0	0	0	12
MAT$_3$	13	1	0	0	0	12

Table 7. Syllable transition analysis with four syllables

label	l	Σ	AA	AB	AC	AD	BA	BB	BC	BD	CA	CB	CC	CD	DA	DB	DC	DD
BAS$_1$	22	4	13	2	1	0	2	0	0	0	0	0	0	0	1	0	0	2
BAS$_2$	25	3	8	3	0	0	2	8	0	0	0	0	0	0	1	0	0	2
BAS$_3$	25	4	18	1	1	0	1	0	0	0	0	0	0	0	1	0	0	2
DAB$_1$	15	2	0	0	0	0	0	8	0	1	0	0	0	0	0	1	0	4
DAB$_2$	14	2	0	0	0	0	0	5	0	1	0	0	0	0	0	1	0	5
DAB$_3$	14	2	0	0	0	0	0	7	0	1	0	0	0	0	0	1	0	4
GAS$_1$	39	4	4	1	3	1	2	0	1	0	2	2	16	2	0	0	2	3
GAS$_2$	36	4	2	2	3	0	2	3	2	0	3	2	12	0	1	0	0	1
GAS$_3$	38	4	2	3	0	0	2	5	4	0	1	3	10	2	0	0	2	2
MAT$_1$	13	1	0	0	0	0	0	0	0	0	0	0	0	0	0	0	0	12
MAT$_2$	13	2	0	0	0	0	0	5	0	1	0	0	0	0	0	2	0	2
MAT$_3$	13	1	0	0	0	0	0	12	0	0	0	0	0	0	0	0	0	0

Table 8. Unsupervised classification results with two-syllables

Species	classification
BAS	100%
DAB	100%
GAS	100%
MAT	92%

Table 9. Unsupervised classification results with four-syllables

Species	classification
BAS	100%
DAB	92%
GAS	100%
MAT	83%

3.7   Species Classification

The collection of feature vectors describing the structure of songs were classified, again, using a simple competitive learning network. Table 8 shows the accuracy of classification obtained for the two-syllable experiment using a four-unit competitive learning network, each representing a different species, with $\eta = 0.1$ and epochs $= 1000$.

Similarly, Table 9 shows the accuracy in classification obtained for the four-syllable experiment using a four-unit competitive learning network each representing a different species, with $\eta = 0.1$ and epochs $= 1000$.

It should be noted that the proposed model has been tested for generalization using an additional test set with similar results.

4   Discussion

Despite its preliminary character, the results shown here seem to indicate that meaningful acoustic categorization of bird species can emerge using hierarchical self-organizing maps. They also show that the accuracy in classification depends on the number of syllables describing the bird songs. This suggests the existence of a particular number of syllables for representing bird songs that is optimum for accurate species classification.

We show that using different abstraction levels for the description of bird song provides a convenient approach for analyzing different aspect of acoustic signals. On the one hand, temporal and spectral features have proven to be useful for the categorization of song segments. On the other hand, compositional features of syllables have proven to be sufficiently informative for species classification.

We think the proposed method could be extended in several ways. For instance, different sources of information could be combined within the same layer (e. g. acoustic localization of signal and the signal itself). Other extensions, such as adding higher layers would combine information from lower-layers for describing more abstract scenarios (e. g. two birds in the same territory at the same time).

It should be noted that the proposed model has only been tested in a simple simulated setting. We will test the proposed model in real settings in the near future.

Acknowledgements

This work was supported by the National Science Foundation under Award Number 0410438 and by Consejo Nacional de Ciencia y Tecnología under Award Number REF:J110.389/2006. Any opinions, findings and conclusions or recommendations expressed in this publication are those of the authors and do not necessarily reflect the views of the sponsoring agencies.

References

Catchpole, C.K., Slater, P.L.B.: Bird song biological themes and variations. Cambridge University Press, Cambridge (1995)

Charif, R.A., Clark, C.W., Fistrup, K.M.: Raven 1.2 user's manual. Cornell Laboratory of Ornithology, Ithaca, NY (2004)

Collier, T.C., Taylor, C.E.: Self-Organization in Sensor Networks. Journal of Parallel and Distributed Computing 64(7), 866–873 (2004)

Estrin, D., Girod, L., Pottie, G.: Srivastava, M.: Instrumenting the world with wireless sensor networks. In: Proceedings of the International Conference on Acoustics, Speech and Signal Processing (ICASSP (2001)

Hertz, J., Krogh, A., Palmer, R.G.: Introduction to the theory of neural computation. Addison-Wesley, Reading (1991)

Kohonen, T.: Self-organizing maps, 2nd edn. Springer, Heidelberg (1997)

Lee, Y., Riggle, J., Collier, T.C., et al.: Adaptive communication among collaborative agents: Preliminary results with symbol grounding. In: Sugisaka, M., Tanaka, H. (eds.) AROB 8th. Proceedings of the Eighth International Symposium on Artificial Life and Robotics, Beppu, Oita Japan, January 24-26, 2003, pp. 149–155 (2003)

Nelson, D.A.: The importance of invariant and distinctive features in species recognition of bird song. Condor 91(1), 120–130 (1989)

Pfeifer, R., Bongard, J.: How the body shapes the way we think. MIT Press, Cambridge (2007)

Rabiner, L., Juang, B.H.: Fundamentals of speech recognition. Prentice-Hall, Englewood Cliffs (1993)

Stabler, E.P., Collier, T.C., Kobele, G.M., et al.: The learning and emergence of mildly context sensitive languages. In: Banzhaf, W., Ziegler, J., Christaller, T., Dittrich, P., Kim, J.T. (eds.) ECAL 2003. LNCS (LNAI), vol. 2801, Springer, Heidelberg (2003)

Taylor, C.E.: From cognition in animals to cognition in superorganisms. In: Bekoff, M., Allen, C., Gurghardt, G. (eds.) The Cognitive Animal. Empirical and Theoretical Perspectives on Animal Cognition, MIT Press, Cambridge (2002)

The Prisoner's Dilemma with Image Scoring on Networks: How Does a Player's Strategy Depend on Its Place in the Social Network?

Markus Brede

CSIRO Marine and Atmospheric Research,
GPO Box 284, Canberra 2601, Australia

Abstract. We investigate the evolution of cooperation in the prisoner's dilemma on different types of interaction networks. Agents interact with their network neighbours. An agent is classified by a value $S \in [-1, 1]$ denoting its strategy and by its image score. It will cooperate if the opponents relative image score is above S and defect otherwise. Agents spread their strategies to their network neighbours proportionally to payoff differences. We find that network topology strongly influences the average cooperation rate; networks with low degree variance allowing for the largest amount of cooperation. In heterogeneous networks an agents place in the network strongly influences its strategy. Thus, agents on hub nodes are found to 'police' the population, while being on low degree nodes tends to favour over-generous less discriminating strategies.

Keywords: Evolution of cooperation, prisoner's dilemma, heterogeneous populations, scale-free graphs.

1   Introduction

A fundamental question in sociobiology is how cooperation and altruism have evolved. Despite its importance, understanding the evolution of cooperation remains one of the challenges to current research (for a summary see, e.g., [1]). In a series of articles starting with [3] particularly the question of the evolution of cooperation not based on kin selection as in many examples in the animal kingdom, but on the evolution of indirect reciprocity [3] (and [4] for experiments) and moral systems [2] has found much emphasis.

The prisoner's dilemma is frequently studied as a common framework for the evolution of cooperation between unrelated individuals [5]. In the prisoner's dilemma, individuals are faced with a choice between two strategies cooperation and defection. For mutual cooperation both receive a reward R, for mutual defection P, whereas a defector exploiting a cooperator receives T while the exploited cooperator gets an amount S. The balance between payoffs is chosen as $T > R > P > S$ and $R > (T+S)/2$, such that in a one-shot interaction defection appears the best choice for both players individually whereas mutual cooperation gives the better average reward for both. Repeated interactions in the iterated prisoner's dilemma [5] or taking into account the opponents reputation in more

M. Randall, H.A. Abbass, and J. Wiles (Eds.): ACAL 2007, LNAI 4828, pp. 222–231, 2007.

sophisticated strategies can lead to more cooperative behaviour than in the one-shot game.

Starting from the important observation that most contact networks exhibit very heterogeneous topologies [6], the evolution of cooperation on graphs has found some interest recently [8,9,10,11,12]. For instance, in [8,9,10] it is found that scale-free interaction networks grown after the preferential attachment procedure of Barabási and Albert [14] greatly facilitate cooperation for the one-shot prisoner's dilemma. In this paper, we re-examine the model of Santos et al. [10] in a situation where players have knowledge about their opponents previous behaviour and can evolve more sophisticated strategies than simple defection or cooperation, based on their opponents reputation. We investigate the evolution of cooperation based on image scoring on several types of networks and explain how individuals' strategies vary with their place in the network.

2 Simulations

We consider a set of agents that play a standard prisoner's dilemma, the parametrization of which follows the conventional setting $2 > T = b > 1, R = 1, P = S = 0$ introduced in [7]. Agents are identified with nodes of a network and interact only with their network neighbours. In a typical iteration, in random order, every agent will interact with each of its neighbours once, but again in each iteration in a new randomly diced order. Thus, agents with different numbers of neighbours interact a different number of times. The total number of interactions in one round is given by the number of links in the network.

It is probably worthwhile to stress that the situation even for interactions via a random network is different from interactions in a well mixed population. In the latter case, agents typically have randomly chosen interaction partners, that are different in each round of the evolution. In contrast, in the network scenario agents always interact with the same neighbours (even though they might change their strategies, see below).

In our experiment agents' decisions whether to cooperate or defect are based on memory about their opponents' behaviour towards other agents in the past. More precisely, this is modelled as a variation of the imagescore framework introduced in [3]. Thus, every agent is associated with an image score I that is increased by one if the agent cooperates in an interaction and decreased by one if it defected. Likewise an agent is characterized by a threshold value $S \in [-1, 1]$ that defines its strategy. If an agent with strategy S interacts with another agent with imagescore I who had M previous interactions, the first agent will cooperate if $I/M \geq S$ and defect otherwise. The introduction of relative imagescores and strategies as expectations on relative imagescores of opponents proves necessary because players may have different numbers of interactions.

Clearly, a value of $S = -1$ defines a pure cooperator, $S = 1$ a defector while in the terminology of [3] $S = 0$ is a discriminator. Values in between characterize different levels of discrimination with $S < 0$ describing a tendency to cooperate even against 'bad' opponents while a strategy $S > 0$ will defect more often than cooperate even against 'good' opponents. The average of S over the population also gives an indication whether interactions are typically cooperative. In our experiments below we allow players discrete values of $S = -1, -.9, ..., .9, 1$.

After a round of playing PD against their neighbouring agents' total payoffs π are determined. Because an agent's number of interactions is given by the number of its neighbours k the maximum payoff possible is kb. The evolution of strategies in the population is then carried out in the following way, which is the finite populations analogue of replicator dynamics [10,13]. Each node, e.g. i, on the network randomly selects one of its neighbours, e.g. j. The player at node i then adopts the strategy of player j with probability

$$p = \max\left((\pi_j - \pi_i)/(k_g b), 0\right), \tag{1}$$

where $k_g = \max(k_i, k_j)$, i.e. proportional to the payoff difference. Further, to simulate invasions of new strategies, a player will be replaced by a player with a randomly chosen strategy with probability p_{invade}.

It is to be noted that by having more interactions than average nodes hub nodes a priori have a higher potential fitness giving them a better chance to spread their strategies in the population. Because strategies are spread to neighbours, for a strategy to do well in the population it is required that it receives large payoffs when interacting with itself. This is the reason for the marked preponderance of cooperation in preferentially grown scale-free networks in the single shot experiments — where players either cooperate or defect (without memory) — as described in [10]. Roughly, this effect will also be observable in our experiments. However, since agents are endowed with the ability to discriminate between 'good' and 'bad' opponents it will not be dominant.

In the following section we study the evolution of cooperation on different types of networks. Our main focus lies on the interdependence between degree heterogeneity and the average rate of cooperative interactions C in the stationary state of the evolution dynamics. For this purpose, we investigate the model on three classes of networks, scale-free (SF) networks constructed after the model of Barabási and Albert [14], Erdós-Rényi random graphs (ER) [15] and regular random graphs (REG), see, e.g., [16]. The SF networks exhibit the largest variance in the degree distribution, ER have a small but non-zero degree variance, while in regular random graphs all nodes have the same degree. In all cases we ensure that the networks investigated were connected.

3   Experimental Methodology and Results

Similar to the case of unconstrained interactions in the donor recipient model of [3] for very low rates of invasions ($p_{\text{invade}} \approx 0$) we find that the population quickly evolves to a very high level of cooperation ($C \approx 1$) provided the mean degree $\langle k \rangle$ of the interaction network is large enough. Generally, in accord with observations of [3] a larger number of interactions per round, i.e. a larger number of links in the interaction network, leads to higher levels of cooperation. For the simulations below we chose $\langle k \rangle = 10$ guaranteeing that the networks

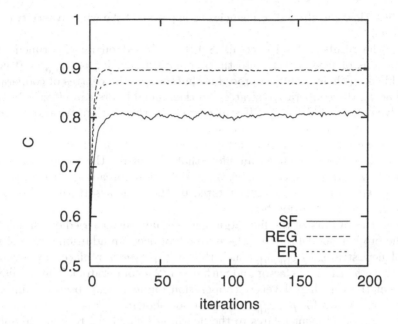

Fig. 1. Evolution of cooperation for different types of networks, SFNW (SF), regular graphs (REG) and Erdos-Renyi random graphs (ER) all with $k = 10$. Otherwise $p_{\text{invade}} = .1$, $b = 1.5$, $N = 10.000$ averaged over 100 Runs.

of $N = 10,000$ players we investigated were sparse, but dense enough that cooperation could evolve in the absence (or for low rates) of invasions. We also systematically varied link densities, observing that lower link densities enhance the results presented below, while much larger link densities reduce differences between different types of networks. The qualitative nature of our observations was, however, unchanged. All following simulation results represent averages over at least 100 different network configurations for each of which a random strategy initialization was chosen.

3.1   Evolution of Cooperation on Different Networks

In the data displayed in Figure 1 the evolution of cooperation in the three different classes of networks is compared. Initially, nodes follow randomly chosen strategies selected from a uniform distribution over $-1, -.9, ..., .9, 1$, resulting in average cooperation rates close to $C = 11/21$, the expected cooperation rate for random strategy encounters. As the evolution progresses more cooperative strategies with $S < 0$ are selected against less cooperative ones with $S > 0$ such that high levels of cooperation can be reached.

In the data one clearly observes that the largest cooperation rates are obtained on regular graphs, followed by ER random graphs and SF networks with a markedly decreased value of C. This is in stark contrast to the results of [10] observing that cooperation is strongly facilitated by SF interaction network topologies. How can these discrepancies be explained? We will advance two main arguments.

First, the result of Santos et al. [8,9,10] is based on an experiment without invasions of new players into the population (i.e. for $p_{\text{invade}} = 0$ in our case). Thus, in their evolution procedure hub nodes, the centres of cooperation, by spreading the cooperative strategy to their neighbours can effectively shield themselves off from defectors. If still present in the stationary state defectors are pushed towards low degree nodes at the periphery of the network. This effect is strongly facilitated in assortative SF networks [6]. The introduction of invasions, however, introduces an effect that can pierce the 'cooperator shield' generated by hub nodes, causing waves of defection spreading through the network. The spreading of these is facilitated by the same assortative SF structure that facilitated cooperation before.

To test the validity of the first argument we have also carried out simulations with the original model of [10], where we introduced an additional rate of invasions of new strategies p_{invade}, such that any player is replaced by a random player (with $p = 1/2$ a defector and with $q = 1/2$ a cooperator). The simulations prove that the very high levels of cooperation found in the absence of invasions quickly break down for $p_{\text{invade}} > 0$ (data not shown).

The second argument relates to the 'learning behaviour' to find the optimal strategy S. Since nodes of different degree have different numbers of interactions, pressures to learn the 'right' strategy are different for the respective players. In general, a player on a hub node can afford less efficiency in terms of payoff per interaction than a player on a low degree node. That is, players on hub nodes can allow themselves to be choosy; relatively a defection against them does not count as heavily as against a player on a low degree node. In contrast, since having only a few interactions to gain payoff from, players on low degree nodes are forced into a different mode of behaviour. Why does this impede and not facilitate cooperation? The reason is that the improved discrimination abilities evolving in hubs (caused by the many interactions they have) are traded-off against much poorer behaviour found on low degree nodes which constitute the bulk of the population. Since having only few interactions players on low degree nodes are forced into overgenerous behaviour, i.e. are expected to have lower average strategy values $\langle S \rangle$ than average nodes.

Further, having only few interactions favours specialization. Thus, being not exposed to a representative mix of the overall population low degree nodes on the periphery are more likely to adapt to the pecularities of a few special neighbours. This delays the evolution into the stationary state and explains the longer transients in SF networks compared to ER random graphs and regular random graphs seen in Fig. 1.

In passing we also note that this decrease of a networks ability to facilitate cooperation with the degree variance strongly resembles synchronization problems on networks. Also in this context the same decrease of synchronizability with degree heterogeneity is observed [17].

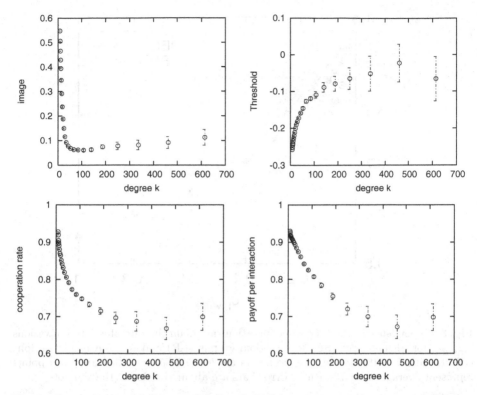

Fig. 2. Dependence of (a) image (b) threshold (c) avg. cooperation rate and (d) avg. payoff per interaction on the degree. SFNW $k = 10$, $p_{\text{invade}} = .05$, $b = 1.5$, $N = 10,000$ averaged over 100 Runs. Avg. cooperation rate is $C = .86$.

3.2 Dependance of the Role of a Player on Its Place in the Network

The above is underlined by the simulation data for the dependence of agents average image, strategy, cooperation rate and payoff per interaction on the number of neighbours shown in Figure 2. Figure 2b underlines the point made in the previous paragraph: there is a clear correlation for increasing average strategy values $\langle S \rangle$ with increasing degree. This reveals different roles for different nodes in the network. Hub nodes with average strategies $\langle S \rangle \approx 0$ work as good discriminators that police the network. Consequently they have lower cooperation rates than average nodes (cf. Fig. 2c). Hub nodes will quickly recognise cheaters and can afford to punish them, even as that means a worse image (cf. Fig. 2a) and a reduction in their payoff per interaction (cf. Fig. 2). However, the relative payoff per interaction decays slower than $1/k$, still retaining their advantage in payoff over other nodes. Thus, they are still strong spreaders of their strategies.

3.3 Stability of Cooperation

Next we investigate the dependence of the average cooperation rate on the rate of external invasions p_{invade}, i.e. the stability of cooperation to external noise.

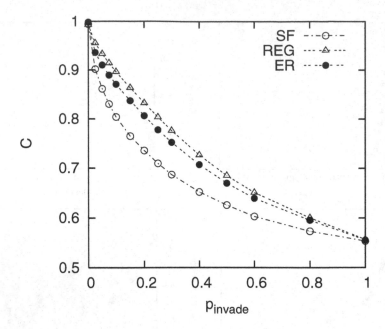

Fig. 3. Dependence of the average cooperation rate C on the rate of external invasions p_{invade} for SF networks (SF), ER random graphs (ER) and random regular graphs (REG) with $k = 10$. The other parameters are $b = 1.5$, $N = 10.000$ and data points represent averages over 100 Runs. Error bars are about the size of the symbols.

Data for this are shown in Fig. 3. Even though for low invasion rates average cooperation rates are close to perfect cooperation $C = 1$ in all three different classes of networks cooperation markedly drops for larger values of p_{invade}. As discussed above, the drop is strongest for SF networks and least for random regular graphs.

We also explored finite size effects, checking whether the results presented in Fig. 3 differ for different numbers of agents N in the population. A systematic investigation shows that the average levels of cooperation quickly saturate at the values displayed in Fig. 3 for network sizes larger than approximately $N = 100$. In smaller systems, however, the advantage of regular interaction network topologies over more heterogeneous topologies is found to be more expressed.

Differences in the cooperation rates for the three different types of networks are of the order of 10%-20% larger for smaller link densities and smaller on more densely connected networks. The maximum difference is typically realized for invasion rates around $p_{\text{invade}} = 1/4$, becoming much smaller as the noise level is further increased.

In summary, our experiments in this section verify that the statement that random regular graphs allow for more cooperation than ER and SF graphs holds over the entire range of invasion rates. This being so indicates that some contact networks facilitate cooperation while others impede it. Naturally the question arises: What is the network that allows for the most cooperation? The absence of a good and easy to calculate indicator of cooperation on a network (like, e.g., the eigenvalue criterion for the synchronizability of networks [18]) renders a direct optimization approach as in [17] unfeasible. Nevertheless our observations seem to encourage a more extensive study in the spirit of the present paper, addressing the dependence of the level of cooperation on specific network properties, like the link reciprocity (for directed graphs), cliquishness or the degree assortativeness.

4   Conclusions

In this paper we have presented simulation results about the evolution of cooperation in a prisoner's dilemma where players' interactions are defined by a contact network. Players' decisions whether to defect or cooperate are determined by a strategy that takes into account their opponents reputation which is determined by the opponent's previous behaviour against other players.

Investigating the evolution on different types of networks we observe that in contrast to previous results for the one-shot prisoner's dilemma [10] cooperation is typically facilitated by very homogeneous contact networks, where no player has more interaction partners than another. The difference between the results for the one-shot situation and the more complicated scenario where players take into account opponents reputation is explained by differences between the learning behaviour of players with different numbers of interaction partners. Whereas players on hub nodes evolve to become very effective discriminators that (by punishing invadors that pursue strategies that prefer defection) effectively police the population, players on low degree nodes become over-generous. The over-generous behaviour of players on low degree nodes results from two sources. One reason is an increased pressure to cooperate (since individual interactions count more heavily players with fewer interactions can less well afford the risk to defect in a generally cooperative environment). The second is rooted in the overspecialization to a small number of neighbours rather than adaptation to a representative sample of the population.

An early criticism against the simple variant of image scoring, that we based our experiments on, was that it hardly appears reasonable to decrease the reputation of a player who justly defects against (and thus punishes) an opponent. A more realistic framework should thus be more elaborate, allowing for a more sophisticated set of values for each individual player like in [19]. Such a scheme, however, is second order. It requires every player to form his own perception about the image of its interaction partners. For reasons of computation time we have avoided such a more complicated setup in this study. Further experiments, to be reported elsewhere, however, suggest that the basic result that cooperation in scenarios with sophisticated strategies that allow for memory is facilitated by homogeneous networks is independant of the detailed setup of the experiment.

In a more speculative vein, our results suggest several interesting patterns that could be tested in social interactions. For instance, it appears worthwhile to investigate whether a correlation between an individual's place in a social network and its likelihood to cooperate can be verified. Second, an interesting sociological study might address the question whether hub node individuals really are more likely to enforce cooperation?

Acknowledgements

The authors wishes to thank Fabio Boschetti and David Newth for carefully reading and commenting on the manuscript.

References

1. Hammerstein, P. (ed.): Genetic and Cultural Evolution of Cooperation. MIT Press, Cambridge (2003)
2. Richerson, P.J., Boyd, R., Henrich, J.: The Cultural Evolution of Human Cooperation. In: Hammerstein, P. (ed.) The Genetic and Cultural Evolution of Cooperation, pp. 357–388. MIT Press, Cambridge (1999)
3. Nowak, M.A., Sigmund, K.: Evolution of Indirect Reciprocity by Image Scoring. Nature 393, 573–577 (1996)
4. Wedekind, C., Milinski, M.: Cooperation Through Image Scoring in Humans. Science 280, 850–852 (2000)
5. Axelrod, R., Hamilton, W.D.: The Evolution of Cooperation. Science 211, 1390 (1981)
6. Albert, R., Barabási, A.L.: Statistical Mechanics of Complex Networks. Rev. Mod. Phys. 74, 47–97 (2002)
7. Nowak, M.A., May, R.M.: Evolutionary Games and Spatial Chaos. Nature 359, 826–829 (1992)
8. Santos, F.C., Pacheco, J.M.: Scale-Free Networks Provide a Unifying Framework for the Emergence of Cooperation. Phys. Rev. Lett. 95, 098104 (2005)
9. Santos, F.C., Pacheco, J.M., Lenaerts, T.: Evolutionary Dynamics of Social Dilemmas in Structured Homogeneous Populations. Proc. Natl. Sci. USA 103, 3490–3494 (2006)
10. Santos, F.C., Rodrigues, J.F., Pacheco, J.M.: Graph Topology Plays a Determinant Role in the Evolution of Cooperation. Proc. R. Soc. B 273, 51–55 (2006)
11. Ohtsuki, H., Hauert, C., Nowak, M.A.: A simple rule for the evolution of cooperation on graphs. Nature 441, 502–505 (2006)
12. Taylor, P.D., Day, T., Wild, G.: Evolution of Cooperation in a Finit Homogeneous Graph. Nature 447, 469–472 (2007)
13. Gintis, H.: Game Theory Evolving. Princeton University Press, Princeton NJ (2000)
14. Barabási, A.L., Albert, R.: Emergence of Scaling in Random Networks. Science 286, 509–512 (1999)
15. Erdös, P., Rényi, A.: On Random Graphs I. Publicationes Mathematicae. Debrecen 6, 290–297 (1959)
16. Wormwald, N.D.: Models of Random Regular Graphs. In: Lambd, J.D., Preece, D.A. (eds.) London Mathematical Society Lecture Note Series, Cambridge University Press, Cambridge (1999)

17. Newth, D., Brede, M.: Fitness Landscape Analysis and Optimization of Coupled Oscillators. Complex Systems 16, 317–331 (2006)
18. Pecora, L.M., Carroll, T.L.: Master Stability Function for Synchronized Coupled Systems. Phys. Rev. Lett. 80(10), 2109–2112 (1998)
19. Ohtsuki, H., Iwasa, Y.: The leading eight: Social norms that can maintain cooperation by indirect reciprocity. Journal of Theoretical Biology 239(4), 435–444 (2006)

Population-Based Ant Colony Optimisation for Multi-objective Function Optimisation

Daniel Angus

Complex Intelligent Systems Laboratory
Centre for Information Technology Research
Faculty of Information and Communication Technologies
Swinburne University of Technology
Melbourne, Australia, 3122
dangus@ict.swin.edu.au

Abstract. Ant inspired algorithms have recently gained popularity for use in multi-objective problem domains. The Population-based ACO, which uses a population of solutions as well as the traditional pheromone matrix, has been demonstrated as an effective problem solving strategy for solving combinatorial multi-objective optimisation problems, although this algorithm has yet to be applied to multi-objective function optimisation problems. This paper tests the suitability of a Population-based ACO algorithm for the multi-objective function optimisation problem. Results are given for a suite of problems of varying complexity.

1   Introduction

Ant Colony Optimisation (ACO) [10, 13], an optimisation methodology based on the foraging behaviour of Argentine ants, has been shown to be useful in the location of optimal or near-optimal solutions to optimisation problems. The first applications of ACO algorithms were to combinatorial optimisation problems such as the Travelling Salesman Problem (TSP) and Quadratic Assignment Problem (QAP), although in recent years the paradigm has been applied to a much wider range of problem domains. A problem domain that remains relatively unexplored by ACO though is Multi-objective Function Optimisation (MOFO).

Multiple Objective Optimisation (MOO) is concerned with finding multiple 'trade-off' solutions in order to optimise many (in most cases conflicting or orthogonal) objectives. MOO problems are found frequently in real-world applications and a commonly cited example is designing an automobile, where a designer may be attempting to simultaneously decrease cost and increase safety and comfort. In this example it is fairly clear that the designer must be willing to make trade-offs between all of these objectives because they are non-complimentary. For all MOO problems there is a set of optimal trade-off solutions which are referred to as the Pareto set, after the economist Vilfredo Pareto. To be classified as Pareto optimal a solution must not be worse then any other valid solution *in all objectives*. A Pareto optimal solution cannot increase its quality in any objective without simultaneously decreasing its quality in another objective.

M. Randall, H.A. Abbass, and J. Wiles (Eds.): ACAL 2007, LNAI 4828, pp. 232–244, 2007.

This study is concerned with the evaluation of a new Population-based ACO algorithm: Population-based ACO for Multi-objective Function Optimisation (PACO-MOFO). The PACO-MOFO algorithm is an extension of previously published work which includes the application of two novel Niching PACO algorithms to a suite of Single-objective Function Optimisation Problems [2] and an improvement of an existing PACO algorithm with a crowding replacement operation for the Multi-objective Travelling Salesman Problem (CPACO) [3]. The PACO-MOFO uses a similar approach to those of the previous works which is primarily the use of niching techniques such as crowding [9, 27] and fitness sharing [19] to maintain solution diversity, taken from the field of Evolutionary Computation (EC). The purpose of maintaining diversity in MOO is to encourage an algorithm to cover the entire Pareto front rather than converging to one specific location [7]. As such, the performance of the PACO-MOFO algorithm is measured by its ability to not only locate good areas of the Pareto front but to spread the population uniformly across it. To determine this, attainment surface comparison described in Sec. 4.2 is used.

2   Population-Based Ant Colony Optimisation

The Population-based Ant Colony Optimisation (PACO) algorithm was introduced in [21, 22] and later extended for a multi-objective optimisation problem, the single machine total tardiness problem with changeover costs (PACO-MO) [20, 23]. The defining difference between PACO and canonical ACO algorithms is the method of solution storage. Whereas most traditional ACO algorithms (Ant System [12], Ant Colony Systems [17], Max-Min Ant Systems [34]) store solution information from an (artificial) ant in a pheromone matrix only, PACO stores solutions in a population and then uses this population to make adjustments to the pheromone matrix. At any time the individual pheromone values contained in the pheromone matrix will reflect the state of the population. PACO still uses the core principles of ACO which include stepwise construction (solutions are constructed one piece at a time) and the use of global information in solutions' construction, i.e. unlike a Genetic Algorithm, all solutions influence the creation of a new solution rather than a few parent solutions. In other respects, there is little difference in the overall algorithmic structure between PACO and ACO as illustrated in Figure 1, and it may be useful to think of the differences that do exist as implementation differences; although these differences can have far-reaching effects on the performance and applicability of the algorithms.

2.1   PACO for Multiple Objective Optimisation

There have been many ant-inspired algorithms proposed for multi-objective optimisation problems and Garcìa-Martínez, et.al. [18] published an excellent review and analysis paper on the subject. In that paper eight major ant-inspired algorithms along with two state-of-the-art Genetic Algorithms (NSGA-II & SPEA2) were implemented and compared. For the particular test cases used (instances

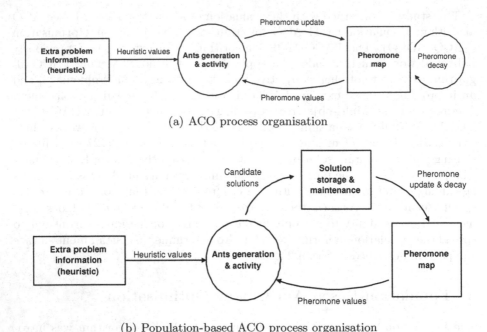

(a) ACO process organisation

(b) Population-based ACO process organisation

Fig. 1. Process organisation of a traditional ACO algorithm versus a population-based ACO algorithm using terminology defined in defined in [1, 5, 11] (taken from [1])

of a bi-criteria TSP) the ant-inspired algorithms performed well. The single Population-based ACO algorithm included in the study, PACO-MO (mentioned previously), performed consistently within the top three ant-inspired algorithms. The PACO-MO algorithm was the subject of a later study that improved not only the quality of result but reduced the overall computational complexity [3].

A trait common to ACO algorithms applied to MOO problems is the presence of multiple pheromone matrices (usually one per objective). These methods tend to store solution quality information in these multiple pheromone matrices and then recombine it according to a unique weight function associated with each artificial ant. A disadvantage of such a technique is that it assumes that information about one objective combined in even proportion with information about another objective will provide good information about the 50/50 trade-off point between these objectives. If this assumption is not true then it is probable that these algorithms may have difficulty locating specific areas of the Pareto front. In the case of PACO-MO and CPACO this assumption is not made since by maintaining a population of solutions it is quite simple to store solutions from the entire extent of the objective space and thus aim to achieve even coverage of the Pareto front. This is similar to other multiple objective Genetic Algorithms such as NSGA-II, which also use a single population of solutions. Unlike other

Multi-objective ACO algorithms CPACO uses a single temporary pheromone matrix. Each iteration a new pheromone matrix is calculated as follows:

1. All solutions (s) in the population (S) are assigned an integer rank according to the dominance ranking procedure used by the NSGA-II algorithm, see [8] for details.
2. All elements in the temporary pheromone matrix are initialised to some initial value (τ_{init}).
3. All solutions in the population increment the corresponding elements in the temporary pheromone matrix according to the inverse of their rank, i.e. $\Delta\tau_{ij}^s = 1/\left(s_{rank}\right)$.

2.2   PACO for Continuous Domains

To date, many ant-inspired approaches for application to function optimisation problems have been proposed, with the Ant Colony Metaphor for Continuous Design Spaces [4] being the first. That algorithm starts by placing a 'nest' somewhere in the n-dimensional search space. after which it projects a group of vectors (ants) into the search space around the nest. Over successive iterations it gradually adjusts the direction of these vectors towards promising areas of the search space. Other approaches include ACO for Continuous Domains with Aggregation Pheromones Metaphor (APS) [35], Continuous Interacting Ant Colony (CIAC) [14] and Continuous ACO (CACO) [29].

ACO for Continuous Domains (ACOCD) [32, 33], an extension of PACO, maintains a population where every population member represents a single point in the n-dimensional search space. Each population member is also assigned a quality which is used for selection purposes. Solution construction is achieved by way of sampling each dimension in turn (stepwise) using a combination of the population's Probability Density Functions to resolve each new point. Newly constructed solutions are evaluated if they fall within the bounds of solution space (otherwise they are discarded) and, if valid, inserted into the population using a quality based replacement strategy.

It is worth mentioning that, to date, there has been one other ant-inspired approach for the MOFO problem [30], however, since this approach was based on the Ant Colony Metaphor for Continuous Design Spaces it is strictly not an ACO algorithm per se [32]. While the algorithm was demonstrated as a good approach for the problems tested, it resembles something closer to a Genetic Algorithm since it uses crossover and mutation to generate new solutions rather than stepwise construction and as such is not included here for comparison.

3   PACO-MOFO

PACO-MOFO reuses components of both ACOCD and CPACO. Considering the problem domain, a pheromone matrix comprised of discrete values is not required, instead Probability Density Functions are used in a similar fashion to

ACOCD. A discrete population is utilised that is modified using a crowding distance comparison operation. The crowding replacement operation ensures that the population retains diverse solutions from the objective space, while a fitness sharing selection strategy encourages an even sampling of the objective space. PACO-MOFO is an *a posteori* preference articulation method as defined in [25], i.e., no preference information is provided to the search process before it begins, rather the search process generates a set of good solutions for the decision maker who then selects from this set. Algorithm 1 outlines the PACO-MOFO algorithm details.

Algorithm 1. Population-based ACO for Multi-objective Function Optimisation (PACO-MOFO)

1: Initialise population with uniform random solutions (*size* = *number of ants*)
2: **while** stopping criterion not met **do**
3:     Rank population according to NSGA2 fitness ranking procedure
4:     Set fitness of each population member as inverse of rank
5:     Adjust fitness of population by applying fitness sharing
6:     **for** $i = 1$ to *number of ants* **do**
7:         Create new empty solution s_i^{new}
8:         **for** $j = 1$ to *number of dimensions* **do**
9:             Probabilistically select a solution (s) from the population based on the adjusted fitness raised to a history exponent power (fitness$^\alpha$), using a biased roulette wheel selection strategy with replacement.
10:            $\mu = s_j$ ▷ Calculate mean
11:            r = Dimension j's range
12:            $c = (\sin(\pi/2 \times$ remaining evaluations/maximum evaluations$))^2$
13:            $\sigma = r \times c/6$ ▷ Calculate standard deviation
14:            **repeat**
15:                $s_{i,j}^{new}$ = Gaussian weighted random value using calculated σ and μ.
16:            **until** $s_{i,j}^{new}$ is within bounds of dimension j
17:        **end for**
18:        Evaluate new solution s_i^{new} for all objectives
19:    **end for**
20:    **for** $i = 1$ to *number of ants* **do**
21:        Select random subset of solutions S from population
22:        **if** s_i^{new} is better in all objectives (strongly dominates) than closest matching solution from S (closest match in terms of objective space) **then**
23:            Replace closest matching solution with s_i^{new}
24:        **else**
25:            Discard s_i^{new}
26:        **end if**
27:    **end for**
28: **end while**

As was previously mentioned PACO-MOFO uses two forms of niching, crowding and fitness sharing, for the population replacement and selection mechanisms. The specific crowding technique used is Restricted Tournament Selection (RTS) [24]. RTS works by selecting a random subset of the population (crowding window size) and comparing a new solution against this subset to determine the closest match. If the new solution is better than the closest match (in the case of MOO better means strongly dominating) the closest match is replaced, otherwise the new solution is discarded. In this sense crowding manages the amount of stored diversity in the population. Just using crowding to control diversity is not enough, however, since selection of solutions from the population for new solution construction must utilise the stored diversity. If only certain solutions are sampled from the population then this negates the utility of having diversity in the population in the first place. In PACO-MOFO fitness sharing [19] has been employed to ensure a balanced sampling of the population. It works by grouping population members according to their proximity in the objective space and derating their quality if they belong to a densely packed neighbourhood of solutions. Fitness sharing advantages the reuse of good quality solutions that are located in sparsely populated areas of the objective space.

4   Test Setup

4.1   Test Problems

The problems selected for testing the proposed algorithm are well documented benchmark test functions. They are taken from [36], and the nomenclature from that source reused, although primary sources are also included where relevant.

MOP1. Schaffer's two objective function [31] is a historically significant test function. It has a one dimensional decision space and two objectives. It is usually unbounded or defined over large bounds, we use the bounds: $-10^5 \leq x \leq 10^5$.

$$f_1(x) = x^2$$
$$f_2(x) = (x-2)^2 \tag{1}$$

MOP2. Fonseca's two objective function [15] is a useful test problem since it allows arbitrary scaling of the number of decision variables without changing the shape of the Pareto front which is continuous and concave. The decision variable space is defined for each dimension as: $-4 \leq x_i \leq 4; i = 1, 2, \ldots, n$

$$f_1(x) = 1 - \exp\left(-\sum_{i=1}^{n}\left(x_i - \frac{1}{\sqrt{n}}\right)^2\right)$$
$$f_2(x) = 1 - \exp\left(-\sum_{i=1}^{n}\left(x_i + \frac{1}{\sqrt{n}}\right)^2\right) \tag{2}$$

MOP3. Poloni's two objective function [28] contains two decision variables and is a maximisation problem that has two discontinuous Pareto fronts. Its decision space is bounded as: $-\pi \leq x, y \leq \pi$

$$f_1(x,y) = -\left[1 + (A_1 - B_1)^2 + (A_2 - B_2)^2\right]$$
$$f_2(x,y) = -\left[(x+3)^2 + (y+1)^2\right] \tag{3}$$
$$A_1 = 0.5 \times \sin(1) - 2.0 \times \cos(1) + \sin(2) - 1.5 \times \cos(2)$$
$$A_2 = 1.5 \times \sin(1) - \cos(1) + 2 \times \sin(2) - 0.5 \times \cos(2)$$
$$B_1 = 0.5 \times \sin(x) - 2.0 \times \cos(x) + \sin(y) - 1.5 \times \cos(y)$$
$$B_2 = 1.5 \times \sin(x) - \cos(x) + 2\sin(y) - 0.5 \times \cos(y)$$

MOP6. This last problem was proposed by Deb [6] and is a two objective, two decision variable problem with four discontinuous Pareto fronts. The decision variable space is defined as: $0 \le x, y \le 1$

$$f_1(x,y) = x$$
$$f_2(x,y) = (1 + 10y) \times \left[1 - \left(\frac{x}{1+10y}\right)^2 - \frac{x}{1+10y}\sin(8\pi x)\right] \tag{4}$$

4.2   Performance Metric: Summary Attainment Surface

The metric selected to measure the performance of the PACO-MOFO algorithm is the summary attainment surface analysis technique [26]. This metric has been chosen since it takes into consideration an algorithm's ability to not only find solutions close to the Pareto front, but also to obtain a diverse range of solutions along the Pareto front. The method is an extension of the work of [16] which was developed to determine the median performance of a stochastic multi-objective optimisation algorithm over several experimental runs. All summary attainment surfaces produced in this study have been created with the aid of Knowles' software package available from http://dbkgroup.org/knowles/plot_attainments/.

5   Results and Analysis

5.1   Testing with Two Decision Variables

Both PACO-MOFO and NSGA2 were run 100 times on all test problems defined for two decision variables and two objectives. Each algorithm was allowed 100,000 solution evaluations and the final population was recorded. The algorithm configuration parameters were the same for all trials and are reproduced in Tab. 1. The average attainment surfaces obtained are shown in Fig. 2.

Except for MOP1 (where NSGA2 is better) there is no statistical difference[1] between the attainment surfaces generated for the control algorithm (NSGA2)

---

[1] Verified by comparing the attainment surfaces using non-parametric statistics.

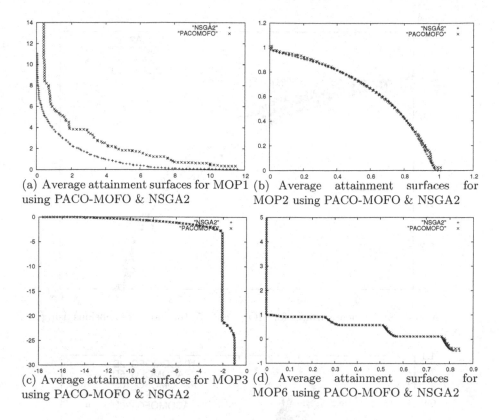

Fig. 2. Attainment surfaces generated from 100 runs of PACO-MOFO and NSGA2 applied to benchmark functions: MOP1, MOP2, MOP3 & MOP6

and the PACO-MOFO algorithm. This is a good result as it demonstrates that on these few benchmark instances the PACO-MOFO algorithm performs on par with an accepted state-of-the-art MOO algorithm in three out of four test cases. Remembering that the intention was not to outperform NSGA2, simply to obtain a result that was somewhat comparable.

5.2 Testing with More Decision Variables

The MOP2 problem was included since it allows arbitrary scaling of the decision space without affecting the shape of the Pareto front. To test the PACO-MOFO algorithms ability to scale (with regard to the number of decision variables) the MOP2 problem was defined in various decision variable dimensions (5,10,15,20) and compared against the NSGA2 algorithm (Fig. 5.1).

The MOP2 problem is extremely difficult in higher dimensions due to the lack of directional information (i.e. the objective functions return a value of 1 for most points in the decision variable space). Given that a random initial population will most likely contain many solutions with objective values of (1.0, 1.0),

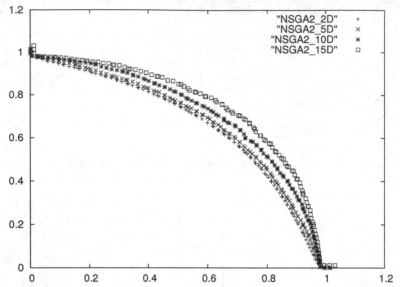

(a) Average attainment surfaces for MOP2 in 2, 5, 10, and 15 dimensions using NSGA2

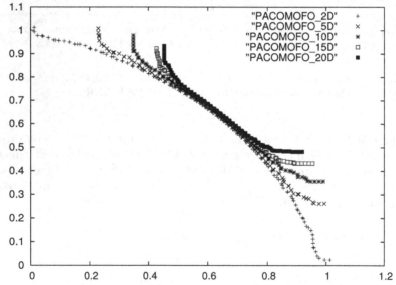

(b) Average attainment surfaces for MOP2 in 2, 5, 10, 15 and 20 dimensions using PACO-MOFO

Fig. 3. Attainment surfaces generated from 100 runs of PACO-MOFO and NSGA2 applied to the benchmark function, MOP2, in 2, 5, 10, 15 and 20 dimensions

Table 1. Algorithm parameter settings

Algorithm	No. of Ants (m) / Population Size	History Exponent	Crowding Window	Fitness Sharing Radius (h = objectives)
PACO-MOCO	50	1.0	0.5	$1/\left((m)^{1/h}-1\right)$
NSGA2	100	n/a	n/a	n/a

Algorithm	Fitness Sharing Power	Crossover Probability	Mutation Probability	Std. Dev. of Gaussian Mutation
PACO-MOCO	1.0	n/a	n/a	n/a
NSGA2	n/a	0.97	0.50	1% of dimension range

the PACO-MOFO algorithm will approach the Pareto front from the top-right corner of Fig. 3(b). Since a crowding replacement scheme is being used, the algorithm will concentrate its search effort on the few good solutions found at the middle of the Pareto front and thus will be slow in exploring outward across the extent of the entire Pareto front, thus explaining the convex Pareto fronts generated. NSGA2 does not suffer from this problem since its population is more volatile and it replaces solutions in a generational manner meaning a quicker uptake of solutions from across the entire Pareto front. The trade-off to this is that in the centre of the Pareto front the solutions found are worse than those found by PACO-MOFO. It cannot be said then which algorithm is better, however, interestingly NSGA2 failed to find any solutions other than those with objective values (1.0, 1.0) above 20 dimensions, while it took PACO-MOFO until 30 dimensions to do the same.

5.3   Computational Efficiency

Not accounting for possible implementation inefficiencies, as a general guide the run times of the NSGA2 and PACO-MOFO were approximately equal. More precisely though the computational complexity of the PACO-MOFO algorithm is derived mostly from two sources, solution ranking (including fitness sharing) and population maintenance (replacement). The solution ranking procedure comprises two steps, the first step is the NSGA2 ranking procedure which has a worst case complexity of $O(hN^2)$ where h is the number of objectives, and N the population size. The second step is the fitness sharing quality adjustment which has a complexity of $O(N^2)$ since every population member has to determine its closest members. It may be possible to mitigate some of this computational effort by combining these two steps, and this is an area of likely future work. The complexity of the crowding replacement operation is dependent on the size of the crowding window (w) since it selects a subset (in this case 1/2 of N) of the population and uses an objective space comparison to find the closest subset member to the new solution and then performs a single non-dominance check, of total complexity $O((N/w)^2 + h)$. The solution creation procedure is relatively lean compared to these other processes and as such has not been included. Compared to NSGA2 the PACO-MOFO algorithm is comparable since both algorithms perform similar sort and crowding distance calculations.

6   Conclusion and Future Work

This paper has introduced a new Population-based ACO algorithm for Multi-objective Function Optimisation, PACO-MOFO. It was never expected that PACO-MOFO was going to be able to achieve a superior result to a second generation Genetic Algorithm such as NSGA2 on first application, however comparable results were obtained on three out of four cases. As the first proof-of-concept study a good starting point for future work has been set and it is expected that with more work, improvements to the solution quality and computational efficiency can be made. Some important results obtained were the insights gained into the sampling behaviour of the algorithm observed in Sec. 5.2. As always, time and space constraints limit the number of test cases able to be included, however an already continuing area of future work is the application of PACO-MOFO to MOFO problems with larger numbers of objectives. Perhaps the most significant contribution of this paper is the expansion of the area of applicability of Population-based ACO algorithms to another problem domain.

Acknowledgement

The author acknowledges the guidance Prof. Tim Hendtlass has provided as well as the rest of the team at CIS.

References

[1] Angus, D.: Niching for ant colony optimization. Tech. rep. Faculty of Information and Communication Technology, Swinburne University of Technology (2006)

[2] Angus, D.: Niching for Population-based Ant Colony Optimization. In: 2nd International IEEE Conference on e-Science and Grid Computing, Workshop on Biologically-inspired Optimisation Methods for Parallel and Distributed Architectures: Algorithms, Systems and Applications (2006)

[3] Angus, D.: Crowding population-based ant colony optimisation for the multi-objective travelling salesman problem. In: 2007 IEEE Symposium on Computational Intelligence in Multi-Criteria Decision-Making (MCDM 2007), pp. 333–340. IEEE, Los Alamitos (2007)

[4] Bilchev, G., Parmee, I.: The ant colony metaphor for searching continuous design spaces. In: Fogarty, T. (ed.) AISB 1995. LNCS, vol. 993, pp. 25–39. Springer, Heidelberg (1995)

[5] Cordón, O., Herrera, F., et al.: A review of the ant colony optimization metaheuristic: Basis, models and new trends. Mathware & Soft Computing 9(2,3) (2002)

[6] Deb, K.: Multi-Objective Genetic Algorithms: Problem Difficulties and Construction of Test Problems. Tech. Rep. CI-49/98, Department of Computer Science/LS11, University of Dortmund, Dortmund, Germany (1998)

[7] Deb, K.: Multi-Objective Optimization using Evolutionary Algorithms, 2nd edn. Wiley-Interscience Series in Systems and Optimization. John Wiley & Son, Chichester (2002)

[8] Deb, K., Agrawal, S., et al.: A fast elitist non-dominated sorting genetic algorithm for multi-objective optimization: NSGA-II. In: Schoenauer, M., Deb, K., et al. (eds.) Parallel Problem Solving from Nature – PPSN VI, pp. 849–858. Springer, Berlin (2000)

[9] De Jong, K.A.: An analysis of the behaviour of a class of genetic adaptive systems. Ph.D. thesis, University of Michigan (1975)

[10] Dorigo, M.: Optimization, Learning and Natural Algorithms. Ph.D. thesis, Politechico di Milano, Italy (1992)

[11] Dorigo, M., Bonabeau, E., et al.: Ant algorithms and stigmergy. Future Generation Computer Systems 16, 851–871 (2000)

[12] Dorigo, M., Maniezzo, V., et al.: The ant system: Optimization by a colony of cooperating agents. IEEE Transactions on Systems, Man and Cybernetics, Part B 26(1), 29–41 (1996)

[13] Dorigo, M., Stützle, T.: Ant Colony Optimization. MIT Press, London (2004)

[14] Dréo, J., Siarry, P.: Continuous interacting ant colony algorithm based on dense heterarchy. Future Generation Computer Systems 20(5), 841–856 (2004)

[15] Fonseca, C.M., Fleming, P.J.: Multiobjective genetic algorithms made easy: selection sharing and mating restriction. In: First International Conference on Genetic Algorithms in Engineering Systems: Innovations and Applications, pp. 45–52 (September 1995)

[16] Fonseca, C.M., Fleming, P.J.: On the performance assessment and comparison of stochastic multiobjective optimizers. In: PPSN IV: Proceedings of the 4th International Conference on Parallel Problem Solving from Nature, pp. 584–593. Springer, London (1996)

[17] Gambardella, L., Dorigo, M.: Solving symmetric and asymmetric TSPs by ant colonies. In: Proceedings of the third IEEE International Conference on Evolutionary Computation (ICEC), pp. 622–627. IEEE Press, Nagoya, Japan (1996)

[18] Garcìa-Martínez, C., Cordón, O., et al.: A Taxonomy and an Empirical Análisis of Multiple Objective Ant Colony Optimization Algorithms for Bi-criteria TSP. European Journal of Operational Research (2006)

[19] Goldberg, D.E., Richardson, J.: Genetic algorithms with sharing for multimodal function optimization. In: Proceedings of the Second International Conference on Genetic Algorithms, pp. 41–49 (1987)

[20] Guntsch, M.: Ant Algorithms in Stochastic and Multi-Criteria Environments. Ph.D. thesis, Universität Fridericiana zu Karlsruhe (2004)

[21] Guntsch, M., Middendorf, M.: Applying population based ACO to dynamic optimization problems. In: Dorigo, M., Di Caro, G.A., Sampels, M. (eds.) ANTS 2002. LNCS, vol. 2463, pp. 97–104. Springer, Heidelberg (2002)

[22] Guntsch, M., Middendorf, M.: A population based approach for ACO. In: Cagnoni, S., Gottlieb, J., Hart, E., Middendorf, M., Raidl, G.R. (eds.) EvoWorkshops 2002. LNCS, vol. 2279, pp. 72–81. Springer, Heidelberg (2002)

[23] Guntsch, M., Middendorf, M.: Solving Multi-criteria Optimization Problems with Population-Based ACO. In: Fonseca, C.M., Fleming, P.J., Zitzler, E., Deb, K., Thiele, L. (eds.) EMO 2003. LNCS, vol. 2632, pp. 464–478. Springer, Heidelberg (2003)

[24] Harik, G.R.: Finding multimodal solutions using restricted tournament selection. In: Eshelman, L. (ed.) Proceedings of the Sixth International Conference on Genetic Algorithms, pp. 24–31. Morgan Kaufmann, San Francisco (1995)

[25] Hwang, C.-L., Masud, A.S.M.: Multiple objective decision making, methods and applications: a state-of-the-art survey. Lecture notes in economics and mathematical systems, vol. 164. Springer, Heidelberg (1979)

[26] Knowles, J.: A summary-attainment-surface plotting method for visualizing the performance of stochastic multiobjective optimizers. In: ISDA 2005. Proceedings of the 5th International Conference on Intelligent Systems Design and Applications, pp. 552–557. IEEE Computer Society, Washington, DC, USA (2005)

[27] Mahfoud, S.W.: Niching methods for genetic algorithms. Ph.D. thesis, University of Illinois (1995)

[28] Poloni, C., Mosetti, G., et al.: Multiobjective Optimization by GAs: Application to System and Component Design. In: Computational Methods in Applied Sciences 1996: Invited Lectures and Special Technological Sessions of the Third ECCOMAS Computational Fluid Dynamics Conference and the Second ECCOMAS Conference on Numerical Methods in Engineering, pp. 258–264. Wiley, Chichester (1996)

[29] Pourtakdoust, S.H., Nobahari, H.: An Extension of Ant Colony System to Continuous Optimization Problems. In: Dorigo, M., Birattari, M., Blum, C., Gambardella, L.M., Mondada, F., Stützle, T. (eds.) ANTS 2004. LNCS, vol. 3172, pp. 294–301. Springer, Heidelberg (2004)

[30] Prakash, B.D.K., Shelokar, S., Jayaraman, V.K.: Ant algorithm for single and multiobjective reliability optimization problems. Quality and Reliability Engineering International 18(6), 497–514 (2002)

[31] Schaffer, J.D.: Multiple objective optimization with vector evaluated genetic algorithms. In: Proceedings of the 1st International Conference on Genetic Algorithms, pp. 93–100. Lawrence Erlbaum Associates, Inc. Mahwah, NJ, USA (1985)

[32] Socha, K.: ACO for Continuous and Mixed-Variable Optimization. In: Dorigo, M., Birattari, M., Blum, C., Gambardella, L.M., Mondada, F., Stützle, T. (eds.) ANTS 2004. LNCS, vol. 3172, pp. 25–36. Springer, Heidelberg (2004)

[33] Socha, K., Dorigo, M.: Ant colony optimization for continuous domains. Tech. Rep. 2005-037, IRIDIA (December 2005)

[34] Stützle, T., Hoos, H.: Improvements on the Ant System: Introducing the $\mathcal{MAX} - \mathcal{MIN}$ Ant System. In: Third International Conference on Artificial Neural Networks and Genetic Algorithms, Springer, University of East Anglia, Norwich, UK (1997)

[35] Tsutsui, S.: Ant colony optimisation for continuous domains with aggregation pheromones metaphor. In: Proceedings of the 5th International Conference on Recent Advances in Soft Computing (RASC 2004), pp. 207–212 (2004)

[36] Veldhuizen, D.A.V.: Multiobjective Evolutionary Algorithms: Classifications, Analyses, and New Innovations. Ph.D. thesis, Department of Electrical and Computer Engineering. Graduate School of Engineering. Air Force Institute of Technology, Wright-Patterson AFB, Ohio (May 1999)

Mechanisms for Evolutionary Reincarnation

Ben Prime and Tim Hendtlass

Complex Intelligent Systems Laboratory
Centre for Information Technology Research
Faculty of Information & Communication Technologies
Swinburne University of Technology
Melbourne, Australia, 3122
ben.prime@gmail.com, thendtlass@swin.edu.au

Abstract. This paper describes the effects of adding gene reincarnation to a biologically inspired evolutionary algorithm. When using the biologically inspired part of the algorithm we are able to draw on experience from real life. Reincarnation capabilities, however, must be constructed without any real life experience to guide us. This paper addresses the question 'can reincarnation be added to a genetic algorithm in such a way as to modify the resulting evolutionary process'? Reincarnation in this context requires that genetic information, saved from earlier generations, be bought back and reintroduced into the population at a later time. A simple algorithm is introduced that selects particular genetic material to add to the storage, performs regular culls of the stored material and inserts some of the stored material back into targeted individuals in later generations. Preliminary experiments show that while much of the reinserted material vanishes without having any obvious evolutionary effect, a small proportion remains for many generations and changes the course of the evolution compared to a genetic algorithm identical in all respects except that it lacks reincarnation.

1 Introduction

Genetic Algorithms (GA) are a family of robust optimisation techniques that are based on Darwinian evolution [1,2]. Although the artificial representation of genetic structure is quite rudimentary, these methods can be used to find solutions to a plethora of problems. The procedure uses Darwinian selection to cause a 'population' of possible answers to evolve towards better answers. This is done through successive generations of evolution in which new potential solutions are 'bred' from the previous generation, with better performing solutions being rewarded with a higher probability of passing on their own genetic material to future generations. Individual solutions consist of a chromosome of genes, and it is the values of these that specify the particular solution represented by this chromosome. Typical GA implementations tend to winnow this genetic information as they progress from one generation to the next.

As pieces of existing solutions are recombined with each other (or copied) and mutated (usually small variations in gene value are applied with an often low

M. Randall, H.A. Abbass, and J. Wiles (Eds.): ACAL 2007, LNAI 4828, pp. 245–256, 2007.
© Springer-Verlag Berlin Heidelberg 2007

probability), the GA causes the population as a whole to 'explore' the problem, particularly concentrating on regions with promisingly high fitness. This concentration of effort is a result of the predominance in the population of particular gene values associated with higher fitness regions. One of the unfortunate drawbacks of the GA is that it is not guaranteed to find the global optimum for every problem. A particular set of gene values may so dominate a population that every individual chromosome describes a very similar solution within a small region of problem space - an inspection of corresponding gene values from the chromosomes of all the individuals will show little diversity in the values stored. When this happens the GA is said to have converged. Premature convergence is said to have occurred if the solution converged on is sub-optimal [3,4]. With no diversity in gene values, standard recombination is highly unlikely to result in exploration outside the current region (unless extreme mutation is allowed) as the new individuals produced will effectively be clones of existing individuals. Since escape from premature convergence is extremely hard, various strategies have been undertaken in order to avoid it. Each of these approaches attempts to preserve diversity in gene values without inhibiting the general focus of the search toward regions of known merit. It has been shown that the performance of the GA can be improved by means of the concept of using 'islands' [5,6]. Island populations can be thought of as dividing an entire population of a GA into physically separate (or caste) parts. It has been noted that under such circumstances each group within the total population will evolve separately. Each population's genetic mix is loosely linked with the others by the occasional transfer of entire individual solutions between groups. The advantage of this method does not become apparent until differing conditions, such as the imposition of varying probabilities of mutation, are applied to each population. For further discussion of this see, for example, [7].

High mutation, that is larger changes applied with a higher than usual probability, has the effect of significantly increasing the diversity of gene values but, being a stochastic process, may interfere with the guided exploration of problem space. It may produce movements that are inconsistent with the (unknown) scale of the significant features in problem space. Cyclic mutation addresses these concerns by the periodic variation of mutation parameters (probability and/or maximum magnitude) throughout the evolution process [8]. High mutation periods force an increase in diversity and are followed by periods of low mutation that allow the exploration of the problem space now occupied. Tabu search maintains a list of areas of the solution space that have already been searched [9]. When used in collaboration with an evolutionary algorithm the breeding process is modified so that only solutions not already on this tabu list are allowed to be bred. This pushes future generations into areas that have been less thoroughly investigated. For a non-quantized problem space, a special neural network may be used to store an ever-evolving series of pointers to regions of known good and bad points. Then the probability of a new solution being accepted into the population is inversely proportional to the distance from the closest of these points. It is important that the probability coefficients differ between good and bad points as both of these directly prevent convergence (premature or otherwise)[10].

The current work takes a different approach to paper [13], using two island populations, one of which is conventional and contains the current population. The other island contains information taken from earlier populations, and holds it for possible reinsertion into the current population, that is old genetic material now extinct may be 'reincarnated' into a later generation. The aim of reintroducing old genetic material is to preserve useful diversity, where useful diversity is that which has a beneficial effect on evolutionary progress by allowing some backtracking to occur along the path to the solution, thus hopefully providing an escape mechanism from evolutionary 'dead-ends'. Acan and Tekol have reported on chromosome reuse [11]. In their work, after the breeding process, individuals from some fitter part of the population that have not been selected as a parent are copied into extra storage. At the creation of the next generation these individuals compete equally with the members of the current conventional population for parenting opportunity. This auxiliary storage has a fixed maximum size, once this is filled only individuals with a fitness higher than the least fit individual currently there can be added and replace that weakest individual. This approach constitutes an extension to the concept of elite [12] individuals.

This paper involves a more extended process in which a random selection is made from a large but not infinite pool taken from all the deliberative stored extinct gene values that have been stored for possible targeted reintroduction into the current population. It represents a substantial development of the ideas first introduced in [13], in particular in the way in which the storage is managed.

2   A Discussion of the Algorithm

There are a variety of reasons for storing information for possible later reinsertion: because it is very good or because it is about to become extinct (disappearing from the population so it is no longer available as a genetic building block). The normal evolutionary process will result in good information playing a part in the evolution of the populations of solutions, reintroducing it is likely to drive the population back towards previously explored areas of problem space (albeit good areas). Storing individuals about to vanish from the population can be of some use as [11] has shown, at least for solutions whose performance is reasonable but which have not had the opportunity to pass on their gene values via normal parenting.

Storing individual genes (rather than whole individuals) when these gene values are about to become extinct should not be restricted to the genes of highly fit individuals: a poor individual performance might be a result of the combining of some 'good - potentially useful' gene values with some 'poor - almost certainly useless' gene values. Since in general interchanging gene values between different gene positions is unhelpful, it will be necessary to maintain a separate storage area for each gene position. Storing every now extinct value for every gene position for all time would obviously lead to excessive amounts of storage being required (as well as the amount of computational overhead) and so a cull has to be performed periodically of the stored gene values. The reinsertion of stored gene values into the population has to be done in such a way as to maintain diversity. When a new population is bred using conventional GA parameters the diversity of the gene values is noted. A gene that is over represented in this new population has some of its instances replaced by values drawn randomly for the appropriate storage area. The actual instances to be replaced are chosen randomly from the set of over represented genes.

3   The Algorithm in Detail

A flow chart of the full algorithm is shown in figure 1. The grey shaded steps are conventional GA steps and will not be discussed further in this paper. The remaining algorithm steps can essentially be summarised into three subsections: Storage of gene values, replacement of over represented gene values with stored gene values and attrition of stored gene values. This process is carried out for every gene position in the solution chromosome.

3.1   Storage of Gene Values

When storing extinct gene values it seems likely to be beneficial to exercise some degree of restraint because of the practicalities of maintaining increasingly large lists. A balance point needs to be found between the practical implementation of the storage mechanism and the desire not to throw away potentially useful gene values. The actual quantity of gene values stored need not be dispropor-tionately large when compared to the actual number of gene values replaced in the population, since a relative few can actually be reused per generation. The exclusivity rule applied to the storage of genes in the algorithm prevents identi-cal values contained in storage from being readmitted. For the real valued genes used presently, a slight variation to this is actually used: A *similarity threshold* is applied so that very minor variations in the floating-point value can be ne-glected in the interests of maintaining a shorter (and thus more manageable) storage list. The similarity threshold must be of reasonable magnitude when compared to the features of the problem space being examined. The smaller this number the more information that is stored.

3.2   Replacement of Overrepresented Genes in the Population

Fundamental to gene reinsertion are the questions of *what genes are to be replaced* and *which genes are they to be replaced with* from the storage. The present series of experiments relies on simple rules for storage and retrieval of genes together with the rate and targeting of attrition within the storage mechanism.

The choice of which genes are to be removed from the population is based upon the desire to thwart premature convergence. In replacing a proportion of the most heavily represented (indeed over represented) gene values with dissimilar gene values from storage the overall tendency of the algorithm to premature

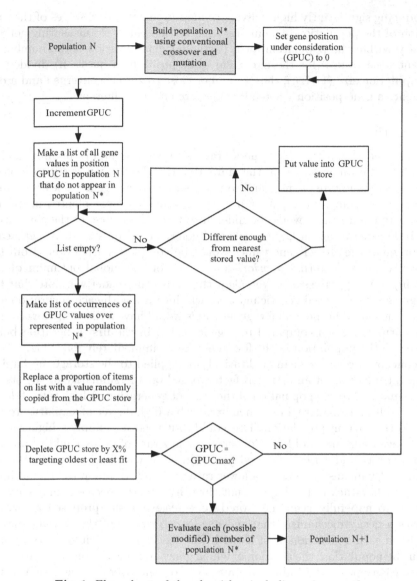

Fig. 1. Flow chart of the algorithm including reincarnation

convergence is lessened. It should be noted that this affects the whole population and that elites are not immune to having genes replaced.

Let the number of occurrences of a particular gene value be A and the (population size / number of different gene values in population) be Avg. Further let (Rf) be the replacement factor (the proportion that are randomly selected to be replaced [0..1]). Then the exact number of instances nominated for replacement, R, is calculated as follows. If $A <= Avg$ then $R = 0$, otherwise $R = Truncate((A - Avg) * Rf)$. The use of this selection method has the result of producing significantly higher diversity measurements at all stages of the progression of the population. The metric used in this paper for measuring genetic diversity within the population for a particular gene position is the number of different gene values (for the particular gene position in question) divided by the population size $(1/Avg)$, thus a number close to zero is converged and a one represents a gene position where all values are entirely different.

3.3 Attrition of Gene Values Within the Storage Mechanism

In Prime and Hendtlass [13], gene storage size was essentially unlimited. The implicit limitation imposed by the similarity threshold upon entry to storage served to slow the rate of entry and to make sure that the gene values contained were of some minimum diversity. Once a gene value had been planted in storage though, its tenure there would be indefinite. Over the course of the simulation (say, 1000 generations) a large number of genes would be stored. A significant percentage of the stored gene values would be completely irrelevant (from the perspective of an external *a priori* knowledge of the global optimum of the problem) to the progression of the algorithm. Also the reader will note that the average age of the stored genetic material will increase with each generation and that a gene stored in the final few generations would have a relatively low chance of reincarnation when compared to a gene stored in the first generation being restored to the population in the few generations immediately following.

A concept of relevancy can qualitatively be applied to the storage mechanism by the introduction of an attrition factor based on the similarity of the stored genetic material to the population of the current generation. A high rate of decay would result in a smaller storage memory, with a high degree of similarity to the genes to the current population. This could, but need not, imply a higher short term relevance to the problem. Conversely, a low rate of decay could result in a larger quantity and variety of stored gene values but with likely lower short-term relevance. By means of analogy, a lower rate of decay results in the maintaining of a larger haystack in the hope of not throwing out the needle it may contain! In order to hopefully keep the stored genes relevant the proposed algorithm employs a decay mechanism that specifies a percentage of the stored genes to be deleted each generation. Just as with the decision of which genes to replace within the population, the decision of which genes to remove from storage during the attrition phase could also have an effect on the overall progression of the algorithm. While similar effects were seen when removing the oldest gene values, a clear performance advantage was achieved by eliminating the genes with the poorest associated fitness in the storage. Only results for this second culling method are presented in this paper. In all methods the rationale is to weight the charitable effects of reincarnation towards better stored gene fitnesses.

4   The Effects of Reincarnation

The basic principle of Darwinian evolution is that features or abilities beneficial to survival in the current environment survive and spread, while those

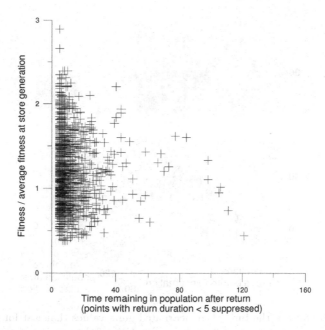

Fig. 2. The time all reincarnated gene values remain in the population (excluding those that remain for four or less generations)

detrimental to survival die out. Thus if reincarnation is to have any effect on evolutionary progress this should first be manifested by the prolonged existence in the current gene pool of some reincarnated features. Figures 2 and 3 show the length of time reincarnated features remained in the population, plotted as a function of the apparent relative fitness (2) or the length of time they had been in storage (3). As can clearly be seen some reincarnated gene values did remain in the population for up to 120 generations. Some qualification must be made, all trace of about 85% of reincarnated values disappeared from the population in 4 or less generations and are not plotted in the two figures. Figure 2 shows the reincarnation history of gene one taken from a single (albeit typical) run solving Ackley's function in 30 dimensions with a (relatively small) population of 200. Since this is a minimisation function, individuals with 'Fitness / average fitness at store generation' less than one were fitter than the population average when they were stored. From figure 2 there is no clear relationship between the relative fitness when stored and the time a reincarnated gene value remained in the population (and hence its apparent usefulness). This is to be expected as the fitness of an individual is calculated from all of its gene values and so will not necessarily correlate well with any one particular gene value.

Figure 2 shows that the longer a gene value had been stored the less time it was likely to remain in the population when returned but that there were some notable exceptions to this. Again this is intuitively reasonable as, in general, when evolution moves into new regions few of the values stored earlier are likely

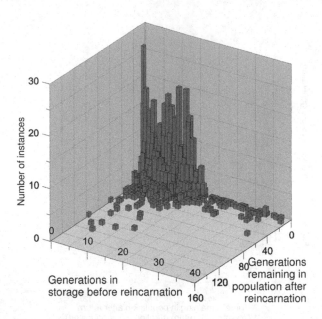

Fig. 3. A histogram of the time all reincarnated gene values that last for five or more generations remained in the population as a function of the number of generations for which they had been stored

to still be relevant. At first sight figure 3 also suggests that few 'long term' storage residents ever last 5 or more generations after reincarnation. More likely the low number of such events on the plot is just a result of the pruning of the stored individuals which results in there being very few 'old' stored gene values to be reincarnated.

Experiments used in this work were conducted using the Conventional and Reincarnation GA's with the Ackley function (simulated in 30 dimensions) as the test problem. The Ackley function [14] is a minimisation problem with a global optimum (minimum) at $(0, 0,...., 0, 0)$ with score of 0. Formally the score $F(\bar{x})$ for any position $\bar{x}$ is given by

$$F(\bar{x}) = -20exp\left(-0.2\sqrt{\frac{1}{n}\sum_{i=1}^{n}x_i^2}\right) - exp\left(\frac{1}{n}\sum_{i=1}^{n}cos(2\pi x_i)\right) + 20 + e$$

For this problem

$$x_i \in (-32.768, 32.768)$$

As this is a minimisation problem a low score represents high fitness. A variety of parameter values have been used to generate results, with each result presented being the average of 100 test runs. The probability of mutation used for all results is 2% per (real valued) gene. The magnitude of mutation as applied to the gene values is a zero-centred Gaussian distributed number added to the previous gene value, with standard deviations of 0.5 and 4.0. The number of elites is also varied with 1 and 30 elites being used. The above parameters were chosen because they represent a relatively favourable choice to one of each of the demonstrated algorithms. 1000 generations were simulated for every test using an attrition rate (K_r) of the least fit 5% of the stored individuals per generation.

For the RGA the replacement factor (Rf) is also varied so that the effect of the aggressive gene replacement can be shown. The four results graphs presented demonstrate the possible combinations of few and many elites (1 and 30) and high and low Rf (1/7th and 1/15th respectively). Within each graph, the magnitude of mutation is varied so as to favour each algorithm in turn. The reader will note that identical results for the conventional GA are replicated in figures 4(a) and 4(b) and also in 4(c) and 4(d).

The low mutation magnitude used ensures the conventional GA is prematurely converged in all cases, this suggesting that the magnitude of mutation change is not compatible with the features of the problem space. For the higher mutation magnitude the conventional GA almost always finds the global minimum. For the RGA the opposite appears to be true: For all settings explored the RGA produces its best outputs for the lower mutation magnitude settings. The RGA seemingly is deriving the required randomness (for coarse search) from the reinserted genes. In all cases the RGA has a more rapid initial fitness improvement than the GA, however upon completion of 1000 generations the conventional GA always has a better average fitness value than the RGA. An increase in the number of elites hastens the final convergence of the GA (and in fact does so for all RGA simulations too), see figures 4(a) and 4(c). Generally, the RGA demonstrated an 'average fitness plateau' suggesting that a certain proportion of tests do not solve the test problem. This trait is not demonstrated by the GA for the high mutation magnitude setting. The fitness level at which this plateau occurs is affected by all of elitism, mutation magnitude and Rf settings and is of interest for further discussion. Figures 4(a) and 4(c) (also figures 4(b) and 4(d)) show that high elitism improves the fitness plateau effect. A more gentle replacement factor similarly improves performance when comparing comparable plots on figures 4(a) and 4(b) (and also figures 4(c) and 4(d)). The best observed performance for the RGA is derived from a combination of low Rf (gentle replacement), high elitism, and low mutation magnitude and is seen in figure 4(c). From figures 4(c) and 4(d) it can be seen that the RGA with low mutation is not equivalent to the GA with higher mutation as the final results differ significantly: while they have some similar effects, reincarnation is not simply another way of performing mutation.

Replacement factor clearly affects the fitness plateau of the RGA. The exact number of gene reinsertions per gene is not precisely fixed by the replacement factor since other characteristics (such as mutation) affect the rate at which the relatively fit begin to dominate the evolutionary process. For more gentle replacement (Rf = 1/15) it is of note that at steady state only about six out of 200 genes are replaced per generation, with a more aggressive Rf of 1/7 this number is close to 11 per generation.

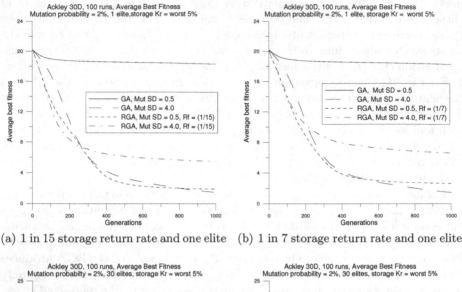

(a) 1 in 15 storage return rate and one elite   (b) 1 in 7 storage return rate and one elite

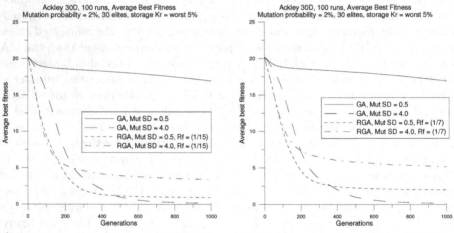

(c) 1 in 15 storage return rate and 30 elites   (d) 1 in 7 storage return rate and 30 elites

Fig. 4. Relative performance of a conventional GA and a reincarnation GA (RGA) for low and high mutation magnitude.

4.1   The New Role of Elites

The real-valued GA as implemented in this work utilises a variety of relatively disruptive parameters such as uniform crossover and tournament parent selection to slow down the rate of convergence in both the conventional GA and in the RGA. The use of such techniques to produce more randomised offspring, as compared to say, single point crossover and roulette parent selection, results in an altered balance of 'normal' optimal parameters for the conventional GA. Often for the conventional GA it can be shown that only a few (for example 2) elites are required to produce steady improvement in best score with extras producing diminishing returns. In this work, significantly more elites also improve the performance of the conventional GA for the above stated reasons. The RGA produces more optimal results (relative to itself) by the use of significantly more elites that might be expected in its working population of 200, in fact even more than 30 elites can be used with only beneficial effects witnessed. Since elite solutions tend to occur more often in the population (due to their high fitness) and are subject to alteration (by gene replacement), the elite section of the population explicitly qualifies itself for the most rigorous gene replacement with the current replacement rule. This tends to produce (structurally speaking) close-to-elite individuals that vary by one or more reincarnated genes. The reincarnation algorithm *needs* more elites because it modifies them so heavily.

5   Conclusions and Future Work

This paper considers if reincarnation would have an effect on an artificial evolutionary process, and clearly shows that it does. While the results reported in this paper only use one function, work on a number of other functions indicates that those results have similar characteristics. While the choice of which gene values in the current population to save is the same as in [13], the addition of attrition of the stored gene values based on the assigned fitness together with a deliberative rather than random method for deciding which gene values are modified by reincarnation has produced an algorithm that clearly shows that reincarnation has an effect on the progress of evolution. In the early stages it seems that reincarnation has similar effects to increasing mutation although not exactly the same. By the time of convergence at the global optimum reincarnation is no longer of use just degrading the final performance.

Further work on a wider range of problems will establish exactly how the effects of reincarnation differ from those of high mutation, knowledge needed to establish the best way to apply reincarnation. Already it is clear that for simple, single optimum, static problems constant reincarnation is unlikely to be the best strategy. Hopefully, at least for a range of problems, careful use of reincarnation will produce a better performance than would be obtained without it.

It has been noted that, particularly for low replacement factors (Rf), that the average assigned quantity of stored genes is much higher than the average assigned quantity of those that actually are placed back into the working population. This implies that the use of a guided choice of the replacement values (e.g. by tournament selection) may prove better than the current random selection.

References

1. Holland, J.H.: Adaption in Natural and Artificial Systems. University of Michigan Press (1975)
2. Davis, Lawrence (eds.): Handbook of Genetic Algorithms, 1st edn. Von Nostrand Reinhold, New York (1991)

3. Mauldin, M.L.: Maintaining diversity in genetic search. In: Proceedings of the National Conference on Artificial Intelligence, pp. 247–250 (1984)
4. Goldberg, D.E.: Genetic Algorithms in Search, Optimization, and Machine Learning. Addison-Wesley, Reading (1989)
5. Tanese, R.: Distributed Genetic Algorithms. In: Proceedings of the Third International Conference on Genetic Algorithms, George Mason University (1989)
6. Levine, D.: A parallel genetic algorithm for the set partitioning problem. Technical report ANL-94/23, Argonne National Library (1994)
7. Copland, H., Hendtlass, T.: Migration Through Mutation Space: A Means of Accelerating Convergence. In: Proceedings of ICANNGA 1997, England (1997)
8. Hendtlass, T.: On the use of variable mutation in an evolutionary algorithm. In: Proceedings of IEA/AIE 1997, Gordon and Breach (1997)
9. Glover, F.: A Users Guide to Tabu Search. Annals of Operations Research vol. 41 J.C Baltzer AG (1993)
10. Podlena, J., Hendtlass, T.: An Accelerated Genetic Algorithm. Applied Intelligence 8(2) (1998)
11. Acan, A., Tekol, Y.: Chromosome Reuse in Genetic Algorithms. In: Cantú-Paz, E., Foster, J.A., Deb, K., Davis, L., Roy, R., O'Reilly, U.-M., Beyer, H.-G., Kendall, G., Wilson, S.W., Harman, M., Wegener, J., Dasgupta, D., Potter, M.A., Schultz, A., Dowsland, K.A., Jonoska, N., Miller, J., Standish, R.K. (eds.) GECCO 2003. LNCS, vol. 2723, pp. 695–705. Springer, Heidelberg (2003)
12. De Jong, K.: An Analysis of the Behavior of a Class of Genetic Adaptive Systems. Doctoral Thesis, Univerity of Michigan (1975)
13. Prime, B., Hendtlass, T.: Evolutionary Computation Using Island Populations in Time. In: Orchard, B., Yang, C., Ali, M. (eds.) IEA/AIE 2004. LNCS (LNAI), vol. 3029, pp. 573–582. Springer, Heidelberg (2004)
14. Ackley, D.H.: A connectionist machine for genetic hill climbing. Kluwer Academic Publishers, Boston (1987)

An Evolutionary Algorithm with Spatially Distributed Surrogates for Multiobjective Optimization

Amitay Isaacs, Tapabrata Ray, and Warren Smith

University of New South Wales, Australian Defence Force Academy, ACT, Australia
{a.isaacs,t.ray,w.smith}@adfa.edu.au

Abstract. In this paper, an evolutionary algorithm with spatially distributed surrogates (EASDS) for multiobjective optimization is presented. The algorithm performs actual analysis for the initial population and periodically every few generations. An external archive of the unique solutions evaluated using the actual analysis is maintained to train the surrogate models. The data points in the archive are split into multiple partitions using k-Means clustering. A Radial Basis Function (RBF) network surrogate model is built for each partition using a fraction of the points in that partition. The rest of the points in the partition are used as a validation data to decide the prediction accuracy of the surrogate model. Prediction of a new candidate solution is done by the surrogate model with the least prediction error in the neighborhood of that point. Five multiobjective test problems are presented in this study and a comparison with Nondominated Sorting Genetic Algorithm II (NSGA-II) is included to highlight the benefits offered by our approach. EASDS algorithm consistently reported better nondominated solutions for all the test cases for the same number of actual evaluations as compared to a single global surrogate model and NSGA-II.

1 Introduction

Evolutionary algorithms (EAs) are particularly attractive for multiobjective problems as they result in a set of nondominated solutions in a single run. Furthermore, EAs do not rely on functional and slope continuity and thus can be readily applied to optimization problems with mixed variables. However, EAs are essentially population based methods and require evaluation of numerous solutions before converging to the desired set of solutions. Such an approach turns out to be computationally prohibitive for realistic Multidisciplinary Design Optimization (MDO) problems and there is a growing interest in the use of surrogates to reduce the number of actual function evaluations.

A comprehensive review on the use of fitness approximation in the context of evolutionary computation has been reported by Jin [1]. The choice of surrogate models reported in literature range from neural network based models like multilayer perceptrons, radial basis function networks, quadratic response surfaces, Kriging and cokriging models. A vast majority of surrogate assisted

M. Randall, H.A. Abbass, and J. Wiles (Eds.): ACAL 2007, LNAI 4828, pp. 257–268, 2007.

optimization methods rely on the use of a single global surrogate model. The surrogate model is either created once and used subsequently throughout the course of search (one shot approach) or created periodically. Algorithms based on the one shot training of the approximation model(s) [2,3] are likely to face the problems when the initial set of solutions generated differ substantially from the final set as in the case of the test function SCH1 [4]. Periodic retraining is necessary as the search proceeds to localized areas. In order to capture local behavior, hierarchical surrogate models have been proposed by Zhou et al [5] and the use of artificial neural network models in the local search strategy have been used by Gasper-Cunha and Vieira [6].

To improve the prediction accuracy with limited samples, multiple surrogates can be used in place of the single surrogate model. Common use of multiple surrogates is in the form of surrogate ensembles where a collection of surrogate models with varying parameters usually trained simultaneously by techniques such as bagging [7], and boosting [8] are used. A survey of neural network ensemble has been reported by Zhao et al [9]. Jin and Sendhoff [10] reported the use of clustering and neural network ensembles to reduce the fitness evaluations. They use k-Means clustering to identify the candidate solutions which need to be evaluated using the actual analysis. Hamza and Saitou [11] have used polynomial regression surrogate ensembles with weighted average response and the most conservative response in the co-evolutionary genetic algorithm for vehicle crash-worthiness design.

Another approach based on multiple surrogates is to use different types of surrogate models simultaneously. Goel et al [3] and Zerpa et al [12] have used a weighted average model resulting from three surrogate types (polynomial response surface model, kriging, and radial basis function). The two approaches differ in the determination of the weights for averaging. Zhou et al [13] reported the use of multiple approximation models in the context of memetic algorithm to perform the local search. They even propose using a surrogate ensemble as one of the approximation models.

In the context of multiobjective optimization, Nain and Deb [14] proposed a multifidelity model (coarse to fine grain) for surrogate assisted multiobjective optimization where a multilayer perceptron was periodically retrained and used in alternation with actual computations to solve a B-spline curve fitting problem. A similar approach of alternating between actual analysis (K) and surrogate models (S) have been reported by Ray and Smith [15]. The study used a RBF model that was trained using the candidate solutions of the population after every K generations. Nain and Deb [16] reported the performance of successive surrogate models on two test functions viz. ZDT4 and TNK. Pareto Efficient Global Optimization (ParEGO) algorithm [17] relies on a kriging based surrogate and the sampling points are generated via design of experiments. However, the method requires knowledge about the limits of the objective function space and cannot guarantee a uniform distribution of the solutions along the nondominated front. Emmerich et al [18] use confidence interval predicted by the kriging model to screen candidates for actual evaluation, reducing the computational cost. A recent paper by Chafekar *et al* [19] reports the use of multiple GAs, each of which uses a reduced model of the objective function with a regular information exchange among GAs to obtain a well distributed nondominated set of solutions.

In this paper an evolutionary algorithm with spatially distributed surrogates (EASDS) is presented. This approach uses multiple surrogates that are spatially distributed in the design space. An archive of the solutions evaluated using the actual analysis is maintained and used to train the surrogate models. The solutions in the archive are split in multiple partitions using k-Means clustering. Using a fraction of the solutions in each partition a Radial Basis Function network surrogate model is trained. The unused points in each of the partition are used to assess the prediction accuracy of the surrogate model. The performance of the EASDS is compared with Nondominated Sorting Genetic Algorithm (NSGA-II) [20] using an equal number of actual function evaluations. The effect of the number of partitions is also studied and the performance is compared with a single global surrogate model.

2   An Evolutionary Algorithm with Spatially Distributed Surrogates

The pseudo code of the proposed Evolutionary Algorithm with Spatially Distributed Surrogates (EASDS) is outlined in Algorithm 1.

Algorithm 1. Evolutionary Algorithm with Spatially Distributed Surrogates

---

Require: $N_G > 1$ {Number of Generations}
Require: $M > 0$ {Population size}
Require: $K > 1$ {Number of partitions}
Require: $I_{TRAIN} > 0$ {Periodic Surrogate Training Interval}
 1: $\mathcal{A} = \emptyset$ {Archive of the Solutions}
 2: $P_1 =$ Initialize()
 3: Evaluate(P_1)
 4: $\mathcal{A} =$ AddToArchive($\mathcal{A}, P_1$)
 5: $\mathcal{S} =$ BuildSurrogates($\mathcal{A}, K$)
 6: **for** $i = 2$ to N_G **do**
 7:    **if** modulo(i, I_{TRAIN}) $== 0$ **then**
 8:       Evaluate(P_{i-1}) {Evaluate parent population using the Actual Analysis}
 9:       $\mathcal{A} =$ AddToArchive($\mathcal{A}, P_{i-1}$)
10:       $\mathcal{S} =$ BuildSurrogates($\mathcal{A}, K$)
11:    **end if**
12:    $C_{i-1} =$ Evolve($P_{i-1}, \mathcal{S}$)
13:    EvaluateSurrogate($C_{i-1}, \mathcal{S}$)
14:    $P_i =$ Reduce($P_{i-1} + C_{i-1}$)
15: **end for**

The basic evolutionary algorithm is on the same lines as that of NSGA-II by Deb *et al* [20]. The algorithm starts with a random initial population and evaluates the population using actual evaluations. Spatially Distributed Surrogate models (using Radial Basis Function network) are created for all the objectives and the constraints. An external archive of actual evaluations is maintained in EASDS and used to periodically train the surrogate models for all the objectives and the constraints. The components of EASDS are described below.

2.1 Initialization

All the solutions in the population are initialized randomly by selecting each variable value from the specified range for that variable.

2.2 Archive of the Actual Evaluations

All the unique candidate solutions that are evaluated using the actual analysis are maintained in an external archive. Every I_{TRAIN} generations, the parent population is evaluated using the actual analysis functions and then added to the archive. New solution is added to the archive only if the normalized distance (using the Euclidean norm) between the new solution and each of the solutions in the archive is more than user defined distance criterion. This condition avoids the numerical difficulties of building the surrogates if the solutions are too close.

2.3 Evolutionary Strategy

The evolutionary strategy of EASDS is the same as that of NSGA-II. Binary tournament is used for the selection the parents undergoing crossover. The simulated binary crossover (SBX) operator [21] and a polynomial mutation operator [22] are used to create an offspring population from the parent population.

2.4 Building Spatially Distributed Surrogate Models

Outlined in Algorithm 2 are the steps involved in building the RBF surrogate models for the objectives and the constraints. A collection of RBF surrogate models is created to approximate the objectives and the constraints. The archive is split into K partitions $(\mathcal{A}_1, \ldots, \mathcal{A}_K)$ using k-Means clustering algorithm where the design variables $x_1, \ldots, x_n$ are used as the clustering attributes.

The solutions in each of the partitions are used to build the RBF surrogate models for the objectives and the constraints. Only a fraction $(0 < \alpha < 1)$ of the solutions are used to train the surrogate model and the rest are used as the validation data set. EASDS uses 80% of the solutions in each partition as the training data and the remaining 20% are used to validate the surrogate models.

If there are very few solutions in a partition (insufficient to build the RBF surrogate model), no surrogate models are built using that partition. If the prediction error on the validation data set in the partition is more than the user defined threshold, the surrogate model on that partition is deemed invalid.

Algorithm 2. Building Spatially Distributed Surrogate Models

Require: $\mathcal{A}$ {Archive of actual evaluations}
Require: K {Number of partitions}
Require: $m \geq 2$ {Number of objectives}
Require: $p \geq 0$ {Number of constraints}
1: $\mathcal{A}_1, \ldots, \mathcal{A}_K = \text{KMeans}(\mathcal{A}, K)$
2: **for** $i = 1$ to K **do**
3:      **for** $j = 1$ to m **do**
4:          $S_{i,f_j} = \text{RBF Train}(\mathcal{A}_i, f_j)$
5:      **end for**
6:      **for** $j = 1$ to p **do**
7:          $S_{i,g_j} = \text{RBF Train}(\mathcal{A}_i, g_j)$
8:      **end for**
9: **end for**

---

k-Means Clustering Algorithm. A k-Means clustering algorithm [23] is used to split given data points into k clusters or partitions. The main idea of k-Means clustering is to define k centroids, one for each cluster, and then assign each data point to one of the k clusters so as to minimize a measure of dispersion within the clusters. A very common measure is the sum of squared Euclidean distances from the centroid of each cluster.

Radial Basis Function Network Surrogate. Radial Basis function networks belong to the class of Artificial Neural Networks (ANNs) and are a popular choice for approximating nonlinear functions. A radial basis function (RBF) ϕ has its output symmetric around an associated centre μ.

$$\phi(\mathbf{x}) = \phi(\|\mathbf{x} - \mu\|)$$

where the argument of ϕ is a vector norm. A common RBF is the Gaussian function with the Euclidean norm.

$$\phi(r) = e^{-r^2/\sigma^2}$$

where σ is the scale or width parameter. A set of RBFs can serve as a basis for representing a wide class of functions that are expressible as linear combinations of the chosen RBFs as shown in Eq. 1.

$$y(\mathbf{x}) = \sum_{i=1}^{k} w_i \, \phi(\|\mathbf{x} - \mu_i\|) \tag{1}$$

Here, k is typically smaller than the number of data points. The coefficients w_i are the unknown parameters that are to be "learned." The training is usually achieved via the least squares solution:

$$\mathbf{w} = \mathbf{A}^+ \, \mathbf{y} \tag{2}$$

where $\mathbf{A}^+$ is the pseudo-inverse and $\mathbf{y}$ is the target output vector.

2.5   Evaluation Using Spatially Distributed Surrogate Models

For accurate prediction of the objectives and the constraints for a new candidate solution, a surrogate model with the least prediction error is chosen from spatially distributed surrogate models. If the new candidate solution is far (using the Euclidean distance measure) from all the solutions in the archive, it is evaluated using the actual analysis.

From the archive of solutions, S solutions closest (using the Euclidean norm) to a new candidate are selected. The values of the objectives and the constraints of these S points are predicted using each of the surrogate models in the collection. For each of the surrogate models, prediction error (RMSE) is computed using the actual and the predicted values of the objectives and the constraints. Surrogate model with the least prediction error is then used to predict the value at the new candidate solution.

2.6   Reduction

The reduction procedure retains the best individuals from the parent and the offspring population (elitism). Combined solutions from the parent population and the offspring population are ranked using the non-dominated sorting and the crowding distance criterion [24]. M elite solutions (better fitness) are retained for the next generation from a set of $2M$ solutions (parent and offspring population). If there are less than M feasible solutions, then infeasible solutions with smaller values of maximum constraint violation are retained.

3   Numerical Examples

3.1   Test Problems

The first two constrained test problems are SRN and OSY [25]. The ZDT test problems [25] are two objective unconstrained problems framed by Zitzler *et al* and they are of the form as shown in Eq. 3.

$$\text{Minimize} \quad f_1(\mathbf{x}),$$
$$f_2(\mathbf{x}) = g(\mathbf{x}) \, h(f_1(\mathbf{x}), g(\mathbf{x})). \tag{3}$$

The description of all the test problems is given in Table 1.

3.2   Experimental Setup

A population size of 100 is used for all the test problems and the algorithm is run for 101 generations. All the test problems are evaluated using EASDS and NSGA-II. For EASDS, the surrogate models were retrained every 5 generations ($I_{TRAIN} = 5$). The probability of crossover is set to 0.9 and the probability of mutation is set to 0.1. The Distribution index for crossover is 10 and the distribution index for mutation is 20. A new solution is added to the archive if the normalized distance between the new solution and closest solution in the

Table 1. Test Problems

Problem	Dim	Objectives & Constraints	Bounds
SRN	2	$f_1(\mathbf{x}) = 2 + (x_1 - 2)^2 + (x_2 - 1)^2$ $f_2(\mathbf{x}) = 9x_1 - (x_2 - 1)^2$ $x_1^2 + x_2^2 \leq 225,\ x_1 - 3x_2 + 10 \leq 0$	$\mathbf{x} \in [-20, 20]^2$
OSY	6	$f_1(\mathbf{x}) = -[25(x_1 - 2)^2 + (x_2 - 2)^2 + (x_3 - 1)^2$ $+ (x_4 - 4)^2 + (x_5 - 1)^2]$ $f_2(\mathbf{x}) = x_1^2 + x_2^2 + x_3^2 + x_4^2 + x_5^2 + x_6^2$ $x_1 + x_2 - 2 \geq 0,\ 6 - x_1 - x_2 \geq 0,$ $2 - x_2 + x_1 \geq 0,\ 4 - (x_3 - 3)^2 - x_4 \geq 0,$ $2 - x_1 + 3x_2 \geq 0,\ (x_5 - 3)^2 + x_6 - 4 \geq 0$	$x_1, x_2, x_6 \in [0, 10]$ $x_3, x_5 \in [1, 5]$ $x_4 \in [0, 6]$
ZDT1	10	$f_1(\mathbf{x}) = x_1$ $g(\mathbf{x}) = 1 + \frac{9}{n-1} \sum_{i=2}^{n} x_i$ $h(f_1, g) = 1 - \sqrt{f_1/g}$	$\mathbf{x} \in [0, 1]^{10}$
ZDT2	10	$f_1(\mathbf{x}) = x_1$ $g(\mathbf{x}) = 1 + \frac{9}{n-1} \sum_{i=2}^{n} x_i$ $h(f_1, g) = 1 - (f_1/g)^2$	$\mathbf{x} \in [0, 1]^{10}$
ZDT3	10	$f_1(\mathbf{x}) = x_1$ $g(\mathbf{x}) = 1 + \frac{9}{n-1} \sum_{i=2}^{n} x_i$ $h(f_1, g) = 1 - \sqrt{f_1/g} - (f_1/g)\sin(10\pi f_1)$	$\mathbf{x} \in [0, 1]^{10}$

archive is more than 0.01. If the prediction error (RMSE) of a surrogate over validation data is less than 20%, then it is considered valid.

To compare the effects of the number of the surrogate models (corresponding to the number of partitions of the archive), each of the problems was run with 3, 5 and 8 partitions. All the test problems are also run with single surrogate model.

The same random seed and hence the same initial population is used for both, EASDS and NSGA-II. Since the number of actual function evaluations in EASDS are much less than 10100 (100×101), NSGA-II is run for fewer generations (with similar number of function evaluations) for performance comparison.

4   Results

Shown in Table 2 are the function evaluations used by EASDS for different number of partitions (K). Traditional evolutionary algorithms with population size of 100 evolved over 101 generations will result in $10,100$ function evaluations. In EASDS, the population is evaluated using the actual evaluations every $I_{TRAIN} = 5$ generations, hence the minimum number of actual evaluations is 2100.

As seen from Table 2, the number of function evaluations for problem OSY decreases as the number of partitions is increased. This shows that the prediction

Table 2. Function Evaluations used by EASDS

	Function Evaluations		
Problem	$K = 3$	$K = 5$	$K = 8$
OSY	7139	4111	3620
SRN	2100	2100	2100
ZDT1	2100	2600	2600
ZDT2	2100	2600	2600
ZDT3	2100	2600	2600

accuracy of the surrogate models with $K = 3$ partitions is poor as compared to the surrogate models with $K = 8$ partitions. If all the surrogates models are invalid (prediction accuracy over the validation data set is more than the user defined threshold), actual evaluations are used.

The non-dominated solutions for problem OSY obtained by EASDS using 3, 5, and 8 partitions are shows in Fig. 1. It is observed that the non-dominated solutions of the EASDS with 8 partitions follow the Pareto front much more accurately than the EASDS with 3 and 5 partitions.

For test problem SRN, the number of function evaluations used are 2100, the minimum possible. This indicates that EASDS with 3, 5, or 8 partitions is able to correctly capture the behavior of the function SRN. As seen in Fig. 2, the non-dominated solutions of EASDS with 3, 5, and 8 partitions are overlapping.

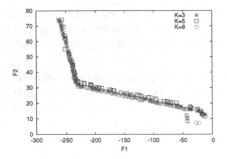

Fig. 1. Effect of number of partitions on problem OSY

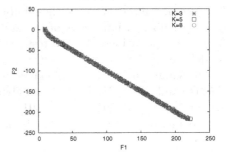

Fig. 2. Effect of number of partitions on problem SRN

Test problems ZDT1, ZDT2, and ZDT3 show a different trend in function evaluations as compared to OSY. The number of function evaluations used by EASDS increase for 5 and 8 partitions as compared to 3 partitions. This can be explained by the fact that the surrogate models in the initial few generations are not very accurate and the actual evaluations are used to evaluate the entire population. In the earlier generations, there are fewer solutions in the archive and those solutions are distributed spatially and split in to multiple partitions to build the surrogate models. Thus each partition might have insufficient number of points to capture the correct behavior of the function and the prediction accuracy is low. As the number of solutions accumulate in the archive, the accuracies of the surrogate models also increase.

Shown in Fig. 3 are the non-dominated solutions for problem ZDT1 obtained by EASDS using 3, 5, and 8 partitions and they are overlapping. It shows that the function ZDT1 is approximated accurately using the surrogate models with 3, 5, and 8 partitions. For the test function ZDT2, the non-dominated solutions are shown in Fig. 4. The surrogate models with 8 partitions are able to achieve better spread of the non-dominated solutions on the Pareto front indicating that surrogate models with 8 partitions have better prediction accuracy than surrogate models with 3 and 5 partitions.

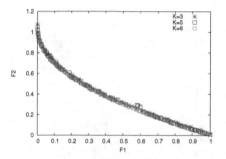 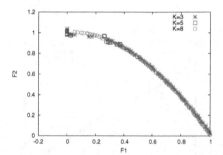

Fig. 3. Effect of number of partitions on problem ZDT1

Fig. 4. Effect of number of partitions on problem ZDT2

Shown in Fig. 5 are the non-dominated solutions for problem ZDT3 obtained by EASDS using 3, 5, and 8 partitions. It is seen that none of the surrogate models are able to completely capture the disjoint Pareto front. Surrogate models with 8 partitions seem to have a better spread than the ones with 5 partitions which are better than the ones with 3 partitions.

Shown in Fig. 6 are the non-dominated solutions for problem OSY obtained by EASDS using 8 partitions (EASDS), single global surrogate model (SGS) and NSGA-II with the same number of function evaluations. The performance of EASDS is better at capturing the Pareto front.

The non-dominated solutions obtained for the problem SRN by EASDS, SGS, and NSGA-II are shown in Fig. 7. Even a single global surrogate is able to capture the behavior of the function adequately and the non-dominated solutions overlap.

The benefit of the spatially distributed surrogate models can be seen from the results of ZDT1, ZDT2, and ZDT3 which are 10-D functions. It can be seen from Figs. 8, 9, and 10 that EASDS captures the Pareto front better than NSGA-II and single global surrogate model (SGS).

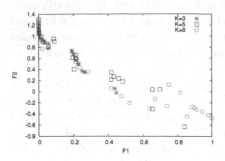

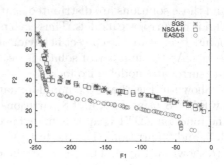

Fig. 5. Effect of number of partitions on problem ZDT3

Fig. 6. Non-dominated solutions for problem OSY

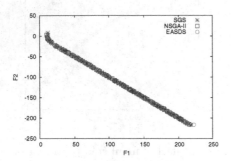

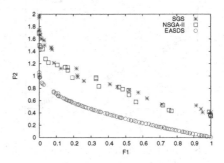

Fig. 7. Non-dominated solutions for problem SRN

Fig. 8. Non-dominated solutions for problem ZDT1

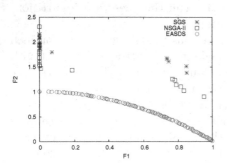

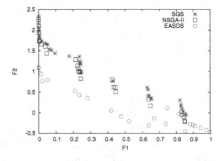

Fig. 9. Non-dominated solutions for problem ZDT2

Fig. 10. Non-dominated solutions for problem ZDT3

5   Summary and Conclusions

In this paper an evolutionary algorithm with spatially distributed surrogates (EASDS) for multiobjective optimization is presented. This approach is an alternative to recent surrogate ensemble proposals, where either multiple types of global surrogates are used or multiple number of same type of surrogates are used for better approximation. EASDS is compared with the non-dominated sorting algorithm (NSGA-II) and single global surrogate model on a set of test functions. Different number of partitions (3, 5, and 8) are used to build the surrogate models and corresponding performance is compared.

It is seen from Figs. 1 - 5 that the surrogate models with more partitions perform better. With more partitions the function behavior is captured better by splitting the design space in multiple regions and approximating each region locally. But as the number of partitions is increased, more number of evaluations are required to populate each partition sufficiently (to be able to correctly capture the behavior of the function locally in the partition). With the computational budget of 1200 evaluations EASDS is able to capture the behavior of 10-D optimization problem with up to 8 partitions. For a smaller computational budget or higher dimensional problem, one may need to use more conservative number of partitions.

Compared to the single global surrogate model and NSGA-II, EASDS performs much better indicating the benefits of the local surrogates built over smaller regions. Effectiveness of EASDS at capturing the Pareto front and the spread of solutions along the Pareto front is clearly seen from Figs. 6, 8, and 9.

References

1. Jin, Y.: A comprehensive survey of fitness approximation in evolutionary computation. Soft Computing - A Fusion of Foundations, Methodologies and Applications 9, 3–12 (2005)
2. Wilson, B., Cappelleri, D., Simpson, T.W., Frecker, M.: Efficient pareto frontier exploration using surrogate approximations. Optimization and Engineering 2, 31–50 (2001)
3. Goel, T., Haftka, R.T., Shyy, W., Queipo, N.V.: Ensemble of surrogates. Structural and Multidisciplinary Optimization 33, 199–216 (2007)
4. Schaffer, J.D.: Multiple Objective Optimization with Vector Evaluated Genetic Algorithms. In: Proceedings of the 1st International Conference on Genetic Algorithms, pp. 93–100. Lawrence Erlbaum Associates, Mahwah, NJ (1985)
5. Zhou, Z.Z, Ong, Y.S, Nair, P.B, Keane, A.J, Lum, K.Y: Combining global and local surrogate models to accelerate evolutionary optimization. IEEE Transactions on Systems Man and Cybernetics Part C-Applications and Reviews 37, 66–76 (2007)
6. Gaspar-Cunha, A., Vieira, A.S.: A hybrid multi-objective evolutionary algorithm using an inverse neural network. In: Hybrid Metaheuristics (HM 2004) Workshop at ECAI 2004, Valencia, Spain (2004)
7. Breiman, L.: Bagging predictors. Machine Learning 24, 123–140 (1996)
8. Abney, S., Schapire, R.E., Singer, Y.: Boosting applied to tagging and PP attachment. In: Joint SIGDAT Conference on Empirical Methods in Natural Language Processing and Very Large Corpora (1999)

9. Zhao, Y., Gao, J., Yang, X.: A survey of neural network ensembles. In: International Conference on Neural Networks and Brain (ICNN&B 2005), vol. 1, pp. 438–442 (2005)
10. Jin, Y., Sendhoff, B.: Reducing fitness evaluations using clustering techniques and neural networks ensembles. In: Deb, K., et al. (eds.) GECCO 2004. LNCS, vol. 3102, pp. 688–699. Springer, Heidelberg (2004)
11. Hamza, K., Saitou, K.: Vehicle crashworthiness design via a surrogate model ensemble and a coevolutionary genetic algorithm. In: Proceedings of IDETC/CIE 2005 ASME 2005 International Design Engineering Technical Conference, California, USA (September 2005)
12. Zerpa, L.E., Queipo, N.V., Pintos, S., Salager, J.L.: An optimization methodology of alkaline-surfactant-polymer flooding processes using field scale numerical simulation and multiple surrogates. Journal of Petroleum Science and Engineering 47, 197–208 (2005)
13. Zhou, Z., Ong, Y.S., Lim, M.H., Lee, B.S.: Memetic algorithm using multi-surrogates for computationally expensive optimization problems. Soft Computing - A Fusion of Foundations, Methodologies and Applications 11, 957–971 (2007)
14. Nain, P., Deb, K.: A computationally effective multi-objective search and optimization techniques using coarse-to-fine grain modeling. In: 2002 PPSN Workshop on Evolutionary Multiobjective Optimization (2002)
15. Ray, T., Smith, W.: Surrogate assisted evolutionary algorithm for multiobjective optimization. In: Proceedings of 47th AIAA/ASME/ASCE/AHS/ASC Structures, Structural Dynamics, and Materials Conference, pp. 1–8 (2006)
16. Nain, P.K.S., Deb, K.: A multi-objective optimization procedure with successive approximate models. Technical Report 2005002, KanGAL, IIT Kanpur (2005)
17. Knowles, J.: ParEGO: a hybrid algorithm with on-line landscape approximation for expensive multiobjective optimization problems. IEEE Transactions on Evolutionary Computation 10, 50–66 (2006)
18. Emmerich, M.T.M., Giannakoglou, K.C., Naujoks, B.: Single and multiobjective evolutionary optimization assisted by gaussian random field metamodels. IEEE Transactions on Evolutionary Computation 10, 421–439 (2006)
19. Chafekar, D., Shi, L., Rasheed, K., Xuan, J.: Multiobjective ga optimization using reduced models. Systems, Man and Cybernetics, Part C: Applications and Reviews, IEEE Transactions on 35, 261–265 (2005)
20. Deb, K., Pratap, A., Agarwal, S., Meyarivan, T.: A fast and elitist multiobjective genetic algorithm: NSGA-II. IEEE Transactions on Evolutionary Computation 6, 182–197 (2002)
21. Deb, K., Agrawal, S.: Simulated binary crossover for continuous search space. Complex Systems 9, 115–148 (1995)
22. Deb, K., Goyal, M.: A combined genetic adaptive search (GeneAS) for engineering design. Computer Science and Informatics 26, 30–45 (1996)
23. Jain, A.K., Murty, M.N., Flynn, P.J.: Data clustering: A review. ACM Computing Surveys 31, 265–323 (1999)
24. Deb, K., Agrawal, S., Pratap, A., Meyarivan, T.: A fast elitist non-dominated sorting genetic algorithm for multi-objective optimization: NSGA-II. In: Proceedings of the Parallel Problem Solving from Nature VI, pp. 849–858 (2000)
25. Deb, K.: Multi-Objective Optimization using Evolutionary Algorithms. John Wiley and Sons, Chichester (2001)

Examining Dissimilarity Scaling in Ant Colony Approaches to Data Clustering

Swee Chuan Tan, Kai Ming Ting, and Shyh Wei Teng

Gippsland School of Information Technology
Monash University
Churchill, Victoria 3842 Australia
{James.Tan,KaiMing.Ting,Shyh.Wei.Teng}
@infotech.monash.edu.au

Abstract. In this paper, we provide the reasons why the dissimilarity-scaling parameter (α) in the neighbourhood function of ant-based clustering is critical for detecting the correct number of clusters in data sources. We then examine a recently proposed method named ATTA; we show that there is no need to use a population of α-adaptive ants to reproduce ATTA's results. We devise a method to estimate a fixed (i.e, non-adaptive) *single* value of α for each dataset. We also introduce a simplified version of ATTA, called SATTA. The reason for introducing SATTA is two-fold: first, to test our proposed α-estimation method; and, second, to simulate ant-based clustering from a purely stochastic perspective. SATTA omits the ant colony but reuses important ant heuristics. Experimental results show that SATTA generally performs better than ATTA on clusters with different densities and clusters that are elongated. Finally, we show that the results can be further improved using a majority voting scheme.

1 Introduction

Ant-based clustering is inspired from the cemetery formation and brood-sorting activities found in real ant colonies. This approach is motivated by the collective problem solving ability found in simple social insects (ants, bees, termites, etc) [2]. Initially proposed by Deneubourg et al. [3], robotic ants have been used to cluster physical objects. Deneubourg's model was later extended by Lumer and Faieta [7] to cluster numerical data. Since then, a number of variants have been proposed; some interesting examples include: (i) combining ant-based clustering with *K-means* ([8], [9]), (ii) using pheromone in ant-based clustering [10], and (iii) specific extensions of Lumer and Faieta's model: Ant-Q [6] and its successor ATTA [5].

Most of the recent studies in ant-based clustering have shown promising results; however, the working principle of this method remains elusive. In this paper, we study a critical parameter in ant-based clustering: the dissimilarity-scaling parameter (α) in the neighbourhood function. Previous works have used a population of heterogeneous ants (e.g., see [5], [7]), in which ants differ by associating with different α-values. However, two issues remain: (i) why is the value

M. Randall, H.A. Abbass, and J. Wiles (Eds.): ACAL 2007, LNAI 4828, pp. 269–280, 2007.

of α parameter critical and (ii) is there a need to use multiple ants to estimate multiple α-values to produce good clustering results? In this paper, we first give the reasons why α is a critical parameter. We also show that a fixed single value of α (for each dataset) is enough to reproduce the existing clustering results.

We propose a method that estimates only a single value of α; this implies that there is no need to use multiple ant-like agents to estimate multiple α-values as in ATTA [5]. Furthermore, our α-value is estimated independently of the clustering process whereas in ATTA the values of α are adapted throughout the entire clustering process; this suggests that there is no need to learn the values of α during the clustering process.

We also present a simplified version of ATTA, called SATTA. SATTA is used to test the estimated α, and it reuses important ant heuristics. However, SATTA differs from ant-based clustering (ATTA) in the following ways: (i) it does not employ a colony of heterogeneous ants, but uses only random sampling to emulate the pick and drop actions of randomly moving ants; (ii) it uses the above-mentioned α-estimation method and also filters noise and outliers in the pre-processing stage. Although our method is simpler than traditional ant-based clustering, it has produced comparable or better clustering results as compared to the recently proposed ATTA model [5].

In the next section we discuss the issues identified. Section 3 presents the proposed method. Section 4 shows that the results of our method are comparable to ATTA and better under some conditions. Section 5 concludes this paper.

2   Issues in Ant-Based Clustering

2.1   Why Is the Value of α Critical?

In the data clustering model proposed by Lumer and Faieta [7], they have generalised the neighbourhood function as:

$$f(i) = \max\left(0.0, \frac{1}{\sigma^2}\sum_j\left(1 - \frac{\delta(i,j)}{\alpha}\right)\right). \tag{1}$$

Here, α scales the distance $\delta(i,j)$ between a currently picked data item i and all the data item j in the neighbourhood of the grid space in which the ants operate. The size of the ant's perception region is σ^2, which is defined as $\sigma^2 = (2r+1)^2$, where r is the perceptive radius.

Handl et al. [5] introduced two modifications to the neighborhood function. First, they kept σ^2 as a constant of 8 regardless of r. Second, they added the following constraint to maintain that any item in a grid-neighbourhood is no more than α-distance apart from the item i at the center of the neighbourhood:

$$f^*(i) = \begin{cases} f(i) & \text{if } \forall j\,(\delta(i,j) < \alpha) \\ 0.0 & \text{else.} \end{cases} \tag{2}$$

While it is generally known that the value of α can somehow affect cluster detection, the criticality of α is not well illustrated in the literature. Take the skewed data in Figure 1(a) as an example, Figure 1(b) shows the distribution of the normalised pairwise Euclidean distances among all data points in the skewed dataset: the left most distribution contains the *intra-cluster* pairwise distances, and the remaining three distributions contain the *inter-cluster* pairwise distances.

In this example, the best α-value is 0.04, which scales the *intra-cluster* pairwise distances to be less than one, and scales the *inter-cluster* pairwise distances to be more than one. Thus, the α-value detects the four underlying clusters in the skewed dataset. If α-value is in [0.15, 1], then only two clusters are detected: the first cluster is C1, and the second cluster contains C2, C3 and C4.

In summary, there are two reasons why α should not be set arbitrarily. First, the range of good α-values can be quite narrow in some cases; and if we set α arbitrarily, then we can miss a good α-value easily. Second, an inappropriate α-value set arbitrarily can lead to misleading results; it can produce wrong number of clusters that appear to be 'correctly' built on the grid, which leads to a false detection of cluster structure.

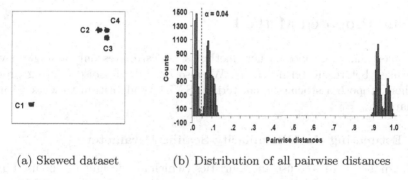

(a) Skewed dataset   (b) Distribution of all pairwise distances

Fig. 1. (a) The skewed dataset contains three nearby clusters and one cluster far from the rest. (b) The best α-value is 0.04 as shown by the dotted line

2.2 Do We Need Heterogenous Ants and Multiple Adaptive α-Values?

In this work, we focus on ATTA [5] since it is well-tested and it shows impressive results. However, our previous study [11] with an earlier version of ATTA (known as Ant-Q [6]) shows that there is no need to use a population of ants. This also implies that there is no need to use heterogenous ants to estimate multiple α-values in order to reproduce the results of ATTA. To show that this implication is plausible, Figure 2(b) shows that after the initial stage of α-adaptation, the α-values of different ants in ATTA converge to an average value of 0.3. This observation is confirmed by running ATTA on VaryDensity dataset using one ant (i.e., only one α-value), and the clustering results (e.g., the F-Measure [12]) are the same as those produced by ten ants (i.e., ten α-values). With the above analysis, we believe that a single value of α can be used to reproduce the results of ATTA.

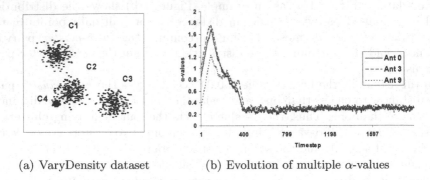

(a) VaryDensity dataset   (b) Evolution of multiple α-values

Fig. 2. (a) This dataset contains three low density clusters (C1, C2, and C3) and one high density cluster (C4). (b) When clustering the VaryDensity dataset, the α-values of all the ten ants in ATTA converge during the process. For clarity, we only show the evolution of α-values for three typical ants.

3   The Proposed Method

In this section, we propose a new method that estimates only a single α-value to be used before clustering starts. We also describe a noise filtering process. We then propose a stochastic clustering (SATTA) algorithm to work with the estimated α.

3.1   Estimating the Dissimilarity-Scaling Parameter

The main idea here is to use ant heuristics (which are designed to be used in the *grid-space*) to estimate an α-value in the *data-space*. First, we use the k-nearest neighbourhood in the data-space to approximate the grid-neighbourhood in the grid-space. Then, we test if a given α-value will render each item i to be *similar* to its nearest neighbours in the data-space. In order to test the *similarity*, we borrow the formulae from ant-clustering. Table 1 presents how each formula in ant-clustering is being adapted for α-estimation in the data-space.

Note that this is not a clustering process; rather, it is a pre-processing step that occurs before clustering begins. Indeed, this procedure takes a negligible time to compute as compared to the runtime of the main clustering process.

Now, we show how to estimate α. First, we compute and store the k-nearest neighbours (*DNeigh*) for each data item i in a data-space (D). Then, for each item i in D, we perform two steps: (i) compute the average similarity of i with its nearest neighbours (using Equations 4 and 5 in Table 1); and (ii) *simulate* a drop event for i using Equation 6 in Table 1. This process is repeated for all items in D, and is based on a given α-value.

If a given α-value is very small, then we expect very few successful drop events to occur in the above process; but as the value of α becomes bigger, then we expect more items to be dropped successfully. When α is increased to a point

Table 1. This table shows how the formulae used by ATTA in the grid-space are being adapted to estimate an α-value in the data-space. Once α is estimated (in the data-space) in this pre-processing stage, it will be used during the actual clustering process in the grid-space. In Equation 6, $Rand(0,1)$ is a random variable in $[0, 1]$.

Equations used by ATTA in the grid-space	Equations adapted for α-estimation in the data-space
Equation (1): $f(i) = \max\left(0.0, \frac{1}{\sigma^2}\sum_j\left(1 - \frac{\delta(i,j)}{\alpha}\right)\right)$	Equation (4): $F(i) = \max\left(0.0, \frac{1}{\sigma^2}\sum_{j\in DNeigh}\left(1 - \frac{\delta(i,j)}{\alpha}\right)\right)$
Equation (2): $f^*(i) = \begin{cases} f(i) & \text{if } \forall j\,(\delta(i,j) < \alpha) \\ 0.0 & \text{else.} \end{cases}$	Equation (5): $F^*(i) = \begin{cases} F(i) & \text{if } \forall j\,(\delta(i,j) < \alpha) \\ 0.0 & \text{else.} \end{cases}$
Equation (3): $p^*_{drop}(i) = \begin{cases} 1.0 & \text{if } f^*(i) \geq 1.0 \\ f^*(i)^4 & \text{else.} \end{cases}$	Equation (6): $DropEvent(i) = \begin{cases} 1 & \text{if } F^*(i)^4 \geq Rand(0,1) \\ 0 & \text{else.} \end{cases}$

that it first renders every item to be successfully dropped, then this is an α-value that we can use for clustering on the grid.

Given a fixed α-value, we can measure the proportion of successful drop events using Equation 7, and this is also the expected drop probability for a dataset D with n items:

$$avgP_{drop}(\alpha) = \frac{1}{n}\sum_{i\in D} DropEvent(i). \tag{7}$$

Figure 3 shows a plot of $avgP_{drop}(\alpha)$ over α in $[0, 0.1]$ for the Skewed dataset shown in Figure 1(a). To make the neighbourhood function discriminative to dissimilar items, we find the smallest α-value that renders an expected drop probability of one. Figure 3 shows that a good α-value is about 0.03.

It is desirable to estimate the value of α automatically from a data source, and we can do so using the following recursive binary search:

$$\alpha_t = \begin{cases} 0.5 & \text{if } t = 0 \\ \alpha_{t-1} - \frac{\alpha_0}{2^t} & \text{if } t > 0 \wedge avgP_{drop}(\alpha_{t-1}) \geq 1.0 \\ \alpha_{t-1} + \frac{\alpha_0}{2^t} & \text{if } t > 0 \wedge avgP_{drop}(\alpha_{t-1}) < 1.0 \end{cases} \tag{8}$$

The search involves $t = 0, 1, ..., 6$. We end at $t = 6$ because $\frac{\alpha_0}{2^6}$ is $\frac{1}{2^7}$, which makes the estimation precise enough in general cases.

Since α is obtained based on neighbours in the *data-space* rather than neighbours in the *grid-space*, the estimated α-value is conservative. This is because each item on the grid is unlikely to be surrounded by their actual neighbours in the *data-space*. Thus, we expect the actual α used on the grid to be higher than our estimated α. To improve the estimated α-value, we repeat the binary search twenty times, and choose the highest α-value found in the process.

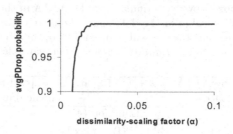

Fig. 3. Plot of $avgP_{drop}(\alpha)$ over α; the estimated α-value is about 0.025 for the Skewed dataset, and this is close to the ideal α-value in Figure 1(b).

3.2   Dealing with Noise and Outliers

The value of data-neighbourhood function (Equation 5), which we denote as F^*-value, is associated with data density: data points with high F^*-values are usually located near the core of each cluster, while low F^*-valued points are usually noise or outliers. To reduce uncertainty errors, we can remove noise and outliers by removing the low F^*-valued points. In the experiments reported in section 4, we remove 10% of the bottommost F^*-valued points from each dataset.

3.3   Main Clustering Algorithm

Algorithm 1 presents our proposed stochastic clustering (SATTA) algorithm. Since we do not use any ant-like agents, the random visitations of items (by the ants) are now simulated using random sampling. First, we randomly select an item i from the dataset; we repeat this random sampling step until i is picked up (using pick probability proposed by Handl et al.[5]). In Algorithm 1, when an item i is being selected, it simulates an event that a free ant A finds and picks up an item i from the grid; when an item j is being randomly selected, it simulates an event that the (randomly moving) loaded ant A encounters another item j on the grid. If i is similar enough to the items in the surroundings of item j (i.e., the probability of dropping i onto j's neighbourhood is high), then i is moved to an empty grid cell near to j using a random search used by [5]. If i is not dropped at the location near j, then the process is repeated for a different selected item. If item i still cannot be dropped even after a maximum number of trials ($maxTrial$), then i is returned to its original position. Algorithm 1 ends when the maximum number of iterations is reached.

Clustering process. First, the pre-processing consists of three steps: (i) store all the pairwise dissimilarities in a matrix, (ii) automatically estimate the value of α using the method described in section 3.1, and, (iii) filter noise and outliers as described in section 3.2. Then, the main clustering process, which has two phases, begins. Phase one contains $30n$ iterations (where n is the data size). To build up the clusters quickly, we use a perceptive radius (r) to one. In phase two, the purpose is to increase spatial separation among clusters (this is required

Algorithm 1. Stochastic Clustering (SATTA)

Begin
INITIALIZATION
Randomly distribute the data on a two-dimensional toroidal grid
MAIN LOOP
for $iteration = 1$ to $maxIteration$ **do**
  repeat
    Randomly select an item i from the data source with n items
  until i is picked up
  Let the location of i be $origLoc$
  Let $Trial$ be 0
  while $Trial < maxTrial$ **do**
    Randomly select an item j from the data source
    Move i to the location of j
    if i can be dropped at its current location **then**
      Move item i to an empty grid location near j
      $Trial = maxTrial + 1$ // to terminate while loop
    else
      Increment $Trial$ by one
    end if
    if $Trial == maxTrial$ **then**
      Move item i back to its $origLoc$ // cannot find a destination cluster
    end if
  end while
end for
End

for automatic cluster retrieval [5]), so we increase r to two, then three; in each setting of r we run $5n$ iterations. Once the clusters are formed on the grid, we retrieve the clusters using the single-link cluster retrieval method as used by ATTA [5], and then we assign each of the noisy points d (previously removed in (iii) above) to a cluster if that cluster contains an item closest to d. As a result, we group the entire dataset into different clusters.

Algorithm 1 builds preliminary clusters on the grid quickly, but these clusters are not compact. To improve the compactness of these clusters, we modify Algorithm 1 slightly after the first $10n$ iterations: we move i to an empty cell near j before we test if i can be dropped at its current location.

Before we compare the results of SATTA to ATTA in the next section, Table 2 summarizes the similarities and differences between SATTA and ATTA.

4   Experimental Set Up and Results

4.1   Experimental Setup

We reuse the real and synthetic datasets used in testing ATTA. These datasets are detailed in [4] and therefore briefly covered here. The Squares series represent

Table 2. Similarities and differences between SATTA and ATTA. Items marked with * are the pre-processing stages.

ATTA	SATTA
*Stores all dissimilarities in a matrix.	*Same as ATTA.
Uses a two-dimensional grid.	Same as ATTA.
Estimates multiple values of α (each ant stores one α).	Estimates a single value of α.
Adjusts α-values throughout the clustering process.	*Estimates an α-value before clustering begins.
No filtering process.	*Filters noise and outliers.
Uses a population of ten ants to pick and drop data items.	Randomly samples data items.
Increases perceptive radius (r) from one to five at five *equal intervals*. This is to increase spatial separation among clusters.	Increases r from one to three, with a substantially longer interval for $r=1$; then set $r = 2$ & 3, each at a *shorter interval*.
Ants are required to disperse data at the middle of the clustering process.	No data dispersion.
Uses ant's local memory with a look-ahead strategy to search for destination items.	Uses random search for destination items.

clusters that are increasingly overlapped. The Sizes datasets have the ratios between the cluster sizes vary from 2 to 10. In addition to the datasets used by ATTA, we introduce the Skewed, VaryDensity and Triangle datasets to test the algorithms' ability to deal with clusters with different shapes and densities. All the synthetic datasets contain 1000 instances and four clusters of bivariate normal distributions. To demonstrate the algorithm's capability in practice, we also use the real data from UCI Machine Learning Repository [1]. These datasets include: Wisconsin, Iris, Wine, Dermatology, Zoo, Yeast and Digits.

The square grid dimension is $\sqrt{10n}$ where n is data size; this is based on Handl's recommendation [4]. The *maxTrial* in Algorithm 1 is set to 2000. For α-estimation, the number of the k-nearest neighbours used is 24. We also remove 10% noise from each dataset; note that the noise points are assigned back to their core clusters for final evaluation.

Results of each dataset were obtained based on 50 independent runs, and finally evaluated using the *F-measure* [12] and the number of clusters detected by both SATTA and ATTA. For the three additional datasets, we obtained the ATTA's results using the ATTA source codes supplied by Handl [4]. As for the rest of the datasets, the results of ATTA were provided by Handl [4].

4.2   Results

Table 3 presents the F-measure and the number of clusters detected by ATTA and SATTA across each dataset.

Table 3. Clustering Results for SATTA and ATTA: each entry is the mean ± standard deviation over 50 runs. Better results are in boldface.

Data Set	#Clusters	F-Measure		#Clusters Detected	
		SATTA	ATTA	SATTA	ATTA
Square1	4	**.985**±.005	.984±.004	4.04±.198	**4** ±0
Square3	4	.937±.026	**.940**±.009	4.08±.396	4±0
Square5	4	.785±.108	**.790**±.061	3.66±.798	**3.74**±.482
Sizes1	4	**.987**±.004	.984±.004	4±0	4±0
Sizes3	4	**.985**±.014	.984±.018	4.06±.24	**3.98**±.14
Sizes5	4	**.988**±.014	.986±.019	**3.98**±.14	3.9±.196
Skewed	4	1±0	1±0	4±0	4±0
Triangle	4	**.993**±.030	.938±0.040	**4**±.202	4.84±0.842
VaryDensity	4	**.917**±.046	.835±0.015	**3.84**±.422	3.06±0.240
Wisconsin	2	**.975**±0	.968±.001	2±0	2±0
Iris	3	.815±.008	**.817**±.015	2.98±.14	3.02±.14
Wine	3	**.893**±.04	.876±.021	3.48±.614	3±0
Dermatology	6	**.894**±.002	.846±.049	5±0	4.36±.625
Zoo	7	.801±.034	**.819**±.047	**3.9**±.463	3.88±.431
Yeast	10	**.453**±.032	.435±.035	5.28±.809	**5.36**±1.179
Digits	10	**.633**±.033	.504±.031	**9.12**±1.043	5.3±.806

Results from synthetic data. For the Square series data, ATTA performs better than SATTA in the highly overlapped Square3 and Square5 datasets. As for the Sizes series data, both ATTA and SATTA show good performance.

For the Triangle dataset shown in Figure 4(a), ATTA tends to produce more than the four actual clusters whereas SATTA consistently detects the four actual clusters and gives a higher F-Measure. Recall that ATTA increases the perceptive radius (r) from one to five; when r is large, there are more items compared in the neighborhood function. If a cluster is elongated, then there is a high chance that an item is more than α-distance apart from the item at the center of the neighborhood. If this occurs, the neighbourhood function value is set to zero by Equation 2 and this promotes the cluster to split. Figure 4(b) shows an example that the elongated cluster is divided into two and three sub-clusters when r is four and five respectively.

For the VaryDensity dataset shown in Figure 2(a), SATTA usually produces four clusters and gives a higher F-Measure than ATTA. ATTA incorrectly merged cluster C2 and C4 because ATTA tends to adjust its α-values based on majority of the low density data points, which results in higher α-value and it cannot distinguish clusters C2 and C4.

As for the skewed dataset, both ATTA and SATTA perform well in detecting the four clusters. If no α-estimation is used, then SATTA returns two clusters (instead of four clusters) for any α-value between [0.15, 1.0].

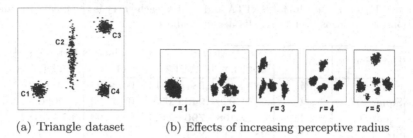

    (a) Triangle dataset   (b) Effects of increasing perceptive radius

Fig. 4. (a) The Triangle dataset contains one elongated and three spherical clusters. (b) When the perceptive radius (r) increases to four or five, the elongated cluster in the Triangle dataset is sub-divided on the grid.

Table 4. Improved clustering results for SATTA* and ATTA*. Each entry is the mean ± standard deviation over ten runs of the majority voting scheme. Better results are in boldface.

Data Set	#Clusters	F-Measure		#Clusters Detected	
		SATTA*	ATTA*	SATTA*	ATTA*
Square1	4	**.986**±.004	.984±.004	4±0	4±0
Square3	4	**.944**±.010	.940±.009	4±0	4±0
Square5	4	**.829**±.044	.820±.024	3.90±.31	**4±0**
Sizes1	4	**.987**±.004	.984±.004	4±0	4±0
Sizes3	4	**.988**±.004	.987±.003	4±0	4±0
Sizes5	4	**.990**±.004	.989±.004	4±0	4±0
Skewed	4	1±0	1±0	4±0	4±0
Triangle	4	**.999**±.005	.931±.042	**4±0**	4.93±.78
VaryDensity	4	**.928**±.037	.833±.001	**3.93±.26**	3±0
Wisconsin	2	.975±0	-	2±0	-
Iris	3	.816±.006	-	3±0	-
Wine	3	.908±.034	-	3.17±.379	-
Dermatology	6	.894±.002	-	5±0	-
Zoo	7	.814±.011	-	4±0	-
Yeast	10	.461±.026	-	5.14±.516	-
Digits	10	.639±.034	-	9.04±.859	-

Results from real data. The last part of Table 3 shows that SATTA performs comparably with ATTA on most of the real datasets. For the Dermatology and Digits datasets, SATTA performs better because our estimated α-value is generally conservatively small, and this makes the neighbourhood function more discriminative on dissimilar items. For the Zoo dataset, both ATTA and SATTA fail to detect some of the small clusters. Handl [4] suggested that this is because the neighborhood function requires clusters to have a minimum size in order to form stable clusters on the grid. For the Yeast dataset, both ATTA and SATTA also perform poorly; previous study [4] has suggested that this is because the cluster structure of Yeast dataset is not easily noticeable.

Improving the results. We can improve the consistency of results for SATTA and ATTA in Table 3. This is done using a majority voting scheme, which involves two steps: (i) examine the results of five sequential runs, (ii) only take the results (F-measure and number of clusters detected) associated with the most frequently occurring **number-of-clusters-detected** in the five runs. For example, if '3 clusters detected' occurs two times, and '4 clusters detected' occurs three times, then '4 clusters detected' is the most frequently occurring result and the three results associated with '4 clusters detected' are used for final evaluation. If a tie occurs then random selection is used.

Since the results of each dataset in Table 3 are based on 50 runs, the above procedure is executed ten times for each dataset. Table 4 shows that the results of SATTA and ATTA have been improved by their respective enhanced versions: SATTA* and ATTA*. In addition, SATTA* now generally outperforms ATTA* on the synthetic datasets. Although the results of ATTA* on real data are not available (this is because we currently do not have enough details of ATTA's results on real data to perform majority voting), the results of SATTA* on real data have generally improved.

5   Concluding Remarks

This work demonstrates a new method to estimate a single value of parameter α, and the estimated α is used with our stochastic clustering method. The proposed stochastic clustering model is a non-ant counterpart of ant-based clustering: it builds upon important ant heuristics contributed by previous works (e.g., [5], [7]). Thus, this work highlights the values of ant-clustering heuristics, and strengthens its most intriguing promise: automatic cluster detection in a dataset with no additional information. On the other hand, our current and previous results [11] continue to challenge a common belief in ant-based clustering that a population of ants is required to achieve good clustering results. We hope that our work can spark further investigation from a swarm intelligence perspective: to understand how collective intelligence can be exploited in ant-based clustering. Although we study ant-based clustering from a purely stochastic perspective, we believe that the additional understanding gained from either perspective will advance the state-of-the-art in ant-based clustering.

Since our focus is on the approach to α-estimation, we have not considered additional heuristics to speed up the overall clustering process. However, our latest preliminary results suggest that, with additional heuristics, SATTA is likely to work faster than ATTA.

Acknowledgments

The first author is supported by Monash Graduate Scholarship and Monash International Postgraduate Scholarship. The authors thank Julia Handl for providing the ATTA's source codes and ATTA's results for the original datasets.

References

1. Blake, C., Merz, C.: UCI repository of machine learning databases. Technical report, Department of Information and Computer Sciences, University of California, Irvine (1998)
2. Bonabeau, E., Dorigo, M., Theraulaz, G.: Swarm Intelligence: From Natural to Artificial Systems. Oxford University Press, New York (1999)
3. Deneubourg, J.-L., Goss, S., Franks, N., Sendova-Franks, A., Detrain, C., Chrétien, L.: The dynamics of collective sorting: Robot-like ants and ant-like robots. In: Proceedings of the First International Conference on Simulation of Adaptive Behaviour: From Animals to Animats 1, pp. 356–363. MIT Press, Cambridge (1991)
4. Handl, J.: Ant-based methods for tasks of clustering and topographic mapping: extensions, analysis and comparison with alternative methods. Master's thesis, University of Erlangen-Nuremberg, Germany (2003)
5. Handl, J., Knowles, J., Dorigo, M.: Ant-based clustering and topographic mapping. Artificial Life 12(1), 35–61 (2005)
6. Handl, J., Meyer, B.: Improved ant-based clustering and sorting in a document retrieval interface. In: Guervós, J.J.M., Adamidis, P.A., Beyer, H.-G., Fernández-Villacañas, J.-L., Schwefel, H.-P. (eds.) Parallel Problem Solving from Nature - PPSN VII. LNCS, vol. 2439, pp. 913–923. Springer, Heidelberg (2002)
7. Lumer, E., Faieta, B.: Diversity and Adaptation in Populations of Clustering Ants. In: Proceedings of the third International Conference on Simulation of Adaptive Behavior: From Animals to Animats 3, vol. 1, pp. 501–508. MIT Press, Cambridge (1994)
8. MacQueen, L.: Some methods for classification and analysis of multivariate observations. In: Proceedings of the Fifth Berkeley Symposium on Mathematical Statistics and Probability, vol. 1, pp. 281–297. University of California Press, Berkeley (1967)
9. Monmarché, N., Slimane, M., Venturini, G.: AntClass: discovery of clusters in numeric data by an hybridization of an ant colony with the K-means algorithm. Technical Report 213, Laboratoire d'Informatique, E3i, University of Tours (1999)
10. Ramos, V., Abraham, A.: Evolving a Stigmergic Self-Organized Data-Mining. In: Proceedings of the fourth International Conference on Intelligent Systems, Design and Applications, Budapest, Hungary, pp. 725–730 (2004)
11. Tan, S.C., Ting, K.M., Teng, S.W.: Reproducing the results of ant-based clustering without using ants. In: Proceedings of Congress on Evolutionary Computation 2006, pp. 1760–1767. IEEE Press, Vancouver (2006)
12. van Rijsbergen, C.: Information Retrieval, 2nd edn. Butterworths, London, UK (1979)

A Framework for the Co-evolution of Genes, Proteins and a Genetic Code Within an Artificial Chemistry Reaction Set

Ken Gardiner, James Harland, and Margaret Hamilton

RMIT University, School of Computer Science and Information Technology,
GPO Box 2476V, Melbourne, Victoria, 3001, Australia
{ken.gardiner,james.harland,margaret.hamilton}@rmit.edu.au

Abstract. We present an artificial chemistry model where genotypic and phenotypic strings react with each other. The model prevents the genome from directly coding for genotype-phenotype mappings or for gene-replication enzymes. Experiments demonstrate the genome can evolve to manipulate reactions of phenotypic strings in such a way as to alter the genotype-phenotype mapping, and produce gene-replication enzymes.

1 Introduction

Artificial chemistries [2] are abstract chemical models, where entities represented as 'molecules' undergo collision reactions or transformations according to a set of rules, inside some specified environment. Artificial chemistries have been applied to the investigation of many life-like phenomena including autocatalytic 'metabolisms' [1,3].

Some artificial chemistry models include both informational (gene) and functional (protein) molecules and permit translation of proteins from genes, as well as gene replication by proteins [4,8]. It has been shown that evolving a genotype-phenotype mapping provides the opportunity to transform a problem representation into a form that is easier to solve [5]. Artificial chemistries with genes and proteins have been designed to evolve a genotype-phenotype mapping, often implemented by permitting genes to code for genotype-phenotype mapping molecules [7,9].

We are interested in developing an artificial chemistry where 'services' required by the genes cannot be directly produced from molecules coded for by the genes. This feature is intended to force the genome to only manipulate *protein-protein* reactions (from an inflow of 'food' proteins) in order to construct molecules for providing those services. While models such as [6,10] have evolved protein-protein metabolisms by manipulating genomes external to the reaction set, we are interested in evolving protein-protein metabolisms under internally produced genetic control.

Our initial artificial chemistry is designed so that molecules for the gene 'services' of gene-duplication and genotype-phenotype mapping can only be implemented by *service* molecules produced from protein-protein reactions.

M. Randall, H.A. Abbass, and J. Wiles (Eds.): ACAL 2007, LNAI 4828, pp. 281–291, 2007.

The artificial chemistry imposes some general syntax restrictions on the structure of genes, proteins and service molecules, but does not specify a priori the effectiveness of any particular (legal) arrangement of atoms. Instead, the semantics of a molecule depends on the set of other molecules that it is in, forcing the genes and service molecules to co-evolve in order to be effective.

This paper presents the model and describes the current implementation. Preliminary results are presented. They demonstrate the model is capable of evolving from initially random molecules, to genetically manipulate protein-protein reactions and produce service molecules for genotype-phenotype mapping and potential gene-duplication.

2   The Model

An artificial chemistry can be described [2] as a triple *(S,R,A)* where S is the set of all possible molecules, R is the set of collision rules describing interactions among molecules, and A is the control algorithm for describing the domain and how the rules are applied to the molecules. The main contribution of our model is the approach to the collision rules.

2.1   The Control Algorithm

The control algorithm determines how the collision rules will be applied to a collection of molecules, manages the 'reactor vessel' environment of the molecules, and implements an evolutionary algorithm based on a reaction-set fitness function.

Each reaction set operates in its own simulated well-stirred (i.e. dimensionless) vessel, which is initially seeded with a food stock of random protein molecules and a random stock of genes. The food stock does not include gene-replication molecules or genotype-phenotype mapping molecules. The food stock has its concentration increased at a steady rate, while the concentration of genes remains constant. The simulated vessel has a total atom-count limit, beyond which molecules are randomly selected for overflow. Future implementations will permit gene replication and gene overflow. Currently we prevent gene molecules from overflowing, and do not implement gene replication.

Once the concentration of a molecule exceeds a user specified threshold, it can potentially take part in chemical reactions. Such reactions may change the concentrations of the molecule and result in new molecular species, which may also take part in reactions if their concentrations exceed the concentration threshold.

The control algorithm implements an evolutionary algorithm. A generation consists of running each reaction set for a user defined number of cycles. A cycle involves adding food stock to the reaction set and stochastically performing molecular collisions until the concentrations of reaction inputs are insufficient to run any more reactions. The simulated vessel then overflows until the number of atoms in the vessel falls below a specified threshold value. The total number of gene-replication molecules produced by a reaction set over the cycles is used as that set's fitness score for that generation.

At the end of a generation, the evolutionary algorithm copies the two highest scoring reaction sets into the next generation. The rest of the generation is populated with copies of reaction sets with some of their genes mutated. The reaction sets are selected for copying by weighted roulette wheel selection based on their fitness score.

2.2   Molecules

We define molecules as consisting of character-based strings, drawn from the set of {F,G,C,P,1,2,3,4,x,y,z}.

Gene molecules can only be composed from {x,y,z}. Gene molecules are capable of undergoing mutation, while other molecules are not.

Protein molecules can be composed of any atoms as long as they contain at least one atom from {F,G,C,P,1,2,3,4}.

Service molecules are a subclass of protein, used to implement a genotype-phenotype mapping, or to perform gene replication. Genotype-phenotype service molecules consist of two atoms drawn from {F,G,C,P,1,2,3,4} followed by two or more atoms drawn from {x,y,z}. Gene-replication service molecules have a mirror-image syntax to genotype-phenotype service molecules: i.e. two or more atoms from {x,y,z} followed by two atoms drawn from {F,G,C,P,1,2,3,4}.

2.3   Collision Rules

The model supports two types molecular interaction: protein-protein interactions and interactions between service molecules (a subtype of protein) and genes.

Protein-protein interactions. If two or more protein molecules collide, one of the proteins is randomly chosen to act as a catalyst in a potential reaction, performing some arrangement of operations on the other molecule(s).

We chose a simple pattern matching operation to determine whether colliding protein molecules interact. Protein-protein reactions are only possible if the molecules bind. A subset of atoms, called *latch* atoms, drawn from {1,2,3,4}, determine if protein molecules will bind, and how they will be aligned if a binding does occur. Latch atoms on one molecule are attracted to latch atoms on another molecule, according to the attraction patterns 1:3, 3:1, 2:4 or 4:2. A protein will bind to a (catalyst) protein if more than a specified number of their latch atoms bind (in the current implementation, this threshold is two). The heterogeneous binding pattern was chosen to reduce the probability of a protein binding with, and possibly destroying, itself.

A protein acting as a catalyst may perform *cut* (ligation) or *paste* (polymerisation) operations on the bound molecule(s). The cut operation is specified by a C atom, and paste by a P atom.

Protein molecules may also contain two inert atoms, indicated by F and G.

Figure 1 illustrates the operation of protein binding, with cut and paste operations. Sites of ligation or polymerisation are shown in grey.

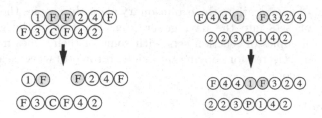

Fig. 1. The *cut* and *paste* operations

Service molecule protein-gene interactions. The model is intended to prevent genes from directly coding for molecules that provide services to genes such as gene replication or a genotype-phenotype mapping. However the model has to permit the production of such molecules from protein-protein reactions. In addition, the protein-protein reactions to produce genotype-phenotype mapping molecules must be simple enough for a reasonable probability that, given a random collection of proteins, genotype-phenotype mapping molecules could be produced by chance. Without this, the system would not have an initial genotype-phenotype mapping and genetic manipulation of the system would be impossible. The service molecule syntax, described in Sect. 2.1 was designed to meet these constraints.

A simple pattern matching operation was used to determine if and where service proteins could interact with genes. Genes are composed solely of *template* atoms, drawn from {x,y,z}. The template atoms of a service protein may bind with gene template atoms according to the patterns x:x, y:y or z:z. Each service protein template atom must bind with a gene template atom, otherwise a reaction will not occur.

To translate a gene into a protein, genotype-phenotype mapping molecules bind to the gene as illustrated in the left-hand side of Fig. 2. The first atom of each mapping molecule then polymerises to produce a new protein (in the example, protein *P2* is produced). The syntax of genotype-phenotype mapping molecules could have been designed to contain only a single atom from {F,G,C,P,1,2,3,4}, instead of two such atoms. The current syntax was chosen due to considerations for future modification of the model, which will not be presented here.

A gene may contain regions that no mapping molecule matches. These regions act to stop translation, permitting a gene to have several protein coding regions, each separated by 'stop' regions.

Gene replication could be implemented by a system similar to that for genotype-phenotype mapping. However in this case, it is the template atom regions of the gene-replication molecules that polymerise into a new gene molecule. The process is illustrated in the right-hand side of Fig. 2. Gene replication is currently not enabled.

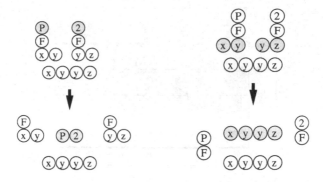

Fig. 2. Gene translation and replication

3   Experiments and Results

Recall that our model was designed to force genes to manipulate protein-protein reactions in order to produce service molecules required by the genes. In this section we describe our series of experiments to generate the expected behaviour of increasing production of potential gene-replication molecules.

The fitness score of a reaction set would increase if genes were able to manipulate protein-protein reactions to increase the production of gene-replication molecules. To test this, we initialised the artificial chemistry system with random molecules and ran it under various settings to see if fitness scores improved over time. None of these initial, unreported, experiments resulted in improvements in fitness score. Analysis showed that genes required more genotype-phenotype mapping molecules than could be produced from the provided concentration of food stock. Therefore genes could not be translated and could not influence protein-protein reactions. Increasing the concentration of each food stock species resulted in an unacceptable run time.

Based on these results, we altered our model to permit unlimited use of any genotype-phenotype mapping molecules produced from protein-protein reactions. The genes still had to evolve to control the protein-protein reactions leading to genotype-phenotype mapping molecule production. The total atom-count was kept constant by counting the number of atoms in gene-translated proteins and adding that number to the atom-overflow cycle of the simulator. The alteration to the model permitted improvements in the fitness score to evolve and various settings of the simulator were investigated (not reported here) resulting in the choice of running the simulator for 600 generations of 300 feed-react-overflow cycles each. Approximately 7% of a reaction set's non-gene contents were replaced with food stock in each cycle. The mutation rate was 0.2% of gene atoms. Figure 3 shows the best-of-generation fitness score for a run using these settings. The model was able to evolve increasingly successful reaction sets, increasing the system's fitness by 223%.

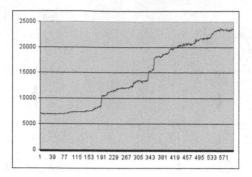

Fig. 3. Best-of-generation fitness verses generation

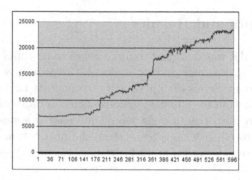

Fig. 4. Fitness verses generation of the ancestors of the highest scoring reaction set from generation 600

The evolutionary algorithm performed duplication and mutation of reaction sets, but not crossover. This meant each reaction set in generation 600 had a single ancestor in each of the previous generations. The ancestors of the winning reaction set from generation 600 were examined. Their scores are shown in Fig. 4. Analysis of these ancestors was undertaken to reveal how their genes evolved to increase the reaction set fitness.

3.1   Genetic Influence over Protein-Protein Reactions

There are two ways genes could influence the production of gene-replication molecules. Firstly, the genes could produce proteins that impede reactions detrimental to gene-replication molecule production. Such genes will be called *impeding* genes. Secondly, they could produce proteins that form part of the reaction path for the production of gene-replication molecules. These genes will be called *production* genes.

Production genes were identified by working backwards through the reactions from gene-replication molecule to food stock, and tagging any genes that produced molecules in the reactions.

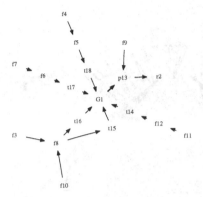

Fig. 5. Generation 59 production gene reaction subset

In order to determine if impeding genes existed, an additional experiment was performed where all genes except production genes in a reaction set were prevented from producing proteins. The results were then compared against running the reaction set with all genes turned off. It was found that turning off the production genes was equivalent to turning off all genes, indicating that the reaction set did not use impeding genes.

Improvements in reaction set scores (and thus gene-replication molecule production) began from generation 59. Prior to then, although genes were producing proteins, those proteins were not influencing reactions that produced gene-replication molecules.

At generation 59, the reaction set contained 9,526 molecular species and 4002 reactions. This was the first reaction set to contain a production gene, *G1*, producing a molecule that lead to the production of a gene-replication molecule *r2*. Figure 5 shows the reactions leading to the production of *r2*. Each molecular species has been given an integer identifier. The identifiers are prefaced as following: *f* indicates a food stock protein, *G* indicates a gene, non-food proteins are prefaced with *p*, *t* indicates a genotype-phenotype mapping molecule and *r* indicates a gene-replication molecule. An example reaction shown in the figure is: food protein *f11* reacts with food protein *f12* to produce genotype-phenotype molecule *t14*. The *t14* molecule is then used by gene *G1* (together with genotype-phenotype molecules *t15*, *t16*, *t17* and *t18*) to produce protein *p13*. Protein *p13* then reacts with food protein *f9* to produce gene-replication molecule *r2*.

Further evolution of the system produced additional production genes, resulting in the production of further species of gene-replication molecule. For example, the major increase in fitness at generation 191 was caused by the evolution of two additional production genes. By generation 600 the reaction set had 13 production genes, assisting the production of 9 species of gene-replication molecule. Figure 6 shows these reactions (a subset of the 21139 reactions occurring in the entire reaction set). In order to reduce the complexity of the figure, the only molecules labelled are genes and gene-replication molecules. Reactions between geneotype-phenotype molecules and genes are shown with dotted lines.

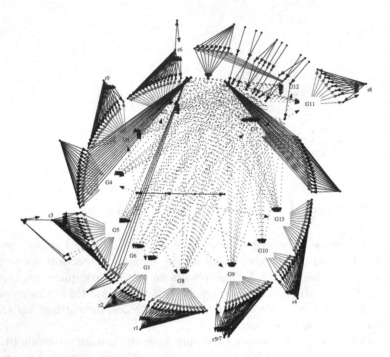

Fig. 6. Generation 600 production gene reaction subset

3.2   Genetic Influence over the Genotype-Phenotype Mapping

The genotype-phenotype mapping produced by the initial random reaction set was a many-to-many mapping between template-atom-pattern and protein-atom. This meant a single segment of gene could be translated multiple ways, resulting in the production of more than one protein. Table 1 compares the set of genotype-phenotype mapping molecules forming the initial genetic code and the set of mapping molecules produced in generation 600. The first atom (underlined) is coded for by the string of template atoms (italicised). This means some genotype-phenotype mapping molecules are functionally equivalent. For example $\underline{G}Pzx$ and $\underline{G}Gzx$ both map zx to G. Gene template atom sequences not included in the genetic code can be considered as stop instructions.

Table 1 shows that most of the genotype-phenotype mapping molecules produced from the initial random reaction set continued to be used in generation 600. However the evolving genes did modify the reaction set to expand the genetic code. The final genetic code included every two-atom template pattern except zy and zz (which therefore formed 'stop translation' codes) although a mapping for zzx was produced. Since genes could use an unlimited supply of any genotype-phenotype mapping molecules produced by protein-protein reactions, there was no incentive to prevent the waste of mapping molecules. This meant a single region of gene could be translated into multiple species of protein without penalty. Therefore there was no incentive to evolve a non-overlapping genetic

Table 1. Genotype-phenotype mapping molecules of generation 0 and generation 600

Generation 0	Generation 600
$\underline{C}P\,xx$	$\underline{C}P\,xx$
	$\underline{3}F\,xy$
$\underline{G}G\,xz$	$\underline{G}G\,xz$
	$\underline{P}C\,xz$
$\underline{F}F\,yx$	$\underline{F}F\,yx$
	$\underline{P}1\,yx$
$\underline{G}4\,yy$	$\underline{G}4\,yy$
$\underline{P}P\,yy$	
$\underline{F}4\,yy$	$\underline{F}4\,yy$
$\underline{F}F\,yz$	$\underline{F}F\,yz$
$\underline{P}F\,yz$	$\underline{P}F\,yz$
$\underline{P}P\,yz$	$\underline{P}P\,yz$
$\underline{G}P\,zx$	$\underline{G}P\,zx$
	$\underline{G}G\,zx$
$\underline{1}F\,zzx$	$\underline{1}F\,zzx$

code, and every incentive to ensure almost every gene template pattern could be translated.

Analysis of the reaction sets showed genes modified the genetic code by evolving to code for proteins that modified the protein-protein reaction set to produce new classes of genotype-phenotype mapping molecule. For example, Fig. 7 shows part of the reaction set for generation 250. Protein $p67$ was produced from translating gene $G20$ (reactions producing the genotype-phenotype molecules used by $G20$ are not shown). Protein $p67$ then catalysed a reaction with food molecule $f59$, leading to the production of genotype-phenotype mapping molecule $t40$ (which was then used by other genes). Such new species of mapping molecule could permit previously un-translatable gene template-atom sequences to be accepted and translated, effectively increasing the number of genes participating in the reaction set.

Analysis of the reaction sets also demonstrated a co-evolution of the genetic code, genes and other reaction set molecules resulting in autocatalytic reaction loops involving genes. For example, Fig. 8 shows part of another reaction subset from generation 250. It shows the genotype-phenotype molecules used by

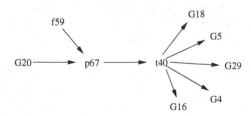

Fig. 7. Genetic control of reaction creating genotype-phenotype molecule

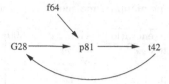

Fig. 8. Gene use of genotype-phenotype mapping molecule produced only from translation of that gene

gene *G28*, and the molecules required to create one of those genotype-phenotype molecules, *t42*. It can be seen that *t42* is produced from the reaction of protein *p81* with food protein molecule *f64*. However protein *p81*, used to create *t42*, is produced by gene *G28*, which requires *t42*. The full reaction set was examined, revealing that no other reaction produced *p81*. This leads to an apparent paradox in that gene *G28* couldn't be translated without genotype-phenotype mapping molecule *t42*, yet the translation of *G28* was required to produce *t42*. Analysis of the reaction set's ancestors showed that there used to be an alternative form of production of *t42* (thus enabling gene *G28* to use it), and later evolution caused the demise of the alternative pathway, leaving the autocatalytic loop between *G28* and *t42*.

4   Conclusions and Future Work

We have presented an artificial chemistry containing informational (gene) and functional (protein) molecules, designed so that gene 'services' of gene replication and a genotype-phenotype mapping can only be produced by genetic manipulation of protein-protein reactions. We presented preliminary results showing the system can and does evolve genetic control of protein-protein reactions in order to produce a genotype-phenotype mapping and increase production of potential gene-replicating molecules. Future work will include further experiments to examine the range of gene control over protein-protein reactions possible under our model.

Currently, a given genetic sequence can lead to the translation of more than one species of molecule due to overlap in the genetic code. If only one of the species from such translations was useful, then the genotype-phenotype mapping molecules used to produce the other translations would have been wasted. A future area of investigation will be to introduce a cost of wasting genotype-phenotype mapping molecules, providing a selection pressure to remove genetic code overlap.

Another area of investigation will be to place the replication of genes under genetic control, via the use of gene-replication molecules. Genes will be permitted to overflow the simulated vessel, providing pressure to copy genes before they are flushed out. The eventual intention is to divide each reaction set into two after a set time period, implementing a simple form of reproduction and further pressuring the reaction set to copy genes and other vital molecules before the division occurs.

References

1. Bagley, R.J., Farmer, J.D.: Spontaneous Emergence of a Metabolism. In: Langton, et al. (eds.) Artificial Life II, SFI Studies in the Sciences of Complexity, vol. X, Addison-Wesley, Reading (1991)
2. Dittrich, P., Ziegler, J., Banzhaf, W.: Artificial Chemistries - A Review. Artificial Life 7, 225–275 (2001)
3. Farmer, J.D., Kauffmann, S.A., Packard, N.H.: Autocatalytic Replication of Polymers. Physica 22D, 50–67 (1986)
4. Hutton, T.J.: Evolvable Self-Reproducing Cells in a Two-Dimensional Artificial Chemistry. Artificial Life 13, 11–30 (2007)
5. Kargupta, H., Ghosh, S.: Toward Machine Learning Through Genetic Code-like Transformations. Genetic Programming and Evolvable Machines 3(3), 231–258 (2002)
6. Lohn, J.D., Colombano, S.P., Scargle, J., Stassinopoulos, D., Haith, G.L.: Evolving Catalytic Reaction Sets using Genetic Algorithms. In: Proc. IEEE Int. Conf. on Evolutionary Computation, New York, pp. 487–492 (1998)
7. Piaseczny, W., Suzuki, H., Sawai, H.: Chemical Genetic Programming - Evolutionary Optimization of the Translation from Genotype String to Phenotypic Trees. In: Sugisaka, M., Tanaka, H. (eds.) Proc. 9th Int. Symposium on Artificial Life and Robotics, vol. 2, pp. 571–574 (2004)
8. Suzuki, H.: Models for the Conservation of Genetic Information with String-Based Artificial Chemistry. In: Banzhaf, W., Ziegler, J., Christaller, T., Dittrich, P., Kim, J.T. (eds.) ECAL 2003. LNCS (LNAI), vol. 2801, pp. 78–88. Springer, Heidelberg (2003)
9. Suzuki, H., Sawai, H., Piaseczny, W.: Chemical Genetic Algorithms - Evolutionary Optimization of Binary-to-Real-Value Translation in Genetic Algorithms. Artificial Life 12, 89–115 (2006)
10. Ziegler, J., Banzhaf, W.: Evolving Control Metabolisms for a Robot. Artificial Life 7, 171–190 (2001)

In-Formation Flocking: An Approach to Data Visualization Using Multi-agent Formation Behavior

Andrew Vande Moere and Andrea Lau

Key Centre of Design Computing and Cognition
Faculty of Architecture, Design and Planning
The University of Sydney, Australia
{andrew,andrea}@arch.usyd.edu.au

Abstract. This paper presents *in-formation flocking*, a novel information visualization technique that extends the original information flocking concept with dynamic and data-driven visual formation behavior generation. This approach extends the emergent swarming properties of a decentralized multi-agent system in order to represent complex time-varying datasets through visually-recognizable formations and motion typologies. In-formation flocking is capable of representing volatile and inherently chaotic time-varying datasets while sustaining a comprehensible representation at a global level as well as revealing more detailed patterns in subsets of the data. This paper demonstrates the capabilities of in-formation flocking to historical stock market data.

Keywords: data visualization, swarming, flocking, boids, motion, emergence, multi-agent systems, self-organization, artificial life.

1 Introduction

Behavioral rule-based *flocking* is a well-known computer graphics technique that provides a conceptual means for visually simulating the natural phenomenon of aggregate motion in birds, fish and other animals. The principle of computational flocking simulation assigns each individual group member or *boid*, short for bird object, with a fixed set of behavior rules [1]. Flocking is an example of an emergent process, which demonstrates complex behavior that arises from a collection of entities that were not individually and explicitly programmed to do so. The recursive interactions of each single entity to those in its immediate environment cause a process which, on a holistic level, can lead to perceivable complex behaviors and an increase in order of the whole collection of entities.

In this paper, the apparent order generated by emergence is exploited to represent patterns reflecting relationships in complex, time-varying datasets. Accordingly, we believe that self-organization principles such as flocking can be used for visualizing abstract data, as it is theoretically possible to group similar data entities without the need for supervision, pre-calculating data similarity matrices or predetermined data mapping algorithms. *In-formation flocking*, an extension of the information flocking [2, 3] approach, aims to generate more readily recognizable flocking motion typologies. Instead of representing data tendencies by separating apparently

M. Randall, H.A. Abbass, and J. Wiles (Eds.): ACAL 2007, LNAI 4828, pp. 292–304, 2007.
© Springer-Verlag Berlin Heidelberg 2007

randomly-moving clusters, in-formation flocking is capable of making dynamic patterns and clusters more apparent by integrating the process of decentralized *formation flying*. Formation flying, as exhibited by birds, describes the synchronized movement of a group in readily distinguishable shapes. Formation flying is believed to be reflective of the underlying internal relationships and energy considerations within the social hierarchy of a flock [4]. Here, the in-formation flocking concept aims to exploits the visually perceivable order as an additional, readily distinguishable visual cue for representing dynamic similarities in complex, time-varying datasets.

We believe in-formation flocking is capable of representing highly volatile, even potentially chaotic, time-varying datasets. By exploiting the concept of emergence and self-organization, this research proposes an alternative data mapping technique that is not predefined or predetermined as in common data mapping techniques within the field of information visualization. In-formation flocking is capable of providing a global and local view of the whole dataset over time based on animating readily recognizable and interpretable motion typologies.

2 Background

The original *information flocking* approach applies emergent spatial clustering behavior of boids to the field of data visualization by assigning a unique data object to each boid [2]. As the three basic behavior rules are extended with an additional data similarity rule, boids with similar data objects tend to flock towards each other. As an emergent result, underlying similarity relationships between data objects are revealed through the formation of separate spatial clusters. More recently, the information flocking concept has been extended to complex and time-varying datasets, and the representation of dynamic data tendencies by distinct dynamic motion typologies [3]. Other research has combined the information flocking algorithm with foraging behavior, enabling clusters of data items to be found according to their spatial position and density [5]. In contrast, our research is not concerned with data mining applications, but focuses on generating more readily discernable, self-organizing information displays.

The boids concept is an example of a *decentralized multi-agent system*. An *agent* is a system situated within an environment, which senses its immediate environment and can act on it autonomously, over time, to achieve a set of objectives [6]. Some agents can collaborate with others, can perceive and respond to changes, and can exhibit goal-directed behavior. A *multi-agent system* consists of multiple agents, mostly because they pursue different goals, or because the environment is too complex for a single agent to observe efficiently. A *decentralized multi-agent system* contains numerous equal agents that have communication links with those in their neighborhood, either directly or through the environment, but always in absence of a centralized coordinator. In visualization, several agent-based approaches have been used to display internal properties, such as the relationships between agents for monitoring and engineering purposes [7]. Multi-agent systems have been implemented to structure the data flow, such as for the generation of information visualizations of complex fuzzy systems [8], or to determine the choice of the most effective visualization method depending on the dataset and user tasks [9].

Self-organizing systems have been used to create emergent spatial organizations which reveal relationships between data objects. The *Narcissus* approach, for instance, aims to integrate behavioral rules into agents in order to aid the comprehension of both high-level and low-level structures using distinctive emergent shapes [10].

Motion is a powerful graphical cue that is capable of attracting attention, maintaining motivation and facilitating comprehension, learning, memory and efficient communication in the contexts of learning or knowledge discovery. Generally, animated objects follow predefined paths or trajectories defined by specific mathematical functions or user-defined control points. Alternatively, motion can be generated by behavior rules, which are inherently unpredictable and more suitable to convey interpretative behavior. Some researchers have demonstrated that even simple motion cues can reveal causal relationships, as launching, entraining and triggering [11]. Ware et al. have demonstrated the rich expressive visual language of motion in the context of information visualization [12]. Lethbridge and Ware [13] used behavior functions based on distance, velocity and direction to model complicated relationships such as pulling, pushing, chasing, escaping, repulsion, collision and anticipation.

Conceptual flocking models reveal that overall group structures in animals are directly affected by transformations at local levels [14]. That is, high-level aggregate movement is dictated by a decentralized system of individuals. This concept has been applied to the decentralized *formation* of robots in space [15-17]. Fredslund and Mataric employ a *neighbor-referenced* approach, which requires that robots attempt to stay at a fixed distance and angle from their so-called robot *friend* [17]. This approach only requires one robot – the friend – to determine the heading of another, rather than more centralized approaches such as *unit-center-referenced* (i.e. robots determine their positions relative to a centre average) and *leader-referenced* (i.e. positions are determined relative to a single leader). Thus, in neighbor-referenced formation, a single *conductor* or *leader*, is able to 'drag' a whole formation forward through the downward filtering of iterative friend relationships [17]. Other researchers have compared the appropriateness and optimization of each of these three techniques to the problem of *obstacle avoidance* [15, 16].

3   In-Formation Flocking Approach

In our in-formation flocking approach, each boid represents a unique data object, retrieved from a time-varying dataset. As illustrated in Figure 1, each boid has a limited field of perception, and is able to communicate only with boids in its immediate vicinity. Each boid is governed by an identical set of behavior rules, which are executed in parallel for the whole boid collection. These rules determine the visual characteristics of a boid, such as its speed and direction. The rules take into account any time-varying changes in the data object which the boid represents, as well as the relative positions, velocities and data values of the boids in its immediate neighborhood. During the visualization, the data values for each boid are updated to match the data values of the next successive iteration in the time-varying dataset, according to a virtual timeline. As a result, each boid is continuously governed by a small set of behavior rules which are directly affected by its own data values as well as those of its immediate boid-neighbors.

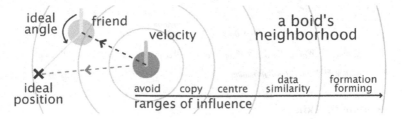

Fig. 1. A boid (center), its view of the neighborhood and its flocking rules ranges of influence

These local, data-driven influences between pairs of boids cause an emergent pattern of visual formations to appear on a global scale. Notably, boids can consider local information only, and have no reference to the global pattern they may be part of. These patterns are able to represent dynamic dataset alterations, as they are essentially formed out of the interactions between pair-wise members according to their relative data values. The visualization is self-organizing and based on the dynamic properties of the underlying data phenomena rather than a traditional, predetermined data mapping rules that directly translate data values into visual form.

3.1 Behavior Rules

Each boid obeys five behavior rules, which are determined by pair-wise comparisons between boids. A behavior rule is only invoked when a boid is in the field of vision of another boid. The fields of vision are ordered by size, where $d_{avoid} < d_{copy} < d_{centre} < d_{similar} < d_{formation}$ so that the behavior rules act as sequential steps and do not overlap.

Rule 1. Collision Avoidance. Each boid avoids any other boid which is within the collision avoid range d_{avoid}. This rule withholds boids to visually overlap, as it causes them to actively move away from each other when nearby.

Rule 2. Velocity Matching. Each boid copies the direction and speed of any other boid which is within the velocity matching range d_{copy}. This rule causes groups of boids to move towards a similar general direction.

Rule 3. Flock Centering. Each boid moves towards the perceived center of gravity of all neighboring boids, present within the centering range d_{centre}. This rule causes localized flocking to occur, so that little internal order occurs over time.

Rule 4. Data Similarity. Each boid moves towards any other boid with a similar data object within a distance range $d_{similar}$ and a data range of $q_{similar}$. This rule groups boids that experience similar data changes [2, 3]. It is proportional to distance, so that boids far away move more quickly towards each other than those nearby.

Rule 5. Formation Forming. Each boid attempts to reach a spot that is positioned at a specific distance and angle from the most similar boid within a formation finding range $d_{formation}$ and a data range of $q_{similar}$ [15, 17]. This rule causes visually distinguishable formations to form containing multiple boid members.

The different weighting factors w_r are applied to the vector outcome v_r of each behavioral rule, depending on the importance of its relative influence. A new velocity v_{new} is calculated using these vectors, and added to the current velocity. d is the distance between a boid and its neighbor in a pair-wise comparison.

$$v_{new} = d.v_{flocking} + \frac{v_{formation}}{d}$$

$$with \begin{cases} v_{flocking} = -w_{avoid}.v_{avoid} + w_{copy}.v_{copy} + w_{center}.v_{center} + w_{similar}.v_{similar} \\ v_{formation} = w_{formation}.v_{formation} \end{cases} \tag{1}$$

3.2 Formation Flocking

The in-formation flocking approach extends the original information flocking algorithm [2, 3] with an additional formation-making rule. This rule generates formations consisting of boids that have experienced similar data value changes between successive time steps. In order to exhibit in-formation flocking (or formation forming) behavior, a boid's data change must be greater than a *minimum relative data value difference threshold* q_{change}, which is the relative change in data values between the current and previous time steps of the time-varying dataset.

$$q_{change} = \frac{(q_{current} - q_{previous})}{q_{previous}} \tag{2}$$

If the difference between a pair of boids' q_{change} is less than the minimum data value difference threshold, normal *information flocking* behavior will be exhibited, that is similar boids will move together in independent flocks. Only if q_{chang} between a pair of boids is more than the predefined minimum threshold value will formation forming be invoked, described by the following steps.

For a predefined data attribute, each boid attempts to find another boid that contains the most similar data values, here called *friend*. Accordingly, once the boid-to-friend relationship is established, the boid becomes its friend's *follower L* (friend and follower terminology is borrowed from [17]). As a restricting rule, each boid may only have one single friend F and one single follower L. More specifically, for a boid X with data change q_X to become the friend of a boid A with a data change q_A, where $q_{similar}$ is the maximum largest difference between the value of two points:

$$\begin{cases} q_X \geq q_A \\ |q_A - q_X| \leq q_{similar} \end{cases} \tag{3}$$

A data splitting rule internally orders resulting groups. If boid X already has a follower F with change q_F, boid A may only split this relationship if it is more similar:

$$|q_A - q_X| \leq |q_L - q_X| \tag{4}$$

As boids split their friend-follower relationships when 'more similar' boids have been detected in its neighborhood in an iterative fashion, formations are ordered by data similarity along the chain of friends and followers, providing a chain-like representation of data similarity. This process of friend- and follower-determination happens continuously, so that the formations constantly change as all data values are continuously updated over time.

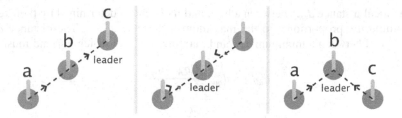

Fig. 2. Reversal of relationships in a friend-follower chain to generate a 2-sided formation

The aim of each boid in a formation is to stay close to an "ideal position" relative to its friend as defined by a distance and an angle, proportional to the data attribute. This relatively simple means of formation-forming would obviously result in a single, straight line. Our approach of formation forming, however, entails that a wedge-like shape with two arms to either side is produced by specifying a single boid as a *leader*, which in turn is followed to the left and right side by a number of boids, as shown in Figure 2. Accordingly, one half of the boids in the formation must reverse their friend-follower relationships, and follow their followers rather than their friends. Because of the data splitting rule, all boids are ordered emergently by data similarity from left to right (or vice versa) along both the wedges. As shown in Figure 3, in the case of a formation with an even number of boids, one of the middle boids must follow the other directly (orthogonally) to the side.

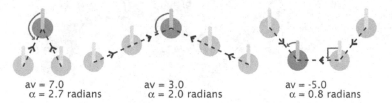

av = 7.0   av = 3.0   av = -5.0
α = 2.7 radians   α = 2.0 radians   α = 0.8 radians

Fig. 3. Wedge and inverted-wedge shapes reflect positive and negative data value averages. A sharper angle between arms reflects larger variation in data change for the whole group.

The angle at which each boid follows its friend is controlled by an averaged data value of the whole group. This average is determined by the boids in a decentralized manner, in which each boid in a formation recalculates and passes on a new average from its friend to its follower. The angle α of a wedge is determined by the *average data value av* and a predefined maximum average data value *max*.

$$\alpha = \frac{\pi}{2} \cdot \left(1 + \frac{av}{max} \right) \tag{5}$$

Figure 3 shows the emergent result of this algorithm: a *negative* numerical average value will construct an inverted-wedge: boids move in the opposite direction than the direction of the wedge. Accordingly, a *sharp* peak (i.e. small wedge angle) indicates a high average data change while an almost *wide* angle or horizontal line (i.e. large wedge angle) conveys close to no variation in the boids' data values.

The ideal distance d_{ideal} between a boid and its friend is determined by their relative data similarity, proportional to the maximum difference $q_{similar}$. The distance d_{step} is interpolated between a minimum d_{min} and maximum d_{max} in which a friend must lie in.

$$d_{ideal} = d_{min} + \frac{|q_A - q_X|}{d_{step}}$$

$$with \ \ d_{step} = \frac{q_{similar}}{d_{max} - d_{min}}$$

$$(6)$$

Using knowledge of a friend's position F_{pos} and velocity vel, in addition to the angle α to follow a boid A calculates its ideal formation position $B_{formation}$. The boid then calculates its new position B_{new} according to $B_{formation}$ in the following manner.

$$\vec{B}_{formation} = (cos\,\alpha \cdot vel_x - sin\,\alpha \cdot vel_y, \ sin\,\alpha \cdot vel_x + cos\,\alpha \cdot vel_y)$$

$$\vec{B}_{new} = \vec{B}_{formation} + \vec{F}_{pos} - \vec{A}_{pos}$$

$$(7)$$

The resulting angles are scaled depending on the relative position of a boid along a wedge to generate a unified curved-like wedge rather than a sharp difference between the two arms, as illustrated in Figure 4. In addition, members of a formation are connected by a continuous spline. Both these visual features emphasize formations as distinct, continuous shapes rather than two separate sequences of objects [18].

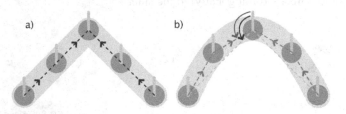

Fig. 4. Altering the relative formation angle from a straight line (a) to a continuous curve (b)

4 Data Analysis Scenario

The in-formation flocking approach was implemented for a large, complex, time-varying dataset: the historical US stock market opening and closing prices, and volume traded, over a one-year period of the 500 leading US companies [19]. Our current prototype application, programmed in Java3D, includes several interactive sliders. These interactive tools were especially required while developing for fine-tuning the weights and threshold values towards the most optimized emergent results.

Each boid represents a single stock market quote company. The formation flocking represents the percentage change in closing price over a day. The *minimum data value change threshold* for a closing quote price change is 3%. The maximum difference $q_{similar}$ between the data value of a boid and its friend is fixed at 0.1%. The volume traded each day is represented by the relative size of the boid.

Several patterns and tendencies can be perceived using the in-formation flocking approach. Firstly, data-similar groups of boids which have experienced relatively large changes in value are highlighted through formation forming. The number of boids involved in each formation shows the extent to which the similarity is common. The shape of the wedge reflects the average across the formation, which can then be compared with other groups. This shape is emphasized by the use of a continuous, underlying curve, in order to link members of a group, and differentiate between the shapes of emerging flocks (for example, formations *a* and *b* in Figure 5).

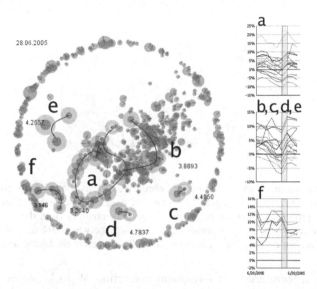

Fig. 5. Components of in-formation flocking, after 250 iterations: separate groups emerge on 28 June 2005, with differentiating features: size, angle, and direction, versus traditional price history line charts. Each formation represents stocks which have experienced almost identical stock price changes (identical / parallel stock price changes, as highlighted in yellow on the line charts). (Charts are based on MSN *MoneyCentral* http://moneycentral.msn.com).

The traditional line charts in Figure 5 highlight the correlations between the data changes and formation representations. For example, chart *f* shows the almost identical, steep drop in price experienced by all the boids in formation *f* at exactly the same time. Although all groups from *a* to *e* have experienced similar, parallel increases in price as can be perceived from charts *a* to *e*, the formations clearly show five separate subsets which cannot be readily seen in the charts. Formation *c* corresponds to the pair of lines at the top of chart *b* to *e*; this is due to the similar starting and ending points of the line for that particular day. Although the line charts might seem more comprehensible in showing stocks which have experienced similar price changes, they only show a small subset of the total of 37 stock quotes in the dataset. In contrast, in-formation flocking is able to depict the whole dataset while highlighting meaningful data patterns as they happen and change over time.

The current prototype allows several different subsets of data similarity to be visually depicted, through the use of motion typology and color coding. Figure 6 shows the in-formation flocking on one of the worst days experienced by the stock market in 2005. Several distinguishable groups emerge: a large group of boids move in formation *c*, two smaller formations *a* and *b*, a green and red cluster, and a large red flock which has separated itself from the others. The visual focus lies on the emerging formation *c*, which correlates with a large group of similarly-changing boids between the values of -3.5% and -2.9% for that day (see histogram). A histogram of the stock market on the day reveals that the spatial separation of the large red group from the others is representative of the dip that occurs around -0.7%.

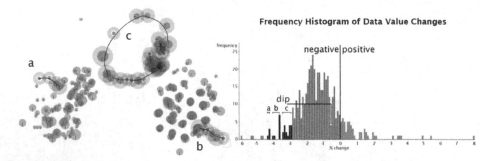

Fig. 6. The state of the stock market on 22 February 2005, as shown by its constituent subsets of data changes, versus a frequency histogram of data value changes. Formations *a* and *b* represent two groups of stocks experiencing large data changes (less than -4%); formation *c* reveals a large emerging group experiencing changes of between -3% and -4%, while two red information flocking clusters highlight the distinct dip as seen in the histogram.

In Figure 7, there are three formations consisting of stock quotes which have all experienced large positive price changes of over 3%. Formation *a* is a flock which has experienced a high average data of 5.6%, represented by a narrow wedge angle. In contrast, formation *b* and *c* convey a wider angle between the wedges, as they experienced an average data change of 3.0% and 3.5%, respectively. The relative distance between boid members in each formation conveys the relative similarity linking the boids and their friends. For instance, in formation *b*, boid PD is much closer to its friend, NSC, than NSC is to its friend UST. This data dependency is also reflected in the data tables, showing how PD versus NSC percentage change (0.007) is smaller than NSC to UST (0.05).

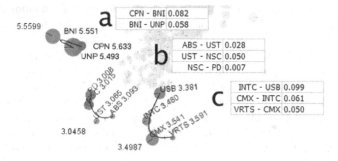

Fig. 7. Shape comparison between different formations on 21 December 2004, as related to the group average. The tables show differences in change between boids and their friends.

Figure 8 shows the difference in formation patterns when altering the weight values for the minimum relative data value difference threshold *mt* and maximum difference *md*. Increasing the minimum threshold causes less boids to satisfy the rule for minimum change thus creating low numbers of groups exhibit in-formation behaviors (Figure 8, *a* and *b*). Decreasing the minimum threshold causes more groups to form (*d*). Increasing the maximum difference creates longer chains of boids (*a* and *c*), whilst decreasing the difference causes shorter chains groups to form (*b* and *d*).

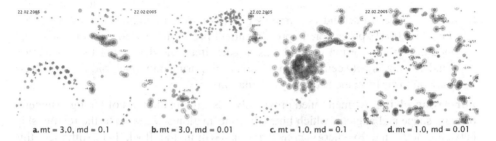

| a. mt = 3.0, md = 0.1 | b. mt = 3.0, md = 0.01 | c. mt = 1.0, md = 0.1 | d. mt = 1.0, md = 0.01 |

Fig. 8. Formations differences by adjusting *minimum threshold (mt), maximum difference (md)*

5   Discussion

The in-formation flocking approach highlights several important characteristics.

Decentralized Multi-Agent System for Data Visualization. The agent-based methodology supports the dynamic nature of time-varying datasets as each individual 'data object' continuously adapts to a changing neighborhood of data values. The decentralized approach is fundamentally different from the normal data mapping method in data visualization, as the resulting visual cues are emergent and inherently unpredictable. It forms the first step towards data visualizations that self-organize, capable of recognizing and highlighting data patterns in an unsupervised fashion

Motion Typology as a Visual Cue. The use of movement enhances the connectedness between similar boids through uniform velocity and direction, and through the formation of shapes and wedges. Dissimilar clusters of boids can also be differentiated by comparing motion typologies. In this work, the use of motion is necessitated by the nature of time-varying data, which can be studied over time in order to create an understanding of complex, dynamic trends that happen in parallel.

Application Domain. We claim that in-formation flocking is most appropriate for representing noisy or highly volatile time-varying datasets with hundreds of data items. Datasets with underlying (but not explicitly-defined) group-structures between data objects could also be effectively represented by both in-formation and information flocking. These methods are specifically useful in recognizing short- and long-term trends and tendencies that were not known before.

Parameter Dependency. Emergent pattern quality is highly dependent on predefined algorithmic parameters, which generally need to be fine-tuned in relation to specific dataset characteristics by a process of trial-and-error. However, even these characteristics generally change over time within a dataset (e.g. volatility of data alterations, data size) questioning the validity of keeping these parameters constant.

Performance. The current implementation has not been optimized for any performance issues in the context of computational efficiency or visual rendering, as we instead focused on demonstrating the in-formation flocking concept. As a worst-case scenario, the performance for n boids is $O(n^2)$ dramatically increasing the number of calculations needed as the number of boids increases. As ascertained through experimentation, a visualization of about a thousand items slows down the frame rate between one and two frames per second on a computer equipped with a Pentium M 2.0GHz processor. Improvements in processing speed could be achieved by updating only a portion of the boids at each iteration, delegating the calculation and rendering tasks between two processors, or requiring only one boid in a pair-wise comparison to perform the necessary calculations.

Formation Flying. As mentioned previously, research in the field of biology suggests that the shape and angle at which birds fly in formation is variable to the relationship between birds and to energy considerations within the flock [4]. Although this research does not aim to create an accurate simulation of natural phenomena, it may be beneficial to integrate knowledge about the physics of and social reasons for formation flying in order to create a truly biologically-valid data visualization. The use of phenomena discovered in nature as a metaphor for information visualization may aid the understandability and learnability of the approach for users. In particular, the use of artificial life insights also demonstrates how interdisciplinary knowledge can enrich the field of data visualization [20].

6 Conclusion

This paper presented a novel approach of visualizing complex, time-varying datasets using a decentralized, multi-agent formation flocking metaphor. It extends the original notion of information flocking [2, 3] with the concept of in-formation flocking, which is implemented as a single, relatively simple, additional behavior rule. As a result, each boid continuously searches and positions itself relative to data-similar friends, resulting in visual formations that can be interpreted in the context of time-varying data tendencies and trends. The relative distance between boids in a formation reflects their degree of similarity, while the wedge angle of the formation visualizes the average data variation experienced by the group. Thus, the shape of each formation in addition to the spatial clustering of boids creates an overall representation of data patterns within time-varying datasets.

With the future integration of algorithmic optimizations, in-formation flocking could be applied in real-time to time-varying datasets consisting of thousands of items. Future developments could integrate additional features to convey underlying data phenomena (e.g. news stories) as flock obstacles or attractors. The application could be enhanced by including dynamic user querying and filtering, and the ability to trace data values or formations. Behavioral rules could be made more flexible to increase the number of emergent characteristics for representing a larger range of data attributes. Further research should focus on user evaluations to analyze the potential for this approach in the context of complex pattern discovery for time-varying datasets and the use of motion typologies for interpreting dynamic data patterns.

References

1. Reynolds, C.W.: Flocks, Herds, and Schools: A Distributed Behavioral Model. Computer Graphics 21, 25–34 (1987)
2. Proctor, G., Winter, C.: Information Flocking: Data Visualization in Virtual Worlds Using Emergent Behaviors. In: Heudin, J.C. (ed.) Virtual Worlds, pp. 168–176. Springer, Berlin (1998)
3. Vande Moere, A.: Time-Varying Data Visualization using Information Flocking Boids. In: IEEE Information Visualization (INFOVIS 2004), pp. 97–104. IEEE, Austin, Texas (2004)
4. Davis, J.: Why birds fly in formation: a new interpretation. Interpretive Birding 5, 30–31 (2004)
5. Folino, G., Spezzano, G.: An Adaptive Flocking Algorithm for Spatial Clustering. In: Guervós, J.J.M., Adamidis, P.A., Beyer, H.-G., Fernández-Villacañas, J.-L., Schwefel, H.-P. (eds.) PPSN VII. LNCS, vol. 2439, pp. 924–933. Springer, Heidelberg (2002)
6. Franklin, S., Graesser, A.: Is It an Agent, or Just a Program? A Taxonomy for Autonomous Agents. In: Jennings, N.R., Wooldridge, M.J., Müller, J.P. (eds.) Intelligent Agents III. Agent Theories, Architectures, and Languages. LNCS, vol. 1193, Springer, Heidelberg (1997)
7. Schroeder, M., Noy, P.: Multi-agent visualization based on multivariate data. In: International Conference on Autonomous Agents, pp. 85–91. ACM, Montreal, Canada (2001)
8. Pham, B., Brown, R.: Multi-Agent Approach for Visualisation of Fuzzy Systems. In: Sloot, P.M.A., Abramson, D., Bogdanov, A.V., Gorbachev, Y.E., Dongarra, J.J., Zomaya, A.Y. (eds.) ICCS 2003. LNCS, vol. 2659, pp. 995–1004. Springer, Heidelberg (2003)
9. Healey, C.G., Amant, R.S., Chang, J.: Assisted Visualization of E-Commerce Auction Agents. In: Graphics Interface 2001, Ottawa, Canada, pp. 201–208 (2001)
10. Hendley, R.J., Drew, N.S., Wood, A.M., Beale, R.: Case study: Narcissus: Visualizing Information. In: IEEE Information Visualization (INFOVIS 1995), pp. 90–96. IEEE, Atlanta (1995)
11. Michotte, A.: The Perception of Causality. Basic Books, New York (1963)
12. Ware, C., Neufeld, E., Bartram, L.: Visualizing Causal Relations. In: IEEE Information Visualization (INFOVIS 1999), pp. 39–42. IEEE, Los Alamitos (1999)
13. Lethbridge, T.C., Ware, C.: Animation Using Behavior Functions. In: Ichikawa, T., Jungert, E., Korfhage, R.R. (eds.) Visual Languages and Applications, pp. 237–252. Plenum Press, New York (1990)
14. Couzin, I.D., Krause, J., James, R., Ruxton, G.D., Franks, N.R.: Collective memory and spatial sorting in animal groups. Journal of Theoretical Biology 2002, 1–11 (2002)
15. Balch, T., Arkin, R.C.: Behavior-based formation control for multi-robot teams. IEEE Transactions on Robotics and Automation 14, 1–15 (1998)
16. Bicho, E., Monteiro, S.: Formation control for multiple mobile robots: a non-linear attractor dynamics approach. In: Proceedings of the IEEE/RSJ International Conference on Intelligent Robots and Systems (IROS 2003), vol. 2, pp. 2016–2022. IEEE, Las Vegas (2003)

17. Fredslund, J., Mataric, M.J.: Robot formations using only local sensing and control. In: IEEE International Symposium on Computational Intelligence in Robotics and Automation 2001, pp. 308–313. IEEE Computer Society, Seoul, Korea (2001)
18. Bertin, J.: Semiology of Graphics. The University of Wisconsin Press, Wisconsin (1983)
19. SWCP: Historical Data for S&P 500 Stocks, vol. 2005 (2005)
20. Lau, A., Vande Moere, A.: Towards a Model of Information Aesthetics in Information Visualisation. In: International Conference on Information Visualisation (IV 2007), IEEE, Zurich, Switzerland (2007)

A Principled Approach to Swarm-Based Wall-Building

Lihan Lai, Jeff Manning, Jeannie Su, and Sanza Kazadi

Jisan Research Institute, Pasadena, CA 91107, USA

Abstract. In this paper, we apply a theoretical swarm-generating technique to a system implementing *cluster-based construction*. The technique, known as *swarm engineering* consists of two stages. In the first stage, which is top down, the global goal is expressed in such a way that specific conditions may be developed which, when satisfied, guarantees achievement of the global goal. The second step, which is bottom up, concerns the design of specific agents. These agents, once built in accordance with the conditions from the top-down step, will provably lead to the global goal. We develop parts of this theory and apply them to the cluster-based construction problem.

1 Introduction

Swarm-based systems have received an increasing amount of attention over the past few years owing to the remarkable abilities of swarms of agents to do things *en masse* that none of the individual agents can do. The phenomenon of *emergence*, in which agents carry out actions that they are not explicitly designed to do, has captured the imagination of many engineers and scientists and this has led to a flurry of work.

A number of researchers have explored the potential use of swarms to carry out construction tasks[11,12,13,14]. Much of this work has centered around trying to understand how animal systems accomplish this task and then to reproduce what animals have done. While many of these studies have produced interesting initial steps, few have actually made the transition from a proof of concept to a useful and competitive construction idea.

Our interest in swarms centers around the careful design of swarms using a reproducible methodology. To date, no standard technique exists which can be used to design swarms with specific global properties. As a result, most practitioners of swarm design must resort to using their own skill as engineers to build swarms of particular design. There is no guarantee that the desired swarm can be built at all. There is no way of knowing whether or not a particular behavior will yield the desired global behavior without running the task.

In this paper, we extend our previous work on puck clustering by applying the formal swarm engineering technique we call the ***Hamiltonian method of swarm design*** to the wall-building subproblem. This problem involves building walls between existing placed clusters. Once we've solved this problem from a theoretical standpoint, we apply it to a simulated swarm of agents.

M. Randall, H.A. Abbass, and J. Wiles (Eds.): ACAL 2007, LNAI 4828, pp. 305–319, 2007.

Though this paper deals only with artificial swarms, this methodology may be used to understand natural systems as well. Each natural system has global properties that must be satisfied in order to accomplish tasks that help keep it alive. Understanding the minimal requirements of that task will help the researcher identify agent behaviors that lead to specific swarm-level behaviors. Moreover, the identification of these behaviors helps to understand how the agents maintain the swarm and how each small behavior contributes to the global behavior of the swarm.

The paper proceeds in the following way. Section 2 reviews swarm-based construction work. Section 3 describes the swarm engineering methodology and its application to this problem. Section 4 gives simulation results. Finally, Section 5 offers some discussion and concluding remarks.

2   Previous Work in Swarm Construction

Swarm based construction deals specifically with the use of swarms of robots in construction. Unlike regular construction, the "workers" in this swarm based system do not use high-level reasoning. The swarm, however, does behave like the insects, exhibiting remarkably dynamic properties that make construction possible. Once developed, swarm based construction applications might range from assisting in civil disasters to remote construction problems in which human labor is impossible or dangerous; robots can be used to build replacement homes in areas struck by disaster, construct levee banks to restrain floodwaters, and build walls to retain chemical spills or nuclear radiation leaks. Swarms have also been envisioned as a solution to building underwater facilities and even structures in space. The main question in dealing with any of these problems lies in determining the role of the swarm, the various castes required, the interaction of the various castes, and the control algorithms for each of the agents in each of the castes.

It is interesting to look at the animal kingdom as a source of inspiration for the design of construction swarms. For instance, bulldozer ants have been observed to build nests by plowing material away from the nest site. These ants, behaving like little "bulldozers", ensure that the construction site is clear of rocks and other obstacles. Parker, Zhang, and Kube [12] explore this collective construction strategy, which they call "blind bulldozing". In their study, the robots, like the ants, carve a nest out of an excess material or rubble. The robots plow a nest by continuing to push the material until the robots cannot exert enough force to move the material. By pushing back the material, the robots build a circular wall structure that encircles the nested area. This version of site preparation can be implemented in the first step of construction. The robots that accomplish this task may be thought of as a specific caste of robots that perform the initial construction task, with subsequent castes required for further construction.

While the blind bulldozing methodology does not lead directly to methods for the construction of larger structures, the simplicity of the method is alluring. Not only is a single individual capable of accomplishing the entire task if given enough time, but other individuals can be added to the task seamlessly, without affecting the original agent's behavior or design. Moreover, additional individuals have a very small effect on the original individual as long as the number of new individuals is relatively small. As a result, the method is both scalable and robust to failure of a single individual. These qualities are desirable, and therefore should appear in the construction task.

This approach has already been used to construct complex structures using modified algorithms "borrowed" from the natural world. Ants can construct complex structures such as arches by first forming small piles of sand and adding onto these piles. Two piles are started near one another. Ants continually place bits of sand on top of the piles, making the overall piles grow upward. The piles grow up, and can bend over as they are buit up. When the piles are built up and curve inward towards one another, eventually meeting at the top. In [1], Bowyer implements a similar method to build arches and walls by adding blobs of polymer foam to piles of foam initially placed near one another. In this instantiation, the robots have legs, which allow them to climb the foam walls and build on the existing structure. The structures developed using this method in the laboratory resembled that found in the natural world, while illustrating one important aspect of the overall task design process - the morphological instantiation of the robot itself.

Our approach to construction is therefore based on this general approach. We are developing a methodology that can be accomplished with groups of individuals or with single individuals, with a graceful improvement in overall performance as new individuals are added to the task. On top of this, we would like predictability in our final results in the sense that we know within given tolerances what type of structure will emerge. This is very different from the requirements of a natural system, as a natural system will have differing overall structure depending on nuances of the environment, and therefore we depart from strict adherence to the designs of a natural world.

The general *puck clustering* methodology [9,10] satisfies the requirements outlined above. This methodology focuses on creating clusters of predetermined size and multiplicity. Special care is taken to provide theory that allows predictions as to the final state of the system to be made. This methodology, which has also been "borrowed" from the natural world, also has the beneficial characteristics described above. Namely, the clustering can be accomplished by a single individual, it benefits from the addition of new individuals, and the addition of new individuals is nearly invisible to the already existing individuals in the system.

Studies have also focused on the movement and correct placement of the clusters formed using the puck clustering methodology. These studies have demonstrated that it is possible to move clusters into predetermined relative positions with precise specificity. This means that a complex cluster grouping can be constructed using agents whose knowledge of the overall structure is extremely limited. However, the final arrangement of clusters is predictable and reasonable. Estimations can be made as to the completion time of such a task. Once again, these algorithms have the general properties described above - the potential for completion even with a single agent, and the seamless addition opportunity for new agents of similar design.

3 Swarm Engineering

Swarm engineering [8] is a method composed of two steps that generates agents and associated algorithms which accomplish predetermined tasks. The first step of swarm engineering consists of the creation of a swarm condition. It is the condition that, when satisfied, leads to the specific completion of the global goal. No specific method exists for this step and the creation of a swarm condition is problem-dependent. The second step is the creation of swarm behaviors that satisfy the given swarm condition.

This two-step process guarantees the achievement of the desired global behavior. However, no general set of techniques has been developed for designing swarms. Because the agents interact independently, they may exhibit unforeseen behaviors. Small deviations in the individual agents' behaviors may cause a large change in the overall system behavior. Therefore, much effort has been expended to devise a rigorous methodology to avoid this potential problem. One of these efforts is [8]; we review the major points from [9].

Our strategy is to begin with an examination of the desired property for the construction of walls using properties that the agent can sense. Once this property has been constructed, an examination of its dynamics will yield a requirement for the behaviors of the agents which causes the system's generation of the desired property.

Our theoretical work is validated using a simulation consisting of a two-dimensional "world" populated by inanimate and animate objects. The inanimate objects are the pucks, which are circular in shape and do not move. The animate objects are the agents. The agents can move around, avoid one another, walk on, pick up, and put down the pucks. Agents are also circular, primarily in order to simplify the simulation. Agents are larger than pucks, with the previous having a diameter of two units while the latter has a diameter of one unit. The simulation is depicted in Figure 1.

The rules of the simulations are meant to mimick conditions that occur in with real laboratory robots. Agents are embodied and so cannot pass over one another. Agents are equipped with collision sensors which allow them to avoid collisions. Moreover, their visual sensors are limited in range, though they've been made to allow a 360 degree visual field. The agents can sense other agents and pucks within four hundred (400) units and avoid collisions with Agents are limited in their ability to pick up or place down pucks, so as to simulate similar limitations for real agents equipped with grippers. Once pucks have been placed, they stay where they've been placed until another agent comes along and moves the puck. We examine the behaviour of our agents within an embodied virtual system.

Initially, the agents do not carry any pucks. The system contains clusters of pucks that have been placed to mark the endpoints of a wall in accordance with

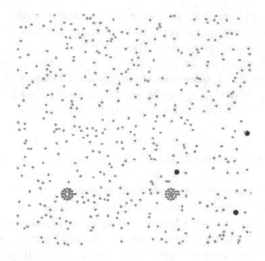

Fig. 1. This depicts the two-dimensional system. The small building materials are scattered about. Two large clusters used as construction markers are evident, as are three agents. The agents' behavior will move the building materials into the space between the two clusters.

the approach of [9,10]. Each of these endpoint clusters is made using pucks of a type that the agents have been programmed to ignore. This way the agents do not move the marking clusters. Pucks of a different type than any of the marking clusters are initially placed somewhere in the arena to be used as building materials. They are either randomly placed or placed in a single large pile, typically much larger than the marking clusters. These pucks may be picked up and put down as the structure being built is constructed.

The system allows agents to carry out construction task by moving pucks in some way from the starting location to other locations. The task is considered completed when all of the pucks that can be placed are placed in the desired region. In this system, we assume that all agents can walk over the pucks, but not over one-another.

3.1   Top Down

In [8], guidelines were generated for writing a global property. The first of these guidelines suggests that the global property be a function of local properties that are themselves accessible to a single agent. Once we write the global property, P_j, then differentiating gives the equation

$$b_j = \frac{dP_j}{dt} = \sum_{i=1}^{n_b} \frac{\partial P_j}{\partial b_i} \frac{db_i}{dt} + \sum_{i=1}^{n_P} \frac{\partial P_j}{\partial P_i} b_i \tag{1}$$

Where $P_1, P_2, \ldots, P_{n_P}$ are the properties of the system, and $b_1, b_2, \ldots, b_{n_b}$ are the behaviors of the system. This equation expresses the golbal property in terms of other, possibly simpler properties that make it up.

The global property for the current work is that the agents move building materials into a region between two already placed clusters. The material must be placed in such a way that it can be used as a foundation for the construction of a larger structure above it. Because this is meant to be a foundation for a larger structure, conditions must be generated which allow for the generation of a foundation with a minimal number of gaps between the building blocks. We assume that the building blocks are all identical. We also assume that they can be placed on top of one another if needed.

We are interested in a system built using agents of minimal complexity. This avoids the potential complications that might arise from the use of sophisticated agents with complex parts and behaviors. As a result, our system utilizes very simple agents which we expect to be able to complete the task in tandem. As a result, we focus on the local actions of the agents. Suppose that an agent encounters a piece of building material. Should the agent pick it up and move it, or should it leave the material where it is? In general, one might expect it to be left where it is unless it is not in the desired region.

It is realistic to expect that a single agent built with current technology on board can tell whether or not an encountered puck is in the desired region. It is also realistic to expect that the agents can determine the range and direction to the nearest region. Thus, our microscopic property can realistically be the minimal distance a puck has to the desired region. This is, of course, zero if the puck is in the desired region.

As a result of this, we can write the global property as

$$P_G = \sum_{p \in \{\text{pucks}\}} d_p. \tag{2}$$

where d_p is the distance the individual puck is from the desired wall region. Differentiating this equation, we obtain

$$b_g = \sum_{p \in \{\text{pucks}\}} b_p. \tag{3}$$

This equation illustrates the idea that the global behavior is a function of the behaviors of the individual pucks. Moreover, these behaviors are functionally independent (they interact with a '+' sign). They can therefore happen without directly interacting agents. Since the behavior of the individual pucks is controlled through the agent behaviors, we have a natural method of manipulating the global property.

Let us examine what the end points of the global property ought to be. Initially, the property has some value determined by the organization of the set of pucks. This value must be reduced by at least

$$\delta P_g = \sum_{p \in \{M \text{ closest pucks}\}} d_p. \tag{4}$$

The reduction of δP_g guarantees that the correct layout of pucks has occurred. Such a layout will be guaranteed if at least M pucks are moved into the region from the exterior region. Thus, we now have a general condition which, when satisfied, will give us the global goal. The next task is to create a behavior that is capable of producing it.

It is perhaps not easy to see how this analysis *guarantees* that the global goal will be met. However, the property so chosen has a unique numerical value at the system configuration desired. The task then, is to find out how to generate this specific numerical value. The specific agents required to achieve this numerical value may be chosen from any number of potential agents whose affect on this property is to move its numerical value from the current system configuration to the desired one.

We also assume, for the moment that the agents need only mark the foundation of the structure being built. More than one property is required for three-dimensional construction, and that is beyond the scope of this paper.

3.2   Bottom Up

Now that a top down condition has been created, we can turn to the bottom up portion of the swarm engineering methodology. Our task is to create a behavior which provably completes the condition from the top-down portion. Thus, this subsection will focus on the development of a behavior set for the individual agents that accomplishes the task in a provable way. This allows the swarm to be designed robustly whilst sidestepping much of the complexity associated with the interactions between agents.

In this study, we've created a global property P_G, which is the sum of all the distances from each puck to the region where the wall is to be built. To manipulate P_G the robots need to have a mechanism for sensing, picking up, transporting, and dropping off pucks. Thus, a basic outline of local behaviors for P_G to reach the desired state is to pick up pucks that are not in the region, move them to the region, and drop them off at an appropriate location within the region. It is necessary that the robots know if they are in the region, in order to ensure that the pucks are dropped off at the right spot. This requires the robots to know where to go if they are not in the region. Many robots are present and moving at the same time in the system. Collisions are extremely likely, particularly as the number of robots increases. In order to avoid collisions, robots must have the ability to sense one another and use this information to robustly avoid collisions.

This gives an idea of what kind of hardware one might need. Clearly some type of sensor array is required which will provide this information along with general obstacle proximity data. Processing is required to determine what direction to go based on sensor data. How much processing is required is not clear, as it will depend on the precise method used to determine the direction. Actuators include both movement mechanisms (i.e. legs, wheels, propellers, floatation devices, etc.) and grippers for grasping and holding the pucks.

Now that we have an idea of what the various requirements of the agents' hardware are, we can proceed to build the behaviors. Note that we can choose any behaviors *as long as they complete the global task*. Thus, the approach allows us to have the same kind of creativeness while guaranteeing that the outcome will be as desired.

The agents in our system have two states: those that are carrying pucks and those that are not. The way each agent moves depends on which of these states it has. Moreover, we assume that the agents are aware of where they are with respect to the construction region (the region between the demarkating clusters). The agents' control algorithm, which utilizes both of these pieces of information, is summarized in the flowchart in Figure 2.

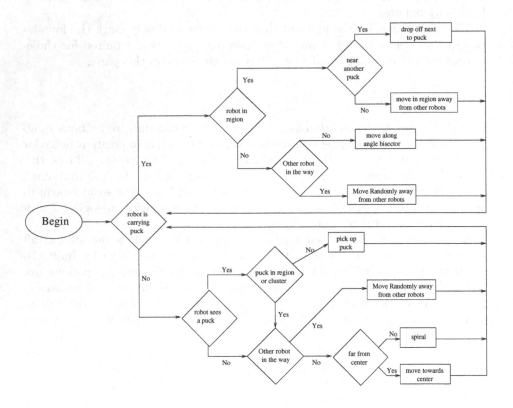

Fig. 2. This flowchart gives an outline of the agent's behavior

If an agent is carrying a puck it will first determine if it is in the region. To do this it first calculates the directions of the tangent segments from itself to each of the clusters. It then calculates the angles between each of the four pairs of

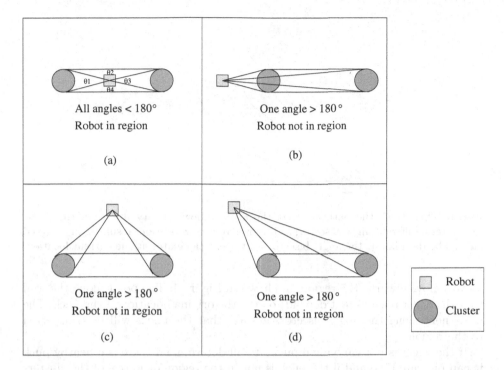

Fig. 3. These figures show different situations in which the agent is either in or out of the region. In (a), the agent is in the region because the all the angles, θ_i, are less than 180°. In (b), (c), and (d), the agent is not in the region because the at least one θ_i is greater than 180° so the agent is not in the region.

adjacent tangent lines, as shown in Figure 3. If all four of these angles are less than 180°, then the agent is in the region; otherwise, it is not.

When a puck-carrying agent determines it is in the region, it will check whether it is near another puck inside the region. If it is, it will drop off its puck next to that puck. If it is not near another puck, it will move a short distance in a random direction.

If a puck-carrying agent is not in the region, it will determine the directions from itself to each of the two wall-demarcating clusters. After this, it will calculate the angle bisector of these two rays. If there are no nearby agents in that direction, it will move a short distance along the angle bisector, as shown in Figure 4. If there is a nearby agent in that direction, it will move in a random direction to avoid a collision.

This way, at each iteration before the agent reaches the region, the agent will re-calculate the angle bisector as before; however, because the locations of the clusters have changed slightly relative to the agent, the agents' directions vary slightly. Thus the agents move along a curve, whose endpoint is always a point on the wall. Because each individual angle bisector intersects the wall, the agent

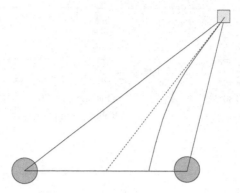

Fig. 4. This shows the pathway that the agent follows. Every iteration, the agent calculates a different angle bisector, as it moves to a new location, resulting in a curved path. The dash line is the angle bisector that agent calculates at the current location.

always move towards the region, as shown in Figure 4. The point along the wall that the agent reaches first depends on the the original location of the puck. The movement along the angle bisectors ensures that the pucks will be transferred to the region.

If the agent is not carrying a puck, it will determine if there is a nearby puck it can pick up. If so and if the puck is not in the region or in one of the clusters it will pick up the puck. If no puck is located nearby the agent which is not in the region or in one of the intial clusters, then the agent will spiral to search for a puck. We say that an agent is lost if it is more than 150 units away from a central location, which in our simulation is defined as the average of the positions of the clusters. Once an agent is lost it will continue to be lost until it is less than 20 units from the central location. If the agent is not lost it will spiral. If the agent is lost it will move toward the central location, this ensures that agents do not just keep spiraling forever and leave the construction area if they do not encounter a puck. In either case if the agent cannot move in the specified manner without hitting another agent, then it will move in a random direction which is not toward any nearby agents.

It is perhaps very easy to see that the effect of the behavior of the agents will be to wander around the arena until they come into contact with the pucks. Once these pucks are picked up, they will be carried by the agents into the building region, and be dropped off at whatever available location the agent carrying them can find. Mathematically, this means that

$$\delta P_g = \sum_{p\in\{\text{M pucks}\}} d_p. \tag{5}$$

which is at least as large a change as that given in (4). Thus, we can be sure that this behavioral set will accomplish the task.

4 Different Geometric Shapes

In the previous section, we discussed how to build single walls. To build rooms with more than one wall, a mechanism must be developed to build different geometric structures. We extend the previously discussed theory to create a multi-walled structure.

To build a multi-walled structure, it is assumed that n clusters, rather than two, have been placed to mark the endpoints of the various walls. Our goal is to build walls between certain specified pairs of clusters, but not other pairs of clusters. This means that there will be multiple construction regions (one for each desired wall), instead of just one. As before, it is realistic to assume that a single agent can determine whether or not an encounted puck is in a region, and that a single agent can calculate the range and direction to any region. So we realisitically make the microscopic property be the distance a puck is from the nearest region. We may write the global property as,

$$P_G = \sum_{p \in \{\text{pucks}\}} d_p \tag{6}$$

where d_p is the distance from p to the desired region. The initial value of P_G is determined by the initial arrangment of the pucks. This must decrease by at least

$$\delta P_g = \sum_{p \in \{\text{M pucks}\}} d_p. \tag{7}$$

where the M pucks are the minimal number of pucks, and the closest ones, required to fill the various regions without unabmiguously.

It should be clear that the agents in this new system must have all the behaviors described in section 3. In addition the agents must be able to determine if a given pair of clusters are required to have a wall between them to ensure that walls are only built in the specified locations. Also, when carrying pucks, agents must be able to choose a specific region, determine its direction, and move towards it.

In our simulation, the pucks in each of these clusters are different. In other words, cluster 1 consists of puck type 1, cluster 2 consists of puck type 2, cluster 3 consists of puck type 3, and so on. The use of differing puck types ensures that the agents do not destroy the clusters and remove the markers by using the pucks in the clusters as building material. A different set of pucks than those in the clusters is used to build the walls.

As an example, the square structure from Figure 5a has four clusters. This means that there are six possible walls - the sides and the diagonals. If all the marking clusters were made up of pucks of the same type, the agents would not have been able to differentiate between them and there would be no way to determine which pairs of clusters to build walls between. In this case, all six walls would be built, resulting in a square with a cross inside. Since this final shape is not what we desire in this case, the new method of using multiple cluster puck types is the obvious choice.

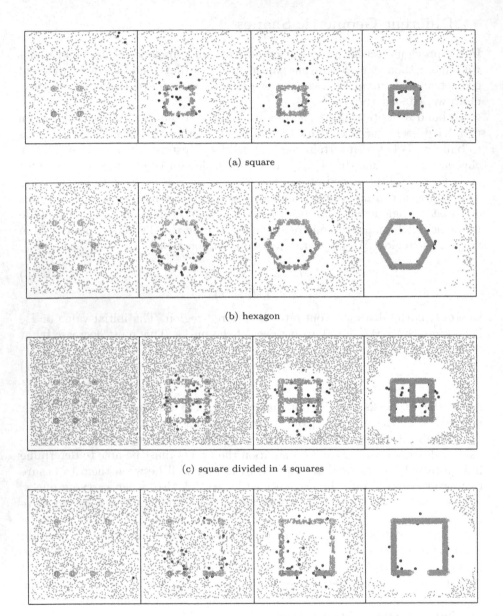

(a) square

(b) hexagon

(c) square divided in 4 squares

(d) This structure can be used as a foundation for buildings with doors.

Fig. 5. The figures above are screen-shots of different steps in our simulation. As the first column illustrates, our simulation is initialized with arbitrarily scattered pucks and specifically placed clusters that mark off the edges of the wall foundation. Each consecutive column is a snap-shot of what the system looked like after a certain amount of time passed. The last column is the finalized structure.

In these simulations, we utilize agents which have identical physical properties to those of the agents used int he previous simulations. However, each of the new agents is provided with a list indicating between which pairs of clusters to drop off the pucks, or in other words where to build the walls.

To implement this new global property, the same local behaviors used to build a one-walled structure are maintained. The agent's behavior after picking up a puck, however, is modified so that it does not move along the angle bisector of the angle between the clusters and itself. There are too many clusters, which complicate this behavior. Instead, the agent randomly chooses one wall to move toward and calculates its direction from that wall as if it were in a single-walled system.

The structures that agents may build in this way can be quite complex. Not only can the structures have many walls, but they can be designed to be practical. As an example, consider the agents laying out the floor plan of a house (Figure 6). The mechanisms described here can be used to carry out practical construction, a natural extension of the work in this paper.

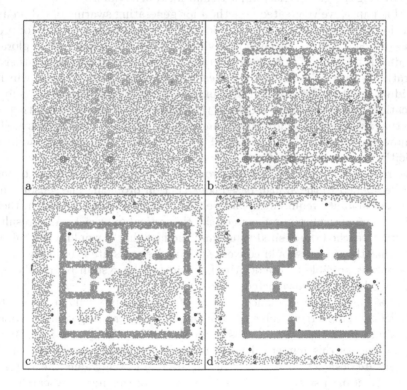

Fig. 6. This algorithm can be used to make a foundation for the outline of a typical house with two bedrooms, a kitchen, and a bathroom

5 Summary and Concluding Remarks

One of the great hopes of swarm systems in the early years was the development of swarms capable of completing construction of large complex structures completely without the need for external intervention [3][11][12][13][14]. Much of this work is based on observations and adaptations of existing systems in the natural world [2] [3][4][5][6][7]. Work continues on this to this day, but has yet to produce a satisfactory solution.

Perhaps one of the reasons this has been so is that most construction swarms have been able to do one or two steps of the whole task, but have not been able to handle the complexity of a true construction task. This may, in turn, be due to the paucity of a theoretical or principalled mechanisms for swarm design. It is extremely difficult to design a swarm of agents that completes the task it is supposed to be completing because the unintended interactions between the agents in the swarm are extremely likely to cause trouble in generating the correct global behavior. In fact, there is no prior theoretical work done that indicates a mechanism for predicting the global effect of particular local behaviors.

In this paper, we've created a method for generating swarms based on an examination of the meaning of the equations that describe the evolution of properties according to their associated behaviors. This is a very powerful exploration, as it affords us the ability to turn the tables on swarm generation. Rather than generating a swarm based on the "guess and check" method during which the individual agent is designed and then examined in simulations or real embodied fabrications, we've designed a method that allows us to say ahead of time what the actual agents will do, individually and in a group. Once these agents are built, we know that the global goal will be able to be achieved, as the interactions are implicitly taken into account.

The remainder of the paper was devoted to demonstrating how the technique could be applied to the swarm-based construction problem. The sensor requirements and mobility requirements as well as behaviors could be shown, then, to satisfy the numerical requirements of the theoretical approach. As a result, the property P reached the desired value, and the fact that it would reach this value meant that the system would converge to the desired state eventually. In other words, it was possible to predict the behavior of the system before the agents themselves had been constructed.

The problem we've examined is part of a larger problem in swarm-based construction. This larger problem can be explored by creating similar properties whose values are unique, given the specific stages that the system is in. As we've seen, as long as this numerical value is specific to the desired state, and the behaviors are designed to generate that desired numerical value, the system will take on the desired states. This is the next stage of this line of research.

In artificial life research, much of the discussion has historically centered around the question of what constitutes a lifelike system. In this particular discussion, we might also ask, what properties of a lifelike system can we expect and how might we generate them? Moreover, we can approach this by utilizing the sensor, processor, and actuator capabilities of the agents to generate the global property. Then, the requirements for the agent behaviors can be constructed. Such an approach might reliably generate agents whose behaviors make the system act more like a living system as a result of generating behaviors of the agents that create the desired system behaviors.

Anecdotally, we found that our method, when properly applied, trimmed the design time from several weeks of trial and error to several minutes of careful planning. If this is the improvement from a single stage of design, one might expect several to have a far greater improvement, possibly making a task that is currently impossible possible. More work is needed to determine whether or not this is so.

References

1. Bowyer, A.: Automated construction using co-operating biomimetic robots. University of Bath Department of Mechanical Engineering Technical Report (November 2000)
2. Bonabeau, E., Theraulaz, G., Fourcassie, V., Deneubourg, J.: Phase ordering kinetics of cemetary organization in ants. Physical Review E 57(4) (1998)
3. Camazine, S., Sneyd, J., Jenkins, M., Murray, D.: mathematical model of self-organized pattern formation on the combs of honeybee colonies. Journal of Theoretical Biology 147, 553–571 (1990)
4. Deneubourg, J., Sendova-Franks, G.S.F.N.A., Chretien, D.C.L.: The dynamics of collective sorting. Robot-like ants and ant-like robots. In: Meyer, J.A., Wilson, S.W. (eds.) Animals to Animats, MIT Press / Bradford Books, Cambridge, MA (1991)
5. Franks, N.R., Wilby, A., Silverman, B.W., Tofts, C.: Self-organizing nest construction in ants: sophisticated building by blind bulldozing. Anim. Behav. 44, 357–375 (1992)
6. Holland, O.: The use of social insect based control methods with multiple robot systems. In: UKACC International Conference on Control 1996 (1996)
7. Holldobler, B., Wilson, E.: The Ants. The Belknap Press of Harvard University Press, Cambridge Massachusetts (1990)
8. Kazadi, S.: On the Development of a Swarm Engineering Methodology. In: Proceedings of IEEE Conference on Systems, Man, and Cybernetics, Waikoloa, Hawaii, USA, pp. 1423–1428 (October 2005)
9. Kazadi, S., Grosz, A., Lim, A., Wigglesworth, J., Vitullo, D.: Swarm-Mediated Cluster-Based Constuction. Complex Systems 15(2), 157–181 (2004)
10. Kazadi, S., Koroleva, O.: Removing Degeneracy From Swarm Mediated Cluster-Based Construction. In: Proceedings of IASTED International Conference Robotics and Applications, Honolulu, Hawaii, USA, pp. 60–66 (August 2004)
11. Leung, H., Kothari, R., Minani, A.: Phase-Transition in a Swarm Algorithm for Self-Organized Construction. Physical Review E 68, 046111-9
12. Parker, C.A., Zhang, H., Kube, C.R.: Blind Bulldozing: Multiple Robot Nest Construction. In: Proc. IROS 2003, Las Vegas, USA (2003)
13. Stewart, R.L., Russell, R.A.: Building a loose wall structure with a robotic swarm using a spatio-temporal varying template. In: Proceedings of the IEEE/RSJ International Conference on Intelligent Robots and Systems (2004)
14. Wawerla, J., Sukhatme, G.S., Mataric, M.J.: Collective Construction with Multiple Robots. In: Proceedings of the IEEE/RSJ International Conference on Intelligent Robots and Systems (2002)

Pattern Extraction Improves Automata-Based Syntax Analysis in Songbirds

Yasuki Kakishita[1], Kazutoshi Sasahara[2], Tetsuro Nishino[1], Miki Takahasi[2], and Kazuo Okanoya[2]

[1] Department of Information and Communication Engineering, Graduate School of Electro-Communications, The University of Electro-Communications, 1-5-1 Chofugaoka, Chofu-shi, Tokyo 182-8585, Japan
kakishita@ice.uec.ac.jp
[2] Laboratory for Biolinguistics, RIKEN Brain Science Institute (BSI), 2-1 Hirosawa, Wako-shi, Saitama 351-0198, Japan
sasahara@brain.riken.jp

Abstract. We propose an automata induction approach to modeling birdsongs on the basis of Angluin's induction algorithm, which ensures that k-reversible languages can be learned from positive samples in polynomial time. There are similarities between Angluin's algorithm and the vocal learning of songbirds; for example, during a critical period, songbirds learn songs from positive samples of conspecific birds. In our previous method, we could not construct song syntaxes for complex songs. In this paper, we introduce a pattern extraction method to improve our previous method and propose a new birdsong modeling method. We estimate the robustness and properness of our method by using artificial song data, and demonstrate that the song syntaxes of the Bengalese finch can be successfully represented as reversible automata. As a result, almost all Bengalese finchs' song syntaxes can be represented with lower k-reversibility; further, one song has 3-reversibility song syntax showing the highest reversibility.

1 Introduction

The Bengalese finch (*Lonchura striata* var. *domestica*) is known as a popular fowl in Japan. They have a complex song structure as compared to other songbirds, such as Zebra finches.

Recent studies on Bengalese finches have reported unique features of their courtship songs. The songs of male Bengalese finches are neither monotonous nor random; they consist of chunks, each of which is a fixed sequence of a few song notes. The song of each individual can be reconstructed by a finite automaton, which we call song syntax (see Fig. 1(a)) [5]. Thus, the songs of Bengalese finches have "double articulation," which is one of the important faculties of human language (i.e., a sentence consists of words, which consist of phonemes). Song syntax is controlled by song control nuclei in the brain. The hierarchy of song control nuclei directly corresponds to the song hierarchy [5]. Due to the structural and functional similarities of vocal leaning between songbirds and humans, the

M. Randall, H.A. Abbass, and J. Wiles (Eds.): ACAL 2007, LNAI 4828, pp. 320–332, 2007.

former have been actively studied as good models of human language [3]. In particular, the song syntax of Bengalese finches sheds light on the biological foundations of syntax.

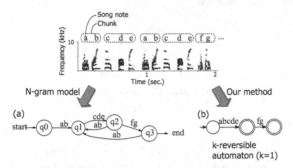

Fig. 1. A courtship song and the song syntaxes of Bengalese finches given in two different ways. The letter string represents the courtship song of a male Bengalese finch. (a) is constructed using N-gram model, and (b) is constructed using our method (1-reversible automaton).

To model birdsong syntaxes, we focus on k-reversible languages, which are subclasses of regular languages. Angluin presented an efficient induction algorithm from positive samples, where $k = 0, 1, 2, \dots$ [1]. Angluin's algorithm provides a finite automaton that accepts the smallest k-reversible language, including the given finite positive sample, within polynomial time. Based on Angluin's algorithm, Berwick et al. proposed a computer model for learning the English auxiliary verb system by using the complete corpus of grammatical sentences [2]. k-reversible automaton A is a type of deterministic finite automata and has following property:

- A is deterministic finite automata
- no two final states of A have a common k-leader
- no two states of A having a common successor have a common k-leader

It should be noted that when A is an automaton, the string u is said to be a k-leader of the state q in A if and only if $|u| = k$ and $\delta^r(q, u^r) \neq \emptyset$, where δ^r denotes the reverse of the state transition function δ, and u^r denotes the reverse of the string u.

So far, Bengalese finches' songs have been analyzed using N-gram models [4]. The N-gram model is effective for the visualization of the structures of song sequences; however, it does not ensure the uniqueness of the resulting automata from given birdsong samples. Therefore, an appropriate N must be selected for each sample through trial and error. On the other hand, if we use the standard minimization algorithm for finite automata, it turns out that simple automata cannot be obtained from the given birdsong samples.

We had proposed a method based on Angluin's inference algorithm to model the syntax of birdsongs [7]. However, it had several problems: it could not represent repeated structures, and it was very difficult to construct an appropriate syntax from complex songs. To resolve these problems, in this paper, we introduce a new pattern extraction method for identifying the chunk structure. Then, we propose a new method to model birdsong syntaxes and demonstrate the extraction of song syntaxes.

2   Method

This section describes the method for representing song syntaxes of Bengalese finches as k-reversible automata and the necessary concepts for the method.

2.1   Song Units

Based on phonetic characteristics, recorded songs are converted from sounds to texts by assigning one symbol to the identical phonemes. A bout refers to a sequence wherein songbirds continue to sing at a draft. The text data of songs is organized as bouts. The following examples are parts of bouts in the songs of a Bengalese finch.

(b1) `abcdbefggabcdbefggabcdbefggabcdbefgghijklibkmgggabcdbefgg...`
(b2) `abcdbefggabcdbefgghijklibkmgggabcdbefggabcdbefgghijklibkm...`

   We can find several identical substrings that frequently appear in a bout (e.g., "abcdbefgg" and "abcdbefgghijklibkmgg"). These substrings are called song units. A bout consists of several song units that appear repeatedly in it. It is the goal of this paper to construct a song syntax from given song units. Thus, it is necessary to extract song units from a bout.

   A typical element pattern frequently appears at the beginning of a bout. When this pattern appears in a bout, we delimit the bout into song units since the pattern can be thought to be the beginning of a song unit. In the above example, the pattern "abcd" appears at the beginning of a bout. If we delimit the bouts into song units by using this pattern, it turns out that the bouts consist of two types of song units as shown below.

(b1) `abcdbefgg`   (b2) `abcdbefgg`
    `abcdbefgg`   `abcdbefgghijklibkmggg`
    `abcdbefgg`   `abcdbefgg`
    `abcdbefgghijklibkmggg`   `abcdbefgghijklibkmggg`

2.2   Chunk Extraction

To construct song syntaxes, we use Angluin's learning algorithm for k-reversible automata. From its formal definition, an input for this algorithm must be a sequence of words. Therefore, a sequence of chunks should be used as an input in song syntax analysis. However, there is no mathematical definition or an appropriate dictionary of chunks. In this paper, we newly introduce a pattern extraction method based on the N-gram model for finding the chunk structure, which is called "chunk extraction method." This method extracts chunks by using statistical information of song note transitions. In this paper, we define a chunk as a substring obtained by this method.

In our previous method, inputs for the algorithm were not sequences of chunks. So, we could not construct appropriate song syntaxes for complex songs. Thus, we try to improve the song syntax analysis by the chunk extraction.

Basic Ideas

Our chunk extraction method is based on the N-gram model. In the N-gram model, the next word in a word sequence is predicted from the previous $N - 1$ words. In our method, chunks are extracted by using the branch structures of the N-gram model of a given phoneme sequence.

Let us consider the case when a sample is {abcdefg,abfg}. Trained researchers can find three chunks {ab,cde,fg} from this sample. In order to obtain these chunks in a simple fashion, we can use a bigram (the case when $N = 2$) (see Fig. 2(a)). If we treat a node with an indegree or outdegree greater than one as the boundary of chunks, we obtain three chunks {ab,cde,fg} from the bigram in this specific example. However, we cannot always use this simple method based on bigrams. For example, in the case of a sample {abcdefag,abfag}, chunks should be {ab,cde,fag}. However, if we also use a bigram for this sample, the chunks {a,b,cde,f,g} are obtained (see Fig. 2(b)). Namely, when different chunks contain the same phoneme, the above procedure fails. To resolve this situation, we use a trigram (the case when $N = 3$) (see Fig. 2(c)) for this sample. As a result, we can represent in the way that a phoneme "a" appears in different chunks. However, we still obtain an inappropriate set of chunks, {ab,cdef,f,ag}. If a larger N is selected, the boundary of chunks tends to shift backwards; however, this is inappropriate in some cases.

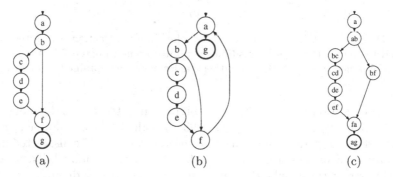

(a)   (b)   (c)

Fig. 2. Examples of the chunk extraction using the N-gram models. Given a sample {abcdefg,abfg}, (a) is constructed by the bigram. Given a sample {abcdefag,abfag}, (b) is constructed by the bigram and (c) by the trigram.

In order to resolve this problem, we propose the following procedure to extract chunks from a sample S.

1. For a large enough N, construct an N-gram model from the sample S.
2. At each node, delete the historical information.
3. Continue to merge states B_1 and B_2 where any of the following is true:
   - There exist transitions from B_1 and B_2 to a common state by a common symbol.
   - There exist transitions from a common state to B_1 and B_2 by a common symbol.

Using this procedure, we can obtain the appropriate set of chunks from the sample set {abcdefag,abfag} mentioned above. In step 1, the N-gram model in Fig. 3(a) is constructed. Here, we set $N = 3$, since N should be less than or equal to the maximum length of a chunk. In step 2, the historical information is deleted from the nodes, and the diagram shown in Fig.3(b) is obtained. Note that Fig.3(b) is a transition diagram among phonemes. In Fig. 3(b), there are two "f" nodes, which have transitions to the node labeled by "a". In step 3, these two nodes are merged and the diagram in Fig. 3(c) is obtained. In Fig. 3(c) , since the boundaries of chunks are "b" node and "f" node, the set of chunks {ab,cde,fag} will be obtained.

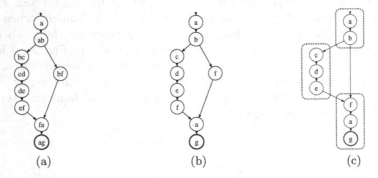

Fig. 3. Diagrams obtained in the process of the chunk extraction. (a) is the trigram obtained in step 1. (b) is the transition diagram obtained in step 2. (c) is the chunk diagram obtained in step 3; the dotted rectangles show the obtained chunks.

By the operation given in Step 3, a bigram-like structure is introduced. By this operation, the obtained diagram, which is called "the chunk diagram," not only has global historical information of the N-gram model constructed in Step 1 but also local historical information of the bigram-like model. Thus, we can correctly extract the chunks {ab,cde,fag} from the chunk diagram.

Notes on Merge Operation

The courtship songs of Bengalese finches may include many types of noises; for example, endogenous ones such as mis-singing and stop-singing by Bengalese finches and exoteric ones created by humans. Since our chunk extraction method and Angluin's algorithm presupposes complete positive samples as inputs, it is necessary to reduce the noises included in our song data before using the algorithms.

In this paper, nodes and edges with occurrence probabilities below a certain threshold value are treated as noises. This threshold value is called the "the transition threshold" and is denoted by TT. During the chunk extraction, the noisy nodes and noisy edges should be carefully treated in Step 3. Before the merge operation of states is executed, there are six possible patterns of situations of noisy nodes and noisy edges, as shown in Fig. 4.

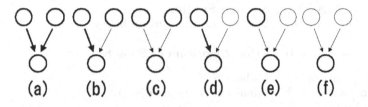

Fig. 4. Six possible patterns of situations of noisy nodes and edges. Noisy nodes and edges are drawn by thin lines.

The ordinary case is shown in (a). In the case of (b) or (c), it is not appropriate to merge the two states into one. This is because each upper node in the figure is not a noisy one, and hence, the node has many occurrences. Thus if we merge these two nodes into one, some global historical information may be lost. In the cases of (d), (e), and (f), a noisy node will be merged into the node that is not noisy. By treating noises in this fashion during the merge operations, we can simultaneously perform noise determination and chunk extraction.

2.3   Extracting Song Syntax

k-Reversibility
It is difficult to constrain the k-reversibility of a song syntax. To obtain reversible automata that are not over-generalized, we adopt the following criteria, which are identical to those proposed by Berwick et al. [2].

– Starting from $k = 0$, increase k, till the constructed song syntax has repeated structures that do not exist in the given samples.

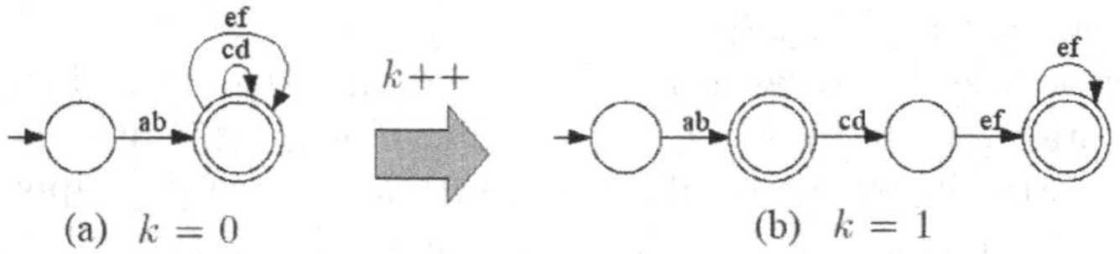

(a) $k = 0$ (b) $k = 1$

Fig. 5. Reversibility and Automata

If the song units are {"ab", "ab cd ef", "ab cd ef ef"} (chunks are de-limited by space), the repeatable chunk is "ef" only. The resulting 0-reversible automaton is shown in Fig. 5(a). This automaton has a loop of "cd" as well as "ef". This indicates the over-generalization of the song syntax, because the song units that are not included in the original data such as "ab cd cd ef" can be accepted. However, if $k = 1$, then we can obtain a suitable reversible automaton that is not over-generalized, as shown in Fig. 5(b). This concept is essential in our method, and thus we can uniquely decide k.

Induction Algorithm of Song Syntax

We show the induction algorithm of song syntax below.

―――――――――――――――― **Induction Algorithm of Song Syntax** ――――――――――――

Input: a song sample with bout S.
Output: a k-reversible automaton A that accepts S.
Procedure:
1. Construct a song unit sample S' from S.
2. Convert to chunk data S'' from S' using the chunk extraction method.
3. Construct a prefix-tree automaton P from S''.
4. Remove the transitions with "!" in P and set the result in P'.
5. Let $k = 0$.
6. Continue to merge states B_1 and B_2 where any of the following is true:
   · B_1 and B_2 have a common k-leader and both are final states.
   · B_1 and B_2 have a common k-leader and transitions to B_3
   with a common input (chunk).
   · There is a transition with a common input (chunk) to B_1 and B_2 from B_3.
7. Set the result in A'. If A' has repeated structures that do not exist in given samples, then repeat step 6 with $k = k + 1$; otherwise, $A = A'$.

2.4 Example

We present an example of the analysis of birdsong using our method. The texts of Fig. 6(a) are bouts of a male Bengalese finch and Fig. 6(b) are song units obtained from Fig. 6(a). Fig. 6(d) shows the chunk diagram constructed from Fig. 6(b). We can extract a sequence of chunks by using the chunk diagram (see Fig. 6(c)).

Using the result of chunk extraction shown in Fig. 6(c) as S'', we construct a prefix-tree automaton P, remove the transitions including "!" in P, and set the result in P' as shown in Fig. 7(a).

Nextly we merge the states in P' using Angluin's algorithm. Fig. 7(b) shows the song syntax as a 0-reversible automaton. This song syntax has a loop of "aabc", indicating its over-generalization. With $k = 1$, we then apply the merge operation to P'. Fig. 7(c) shows the song syntax as a 1-reversible automaton. This automaton has no repeated structures that do not exist in the original

(a) Bouts

faabcdddddefffaabcdddddffffaabcdddddefffffaabcdddddfafffabcdddddd...
faabcdddddefffaabcdddddefffaabcddddddddffffaabcdddddefffffaabcf

(b) Song Units
faabcdddde
fffaabcddddddffffaabcdddddde
fffffaabcdddddfafffabcdddddd...
fffaabcff

(c) Sequence of Chunks
f aabc d d d d e
f f f aabc d d d d d d f f f f...
f f f f f aabc d d d d d f a f!...
f f f aabc f! f !

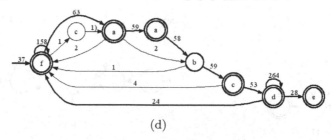

(d)

Fig. 6. An example of the chunk diagram of a male Bengalese finch's songs. The chunk diagram is constructed from 37 song units. By using the obtained chunk diagram, the song unit data is converted into a sequence of chunks in the following fashion. For example, at the beginning, several transitions are possible from the node "**f**"; a blank symbol, which is the delimiter of chunks, is inserted after the first input symbol "**f**". During the transition "**aabc**", the possible transition is uniquely determined, and so it is treated as a single chunk. Note that **a**→**f** is a noisy transition in this example. If this transition occurs, the special symbol "**!**" is printed in the output text. In this text, "**f**" and "**aabc**" are successfully treated as chunks, and the special symbol "**!**" is assigned to noisy sequences.

songs. In this case, the induction algorithm stops, and the resulting 1-reversible automaton is the target song syntax from the song sample S.

Our previous method could construct automata only for simple songs and could not ones for complex songs like the above example. In addition, a repeated structure like "**f**" in Fig. 7(c) could not be represented in our previous method. By using our new method, we can deal with a repeated structure, as shown in Fig. 7(c), and we can construct automata for complex songs (see section 3.3).

3   Result

3.1   Robustness Against Noise

To study the robustness against noise of our method, we developed a song generator — a program for creating artificial song unit data based on the song syntax shown in Fig. 8.

The song generator can add five types of noises, which are based on our observations of real song data.

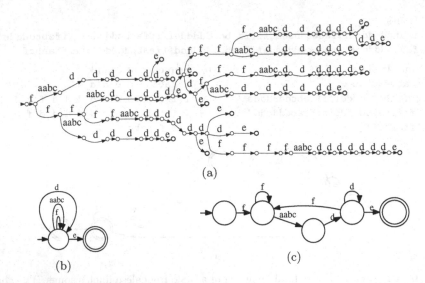

(a)

(b)   (c)

Fig. 7. (a) is prefix-tree automaton P'. (b) is constructed as 0-reversible automaton and (c) is constructed as 1-reversible automaton.

1. **Repetition1:** Repeat a few head symbols
   (e.g., abcdef → ababcdef or aaabcdef)
2. **Repetition2:** Repeat a few tail symbols
   (e.g., abcdef → abcdefef or abcdefff)
3. **Skip:** A song note is skipped
   (e.g., abcdef → abdef)
4. **Simple mistake:** A wrong song note is output
   (e.g., abcdef → abfdef)
5. **Stop:** Stop singing at the non-final states
   (e.g., abcdef → abcd)

With the song generator, we created artificial song data with noises 10%, 30%, 50%, and 70%, respectively. For example, 50% noises means that songs contain any one of above noises per two song units.

In our method, the robustness against noise depends on the chunk extraction method. This section examines the parameter setting of the chunk extraction method for the induction of correct song syntaxes. Fig. 9 and Fig. 10 show the relationship between the amount of noise and TT regarding each song syntax; each point denotes the average value of TT that can provide the correct syntax (N is fixed as 3).

These figures indicate two important features. First, as the noise per song unit increases, TT for constructing the correct reversible automaton increases. Second, even if the song unit size decreases TT per song unit does not change as much. For example, in the case of RA_K0 (Fig. 8(a)), a good setting of TT is approximately 50 when the song unit size = 300; TT is approximately 35 when the song unit size = 200. In both cases, TT per song unit size is approximately

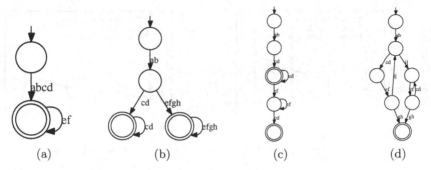

Fig. 8. Artificial song syntaxes. (a) RA_K0, (b) RA_K1, (c) RA_K2, and (d) RA_K3. They belong to 0, 1, 2, and 3-reversible languages, respectively.

17%. We conclude that in the Bengalese finches' songs, the appropriate TT is 5% to 15% of the song unit size.

It should be noted that the automata induction is not always possible; if the song unit size decreases, and the amount of noise increases, it becomes harder to obtain correct reversible automata.

3.2 Size of Required Samples

We estimated the number of samples that are necessary to correctly extract the song syntax. Fig. 11 shows the relationships between the noise ratio and the necessary number of song units to correctly extract the syntax. We randomly selected song units from the artificial song data obtained in the previous section and estimated the number of song units required for correct song syntax induction. We repeated this estimation 300 times and plotted the mean values of the number of song units for correct song syntax induction.

This result suggests that more and more song units are required as the k-reversibility of songs increase. In addition, it is found that the noises affect the induction of song syntaxes seriously in the cases of complex songs. For example,

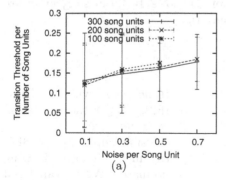
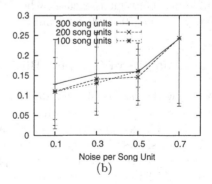

Fig. 9. Results of (a)RA_K0 and (b)RA_K1 in Fig. 8. Horizontal axis represents noise per song unit. Vertical axis represents TT per song unit.

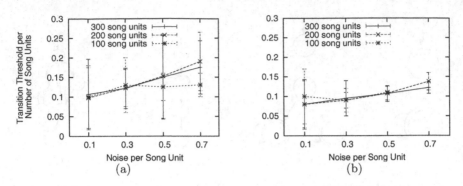

Fig. 10. Results of (a)RA_K2 and (b)RA_K3 in Fig. 8. Horizontal axis represents noise per song unit. Vertical axis represents TT per song unit.

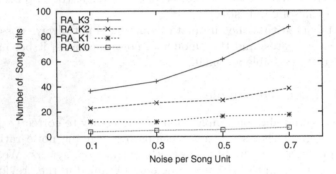

Fig. 11. Relationship between noise rate and song unit size for the induction of correct song syntax

for a simple song syntax like RA_K0 in Fig. 8(a), the correct automaton can be obtained from 10 song units even when the noise rate is 70%. However, in the case of a complex song syntax like RA_K3 in Fig. 8(d), there is an essential difference between the cases when the noise ratio is 10% (40 song units are required) and 70% (90 song units are required).

3.3   Analysis of Complex Birdsongs

We show the results of the induction of Bengalese finches' song syntaxes. These are from the cases of the adult birds (BF1 and BF2). As per the estimation mentioned in section 3.1, we set TT at 10% of the song unit size.

Fig. 12 shows a part of the song unit and the reversible automaton of BF1. BF1's songs are comparatively complex in Bengalese finches' songs. The song syntax of BF1 can be represented as a 1-reversible automaton. We have analyzed more than 20 Bengalese finches' songs using our method, and almost all the songs can be represented as 0-reversible or 1-reversible automaton.

(BF1) bcddefghefghijkfddefghighijkfdefghighijkfdefgh...
       bcbcbcddefghijkfd
       bcbcddefghijkfefghighijkfddefghijkf
       bcbcddefghijkfef

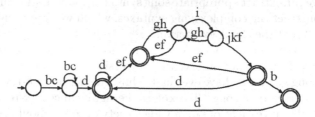

Fig. 12. BF1 (# of song units = 140, $k = 1$, $N = 4$, $TT = 14$)

(BF2) babababacdaabcdefggghiijkaafggghibcdeb
       babababababcdefggghi
       babacdaabcdefggghiijkaafggghibcdeb
       bababababcdefggghi

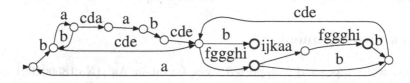

Fig. 13. BF2 (# of song units = 450, $k = 3$, $N = 3$, $TT = 45$)

Fig. 13 shows a part of the song unit and the reversible automaton of BF2. BF2 has the most complex song syntax we have ever obtained, and 3-reversiblity is required to represent the song syntax appropriately.

This result suggests that a complex song syntax tends to have higher k-reversibility and includes many transitions, which correspond to possible song variations.

These songs could not be constructed as reversible automata using our previous method. We now can appropriately construct the song syntaxes by using our new method with the chunk extraction. In addition, our new method enables to represent repeated structures like sequence of "ef gh" in Fig. 12.

4   Conclusion and Discussion

Recently, formal inductive inference has been applied to bioinformatics [6], but it has rarely been applied to ethology. In this paper, we proposed an automata induction approach to the song syntax of the Bengalese finch by introducing a new pattern extraction method and extending Angluin's induction algorithm. Due to the similarity between Angluin's algorithm and the vocal learning of songbirds (i.e., both are learning from positive samples), our method has biological plausibility. Using the proposed method, we analyzed the actual song data of Bengalese finches and demonstrated that song syntaxes could be represented as k-reversible automata. As a result, we can extract appropriate song syntax and chunk structure. Furthermore, we can construct the complex song syntaxes, which we cannot using the previous method without the chunk extraction.

In most cases, simple Bengalese finches' songs can be represented by 0 or 1-reversibility. Further, we observed a 3-reversible automaton, which is the most complex song syntax we have ever had. Thus, the degree of k-reversibility may reflect the complexity of songs. To explore the significance of k-reversibility in song syntax, further investigations on a wide variety of songs should be undertaken.

As mentioned in section 3.1, the setting of the transition threshold(TT) is an important factor for extracting the correct syntax in our method. In this paper, we estimated the robustness and properness of our method using artificial song data and obtained some important results. First, the value of TT should be decided as a ratio of the number of song units; the appropriate TT is approximately 5% to 15% of number of song units. Second, the noises affect to the induction of song syntaxes seriously in the cases of higher k-reversibility songs. This result suggests that more song units are required for the induction of complex songs.

Acknowledgements

This work was partly supported by a Grant-in Aid (No.18500109) from the Ministry of Education, Culture, Sports, Science and Technology of Japan.

References

1. Angluin, D.: Inference of Reversible Languages. Journal of the Association for Computing Machinery 29(3), 741–765 (1982)
2. Berwick, R.C., Pilato, S.F.: Learning Syntax by Automata Induction. Machine Learning 2(1), 9–38 (1987)
3. Doupe, A.J., Kuhl, P.K.: Birdsong and Human Speech: Common Themes and Mechanisms. Annual Reviews Neuroscience 22, 567–631 (1999)
4. Eugene, C.: Statistical Language Learning (Language, Speech, and Communication). MIT Press, Cambridge (1993)
5. Okanoya, K.: Song Syntax in Bengalese Finches: Proximate and Ultimate Analyses. Advances in the Study of Behavior 34, 297–346 (2004)
6. Searls, D.B.: The Language of Genes. Nature 420, 211–217 (2002)
7. Sasahara, K., Kakishita, Y., Nishino, T., Takahasi, M., Okanoya, K.: A Reversible Automata Approach to Modeling Birdsongs. In: 15th International Conference on Computing(CIC-2006), pp. 80–85. IEEE Computer Society Press, Los Alamitos (2006)

A Modified Strategy for the Constriction Factor in Particle Swarm Optimization

Lam T. Bui, Omar Soliman, and Hussein A. Abbass

The Artificial Life and Adaptive Robotics Laboratory, School of ITEE,
UNSW@ADFA, University of New South Wales, Canberra, Australia
lam.bui07@gmail.com, {o.soliman,h.abbass}@adfa.edu.au

Abstract. In this paper, we propose a modification to particle swarm optimization in order to speed up the optimization process. The modification is applied to the constriction coefficient, an important parameter that controls the convergence rate. To validate the proposed strategy, we carried out a number of experiments on a wide range of 25 standard test problems. The obtained results show that the proposed strategy significantly improves the performance of the selected PSO algorithm.

1 Introduction

In recent years, there has been an increasing number of research papers using the particle swarm optimization (PSO) algorithm to solve hard optimization problems. In contrast to other population–based search techniques such as evolutionary algorithms, PSO is inspired by the social behavior of a swarm such as bird flocking or fish schooling [6,2]. Since the introduction of the PSO algorithm in 1995 by Kennedy and Eberhart [6], there have been tremendous efforts to improve the performance of PSO. This effort materialized through studying a number of parameters such as the constriction coefficient [1], the inertia weight [10], adjusting the bound of the velocity [3]; and through the use of cluster analysis [5]. The PSO algorithm has also been used successfully to solve problems in dynamic [9], multi-objective environments [14,8].

In this paper, we modify the constriction coefficient, where it is varied during the optimization process. We call this new PSO version APSO, for *"Adaptive"* PSO. By decaying this coefficient value over time, we can reduce the fluctuation of the particle velocity. Our approach was validated by conducting a series of experiments on 25 standard test problems. The test problems are associated with different difficulties that usually exist in practical optimization problems such as multi-modality, epistasis (rotation), and noise. The proposed scheme is compared against other versions of the PSO algorithm.

The remainder of the paper is organized as follows: an overview of PSO and related work is presented in Section 2. The proposed methodology is introduced in Section 3. The test problems along with a discussion of problem difficulties and experimental studies are presented in Section 4. The last section is devoted to the conclusion.

M. Randall, H.A. Abbass, and J. Wiles (Eds.): ACAL 2007, LNAI 4828, pp. 333–344, 2007.
© Springer-Verlag Berlin Heidelberg 2007

2   Background

The particle swarm optimization algorithm is inspired by the natural phenomena of bird flocking. The principle of bird flocking is applied in optimization problems as follows. A population of solutions (particles/swarm) is generated. The movement of each particle is determined by its location and velocity. By changing the velocities over time, the particles are likely to move towards the global optima. Clearly, the behavior of a particle is dependant on how the associated velocity vector is defined. As originally stated in [6] (p. 1946): Formally, assume $v_i(t)$ and $x_i(t)$ are the velocity and position of a particle i at time t, we calculate the particle's velocity and position at time $t+1$ as follows:

$$v_i(t+1) = v_i(t) + c_1r_1(P_i(t) - x_i(t)) + c_2r_2(P_g(t) - x_i(t))$$
$$x_i(t+1) = x_i(t) + v_i(t+1) \tag{1}$$

in which P_i is the best position the particle has achieved (the local best position) and P_g is the global best position (could be the best for the swarm); coefficients c_1 and c_2 are used to compromise the search towards the best local knowledge of the particle and the best knowledge of the entire swarm; and r_1 and r_2 are random values between 0 and 1.

The strategy for determining the particle's velocity obviously has a strong effect on the optimization process. If $P_g(t)$ is the global best that the swarm has found so far, we get the *gbest* version of the PSO algorithm [2]. On the other hand, if $P_g(t)$ is defined separately for each particle as the best position found within the neighborhood of the local best position, we get the *lbest* version [4]. There are some discussions on the advantages/disadvantages of these versions [13] for controlling the balance between exploitation and exploration. However, in this paper, we consider the case of the *gbest* version only. There is no particular reason for selecting this option. From this point onwards, we use the term PSO for the *gbest* version.

There have been several attempts to control the exploitation and exploration abilities of the PSO algorithm. An obvious approach is to adjust c_1 and c_2 for the balance between the local and global directions; or to limit the range of velocity between $[-V_{max}, V_{max}]$ as follows (this is for the case of positive velocities; for negative velocities, it needs to be adjusted accordingly):

$$v_{t+1} = \begin{cases} v_{t+1}, & \text{if } v_{t+1} < V_{max} \\ V_{max}, & \text{Otherwise} \end{cases} \tag{2}$$

Recently, Kwok et al [7] investigated the performance of the PSO algorithm with different schemes for setting the control coefficients (c_1 and c_2) using scalars, uniform, and gaussian generated values. Their investigation showed a good performance of the PSO algorithm using randomly generated coefficients over the fixed-value ones.

In [10], Shi and Eberhart proposed the concept of inertia weight (w) to control exploration and exploitation (see Eq. 3). One of the interesting aspects of this work is that the inertia weight sets the influence level of the effect of the history on the current velocity. They found that if $w \geq 1$, the velocities will increase over time, the swarm will diverge and therefore it affects the exploitation of the algorithm.

$$v_i(t+1) = wv_i(t) + c_1 r_1 (P_i(t) - x_i(t)) + c_2 r_2 (P_g(t) - x_i(t)) \quad (3)$$

Meanwhile, Clerc and Kennedy [1] introduced the constriction coefficient to ensure the stable convergence of the PSO algorithm (see Eq. 4 and 5). In comparison to Eq. 3, we can see that, for a given χ, we will have $w = \chi$, the first constant is $c_1 \chi$, and the second one is $c_2 \chi$. Therefore, the effect of the constriction factor is somewhat similar to that of the inertia weight.

$$v_i(t+1) = \chi[v_i(t) + c_1 r_1 (P_i(t) - x_i(t)) + c_2 r_2 (P_g(t) - x_i(t))] \quad (4)$$

$$\chi = \frac{2}{|2 - \varphi - \sqrt{\varphi^2 - 4\varphi}|}, \quad \varphi = c_1 + c_2 \quad (5)$$

All the above work recommended fixed values for controlling the parameters; for example, Clerc and Kennedy [1] indicated that $\varphi = 4.1$, $\chi \approx 0.729$, and $c_1 = c_2 = 2.05$. However, it is obvious that when the algorithm converges, the fixed values of the parameters might cause the unnecessary fluctuation of particles. The question is whether we can allow some of parameters changed over time following either fixed or adaptive patterns?

Fan [3] introduced an approach to adapt V_{max} for the PSO algorithm in which the value of V_{max} is reduced over time as in Eq. 6. This approach does not directly change the value of the velocity; however, it will adjust the bound for the velocity over time.

$$V_{max}(t) = \left[1 - \left(\frac{t}{T}\right)^h\right] V_{max}(0) \quad (6)$$

where T is the maximum number of iterations, h is a positive constant which is chosen by "trial and error", and $V_{max}(0)$ also depends on test problems. In the paper, the author showed an excellent performance of this approach in comparison with the standard version of the PSO algorithm.

Another approach is to adapt the inertia weight over time. An example is referred to in Suganthan's work [11] where the author used linear decay of the inertia weight (Eq. 7)

$$w(t) = \left[w(0) - w(T)\right]\left(1 - \frac{t}{T}\right) + w(T) \quad (7)$$

where $w(0)$ is the predefined initial weight and $w(T)$ is the predefined weight at time T.

From the previous analysis, we can see that although χ is an important factor to control the convergence of PSO algorithm, there is no research on how to adapt χ. Therefore, it is the motivation for this paper to look at this issue. In the next section we will introduce a method to adapt χ and also discuss how it is different from previous work.

3 The Modified Strategy for the Constriction Factor

3.1 Time-Dependent Strategy

We first clarify that this paper is about adapting a parameter rather than self-adaptation of the parameter. The first refers to the process of allowing the parameter to change in response to changes in one or more of the search algorithm's states. Therefore, we can adapt a parameter by introducing a control parameter for it. However, self-adaptation normally refers to a process by which the change in the value of a parameter does not introduce additional control parameters; hence, the word "self".

We propose to use both the inertia weight and the constriction coefficient. We still keep a relatively small and fixed value of the inertia weight (0.729, as pointed out from Maurice Clerc's analysis and also discussed in [2]) and allow χ to be adapted ($\chi_0 = 1.0$) through decaying its value over time. The reason to employ the inertia weight is that, early in the search, the value of χ is around 1.0, the inertia weight will be responsible for keeping the algorithm exploring. As the value of χ gets smaller, χ will take over the role of the inertia weight.

Under our strategy, updating the velocity is performed as in Eq. 8 where χ acts as a function of the time-step.

$$v_i(t+1) = \chi(t+1)[w * v_i(t) + c_1 r_1(P_i(t) - x_i(t)) + c_2 r_2(P_g(t) - x_i(t))] \quad (8)$$

We use two different nonlinear decay rules to adapt χ. For the first one (Eq. 9), χ is annealed by a factor of $(\frac{t}{T})^h$ at each time step t (T is the maximum number of time steps). The parameter h is used to adjust the reduction rate of the velocity. For the second rule (Eq. 10), we add a small random effect to the first rule.

– **The first rule: APSO1**

$$\chi(t) = 1 - \left(\frac{t}{T}\right)^h \quad (9)$$

– **The second rule: APSO2**

$$\chi(t) = \left[1 - \left(\frac{t}{T}\right)^h\right] rand(0, 1) \quad (10)$$

Clearly, the main difference between our proposed strategy and the adaptation of V_{max} (we call it DPSO) is that we directly make changes to the velocities of the particles, instead of adapting the bound of the velocity as in DPSO. It is worthwhile to note that the method of using the constriction coefficient has a stable convergence towards the global optima, while this is not guaranteed in the case of using V_{max} (see [1]).

3.2   Non-linear Reduction of the Constriction Coefficient

From Eq. 8, we can see that the rate of reduction of the constriction coefficient has an important effect on the convergence of APSO. A fast reduction of χ might not be good for converging to the optima in the multi-modal problems where APSO can be trapped easily in local optima. However, if χ decays too slowly, it might affect the convergence speed of the algorithm. From Equations 9 and 10, it can be realized that the rate of reduction can be controlled by adjusting the parameter h. For unimodal problems, the effect of h is obvious. If h increases, it will slow down the reduction of χ, and therefore reduces the speed of convergence. However, note that real-world problems are rarely unimodal.

To investigate this matter, we carried out an experiment in which APSO1 was used to test a well-known multi-modal Rastrigin problem (the detailed description of this problem is given in Section 4), and with different h (each was tested with 25 different runs). The average error values (the difference between the obtained best solution and the optima) was recorded after 1e5 evaluations. These values are plotted in Figure 1.

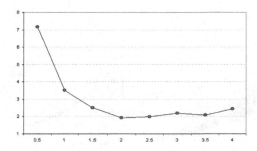

Fig. 1. The error values recorded after 1e5 for different h. The horizontal axis is for h, while the vertical one is for the error.

The figure shows an obvious tendency that if the value of h is too small, it will slow down the convergence of the algorithm, while the large value of h also deteriorates the performance of the algorithm. From Eq 9, we can easily explain this phenomena. Since $\frac{t}{T} \leq 1$, if we reduce h, the rate of reduction in the function of the constriction coefficient (Eq 9) will be faster. Therefore, in multi-modal problems, the side effect might be that the algorithm pre-maturely converges, and hence gets trapped at local optima. Meanwhile, if we increase h, the rate will be slower and the algorithm may over-emphasize explorations, therefore, losing the balance between exploration and exploitation.

3.3   Behavior of APSO

In this section, we will investigate the behavior of APSO (we use APSO1) by considering the change of the velocity over time. We also make a comparison with the method of adapting V_{max} (DPSO). For DPSO, all velocities of the particles are restricted by the bound of V_{max}. Since V_{max} decays over time, the velocities of the particles in the population will quickly become identical and approach zero. This causes a considerable reduction of the diversity of the population. Therefore, in multi-modal, rotated, or noisy problems, it will affect the performance of DPSO. Meanwhile, for our strategy, all particles are allowed a certain level of freedom since their velocities are adapted independently. Therefore, we expect the variance of the velocities in the population to be higher than that of DPSO.

To verify this issue, we tested both APSO1 and DPSO on the Rastrigin problem to measure the variance (or the standard deviation) of the velocities in the population over time. The parameter h was kept the same for both algorithms and set as recommended in [3]. In Figure 2, we plot the standard deviation of the velocities of the particles over time. For this problem, it has a 10-D search space, so that we will have 10 continuous lines on each figure where each line is for the standard deviation of the velocity in each dimension.

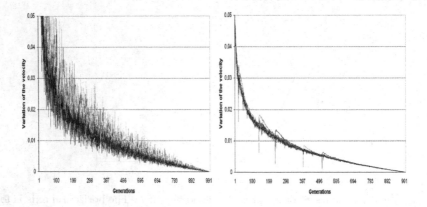

Fig. 2. The variation of the velocity in Rastrigin problem. The left graph is for APSO1, and the right one is for DPSO.

Form the figure, we can see clearly that the variance of the velocity for DPSO is quickly reduced as expected. Further, its reduction is quite smooth. Whereas, for APSO1, the reduction rate of the velocity is much slower and zigzagged. These results is consistent with our explanation.

4   Experimental Studies

4.1   Test Problems and Experiment Settings

In order to examine the performance of APSO, we used a number of standard problems such as Sphere, Griewank's, Schwefel's, Rosenbrock's, Rastrigin's, etc.

They cover both uni-modal and multi-modal functions. However, in our experiments, we want to add to these problems a wide range of difficulties that usually exist in practice:

- **Shifting:** Most test problems have an optima at the origin. For those algorithms that have a bias of approaching the origin, they will easily solve these problems. Therefore, we use shifting to examine the existence of such a bias.
- **Rotation:** Rotation adds a high level of linkage (interaction). This makes the problem become non-decomposable.
- **Noise:** Noise is unavoidable in all practical optimization problems (especially engineering design problems). It usually deceives the optimizers.
- **Narrow basin of the optima:** This causes an algorithm to spend more time to obtain information about the optima; especially in multi-modal problems. It increases the chance that the algorithm gets stuck in a local optima.
- **Problem composition:** This makes artificial test problems closer to the practical ones, which are usually black-box. With composition, the problem structure becomes more complicated or un-clear.

For this, we used a wide range of 25 test problems which were introduced at a special session of the IEEE Congress on Evolutionary Computation 2005 [12]. They all have 10 variables. All important features of the problems are listed in Table 1. However, for a detailed description, readers are referred to [12].

For each test on a particular problem, we run the algorithm 25 times with different seeds. For APSO, the parameter h was set to 2.0 as recommended in Section 3.2, while w, c_1, and c_2 followed the standard settings: 0.729, 2.05, 2.05

Table 1. Test problems and Descriptions: (F1 ÷ F5) are uni-modal, F6 ÷ F25 are multi-modal, F3 ÷ F25 are rotated, and F15 ÷ F25 are hybrid

Probs	Descriptions
F1	Shifted Sphere Function
F2	Shifted Schwefels Problem 1.2
F3	Shifted Rotated High Conditioned Elliptic Function
F4	Shifted Schwefels Problem 1.2 with Noise in Fitness
F5	Schwefels Problem 2.6 with Global Optimum on Bounds
F6	Shifted Rosenbrocks Function
F7	Shifted Rotated Griewanks Function without Bounds
F8	Shifted Rotated Ackleys Function with Global Optimum on Bounds
F9	Shifted Rastrigins Function
F10	Shifted Rotated Rastrigins Function
F11	Shifted Rotated Weierstrass Function
F12	Schwefels Problem 2.13
F13	Expanded Extended Griewanks plus Rosenbrocks Function
F14	Shifted Rotated Expanded Scaffers F6
F15	Hybrid Composition Function
F16	Rotated Hybrid Composition Function
F17	Rotated Hybrid Composition Function with Noise in Fitness
F18	Rotated Hybrid Composition Function
F19	Rotated Hybrid Composition Function with a Narrow Basin for the Global Optimum
F20	Rotated Hybrid Composition Function with the Global Optimum on the Bounds
F21	Rotated Hybrid Composition Function
F22	Rotated Hybrid Composition Function with High Condition Number Matrix
F23	Non-Continuous Rotated Hybrid Composition Function
F24	Rotated Hybrid Composition Function
F25	Rotated Hybrid Composition Function without Bounds

respectively. We also tested the performance of the PSO algorithm using the inertia weight and with the above settings, called IPSO as well as DPSO - the modified version using the technique of adapting V_{max} with $V_{max}(0) = 0.2$ and $h = 0.05$ (as recommended by its author). These are also the best settings we found for DPSO after a series of experiments. In all cases, the population size was 100. The criteria for comparison include error rate after 1e5 evaluations, and the speed of convergence.

4.2   Performance Analysis

We start our analysis on the performance of APSO by recording the error value (the difference between the obtained objective value and the optima's) after 1e5 evaluations. The mean and standard deviation values of the 25 different runs are reported in Table 2 for all problems.

Table 2. Error values (mean and standard deviation) that the different algorithms achieved after 1e5 evaluations for all problems: all values in bold are the best ones among the different algorithms. We use † to indicate the difference between the results of the algorithm and APOS1 is statistically significant (using t-test with the significance level of 0.05).

	APSO1	APSO2	IPSO	DPSO
F1	**0.000+0.000**	**0.000+0.000**	29.402+43.209†	**0.000+0.000**
F2	**0.000+0.000**	9.365+18.933†	333.338+159.758†	**0.000+0.000**
F3	**201517.228** +141658.655	520624.004 +350540.389†	1746681.800 +1081941.056†	234706.324 +144922.625
F4	2.243+6.201	802.939+1214.050†	529.214+273.340†	**0.000+0.000**
F5	1129.207+949.066	312.626+591.144†	4677.125+879.774†	**0.006+0.002†**
F6	74.916+99.698	83.515+167.631	35447.754+78082.281†	**27.827+73.101**
F7	**0.230+0.138**	4.344+3.810†	0.656+0.303†	1.919+0.730†
F8	20.302+0.051	20.303+0.070	20.362+0.041†	**20.183+0.077†**
F9	2.667+1.268	12.934+4.989†	7.045+3.955†	**0.637+0.938†**
F10	**14.407+5.872**	22.207+9.337†	35.593+10.700†	22.848+9.273†
F11	4.741+1.363	5.777+1.762†	6.841+2.193†	**2.006+1.262†**
F12	**94.029+201.318**	3040.124+3556.954†	1218.953+1162.912†	373.896+583.386†
F13	**0.643+0.229**	1.099+0.571†	1.333+0.545†	0.719+0.202
F14	2.926+0.345	3.485+0.437†	3.378+0.301†	3.139+0.436
F15	**119.126+120.629**	324.316+145.824†	182.053+70.153†	240.365+192.898†
F16	**135.310+28.141**	144.635+19.650	179.893+32.797†	151.012+28.133
F17	**130.347+18.753**	143.213+27.281	207.949+26.806†	160.943+40.726†
F18	**808.528+179.442**	917.157+122.130†	911.298+104.314†	883.394+203.050†
F19	**803.662+178.977**	917.730+120.888†	918.716+111.619†	898.986+200.665†
F20	**784.674+188.930**	917.738+120.894†	908.786+108.523†	897.091+199.256†
F21	633.026+279.892	978.696+234.801†	**613.081+285.171**	906.711+291.857†
F22	**765.011+105.418**	812.280+46.952†	823.703+37.271†	797.034+47.880
F23	769.328+210.998	1036.789+266.986†	**768.038+209.080**	877.885+265.022
F24	**236.000+99.216**	438.868+368.698†	327.705+125.475†	418.217+348.385†
F25	**406.635+19.698**	1029.979+239.062†	411.878+19.126†	454.104+186.589

For the two simple uni-modal problems (F1 and F2), we expect both APSO1 and APSO2 to converge quickly to the optima. However, APSO2 converged only in the case of F1. To explain why APSO2 did not converge in F2 after 1e5 evaluations, recall that for APSO2, we multiplied the component of $rand(0,1)$ by the value of χ (that is also between 0 and 1). The multiplication of these two components resulted in a faster reduction of the velocity. This resulted in a premature convergence for APSO2 and therefore a faster convergence in comparison with APSO1. We will look at this matter later on in the section. Note that, the difficulty of shifting in these experiments does not have much effect on the performance of both versions of APSO.

The difference in performance between the APSO1 and APSO2 is obvious in harder problems starting with F3 to F25. In almost all problems, multi-modality, rotation, or noise caused both APSO1 and APSO2 more difficulties. However, APSO2 was inferior to APSO1. The difference between the obtained results from 25 problems are significant (except for F1, F16 and F17). This implies that the use of a random factor is not useful for our strategy.

We now shift our attention to the performance of APSO (especially APSO1) in comparison with other versions of the PSO algorithm: IPSO and DPSO. It is clear that IPSO (the standard one) had the worst performance amongst the contestants. It did not reach the optima in all 25 problems, even for F1, a shifted unimodal problem. The t-test also supports the findings where the results achieved by IPSO are almost significantly worse than that of APSO1. However, IPSO achieved better performance on F21 and F23. These differences are not statistically significant. These results indicate that the use of fixed coefficients does not give the PSO algorithm the best performance in comparison to parameter-variable strategies.

In comparison to DPSO (adapting V_{max}), APSO1 had smaller error values in most problems. As shown in Table 2, DPSO is significantly better than APSO1 only in four (out of 25) problems including: F5, F8, F9, F11. Especially, DPSO performed badly in almost all rotated problems. The adaptation of the boundary causes the velocities of particles to quickly decay and become identical.

Interestingly, DPSO's performance was inferior to APSO1 in all composition problems (although there are some problems where the results are not significantly different). Recall that composition problems were used in order to represent a class of black-box problems with different difficulties.

The last consideration is given to the convergence rate of all test versions of PSO algorithms. We visualized the error values over time for some problems in Figures 3, 4. Once again, IPSO shows a very slow convergence in comparison with the others; even after 1e5 evaluations, it still did not converge for all problems.

In all cases, DPSO and APSO2 converged quickly at early generations. However, they made very little improvement in later ones. This might be the cause for the inferior performance shown in Table 2. For DPSO, the stagnation is perhaps because of the quick reduction of V_{max}; so that in special situations such as rotation or composition, it did not have enough diversity to recover from the effect of difficulties.

In general, the results indicated that under the proposed strategies, the PSO algorithm becomes much more reliable and effective in searching for optima. Furthermore, the adaption of the constriction coefficient showed a better capability of the PSO algorithm in solving practical black-box problems than its counterpart of adapting V_{max}.

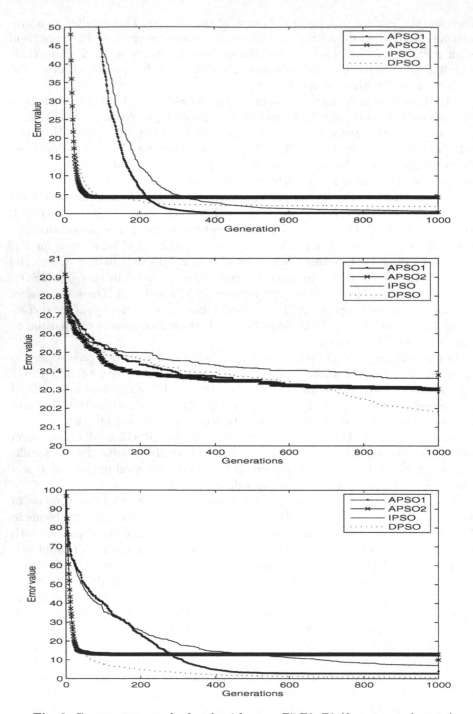

Fig. 3. Convergence graphs for algorithms on F7,F8, F9 (from top to bottom)

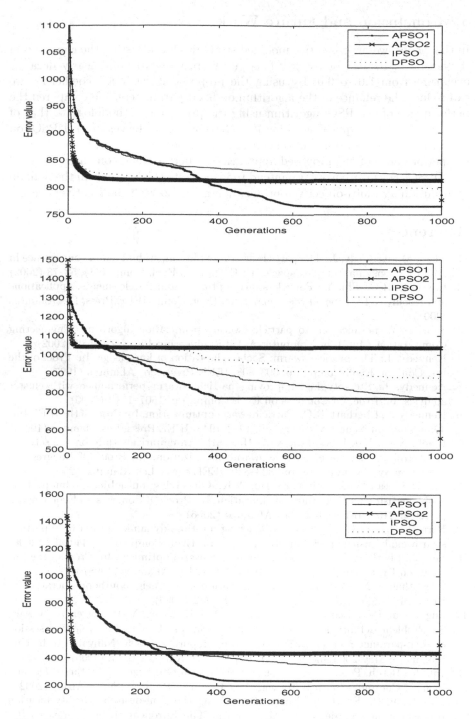

Fig. 4. Convergence graphs for algorithms on F22, F23, F24 (from top to bottom)

5   Conclusion and Future Work

In this paper, we proposed two modified strategies for adjusting the constriction coefficient χ. Instead of using a fixed χ, we propose to non-linearly decay χ over time from 1.0 to 0.0. By using the proposed constriction coefficient, we can reduce the reliance of the algorithm on fixed parameters. We compared the performance of the PSO algorithm using the proposed methodology to that of IPSO and the technique of adapting V_{max} (the bound of the velocity) (DPSO) on 25 test problems with different difficulties. The experimental results showed that the performance of the proposed approach was better when compared to IPSO and DPSO. For future work, we intend to investigate this strategy in the domain of evolutionary multi-objective optimization and also with the *lbest* version.

References

1. Clerc, M., Kennedy, J.: The particle swarm-explosion, stability, and convergence in a multidimensional complex space. IEEE Tran. on Evol. Comp. 6(1), 58–73 (2002)
2. Eberhart, R.C., Shi, Y.: Particle swarm optimization: developments, applications and resources. In: Proc. of the Congress on Evol. Comp. IEEE Press, Los Alamitos (2001)
3. Fan, H.: A modification to particle swarm optimization algorithm. Engineering Computations: Int J for Computer-Aided Engineering 19(8), 970–989 (2002)
4. Kennedy, J.: The particle swarm: Social adaptation of knowledge. In: Proc. of the Int. Conf. on Evol. Comp., pp. 303–308. IEEE Press, Los Alamitos (1997)
5. Kennedy, J.: Stereotyping: improving particle swarm performance with cluster analysis. In: Proc. Of Congress on Evol. Comp., pp. 1507–1512 (2000)
6. Kennedy, J., Eberhart, R.C.: Particle swarm optimization. In: Proc. of the IEEE Int. Joint Conf. on Neural Networks, pp. 1942–1948. IEEE Press, Los Alamitos (1995)
7. Kwok, N.M., Liu, D.K., Tan, K.C., Ha, Q.P.: An empirical study on the settings of control coefficients in particle swarm optimization. In: Proc. Of Congress on Evolutionary Computation, pp. 823–830. IEEE press, Los Alamitos (2006)
8. Liu, D.S., Tan, K.C., Goh, C.K., Ho, W.K.: On solving multiobjective bin packing problems using particle swarm optimization. In: Proc. Of Congress on Evol. Comp., pp. 2095–2102. IEEE press, Los Alamitos (2006)
9. Parrott, D., Li, X.: Locating and tracking multiple dynamic optima by a particle swarm model using speciation. IEEE Tran. on Evol. Comp. 10(4), 440–458 (2006)
10. Shi, Y., Eberhart, R.C.: A modified particle swarm optimizer. In: Proc. of the Int. Conf. on Evol. Comp., pp. 69–73. IEEE Press, Los Alamitos (1998)
11. Suganthan, P.N.: Particle swarm optimisation with a neighbourhood operator. In: Proc. Of Cong. on Evol. Comp., pp. 1958–1962 (1999)
12. Suganthan, P.N., Liang, J.J., Hansen, N., Deb, K., Chen, Y.-P., Auger, A., Tiwari, S.: Problem definitions and evaluation criteria for the CEC 2005 special session on real-parameter optimization. Technical Report 2005005, Nanyang Tech. Uni. Singapore, May 2005 and KanGAL Report, IIT Kanpur, India (2005)
13. van den Bergh, F.: An Analysis of Particle Swarm Optimizers. Phd thesis, Department of Computer Science, University of Pretoria, Pretoria, South Africa (2002)
14. Yapicioglu, H., Smith, A., Dozier, G.: Solving the semi-desirable facility location problem using bi-objective particle swarm. The European J. of Operations Research 177(2), 733–749 (2007)

A Differential Evolution Variant of NSGA II for Real World Multiobjective Optimization

Chung Kwan, Fan Yang, and Che Chang

Electrical and Computer Engineering, National University of Singapore

Abstract. This paper proposes the replacement of mutation and crossover operators of the NSGA II with a variant of differential evolution (DE). The resulting algorithm, termed NSGAII-DE, is tested on three test problems, and shown to be comparable to NSGA II. The algorithm is subsequently applied to two real world problems: (i.) a mass rapid transit scheduling problem and (ii.) the optimization of inspection frequencies for power substations. For both the real world problems, NSGAII-DE is found to have generated better results based on comparative studies.

Keywords: differential evolution (DE), NSGA II, real world problems.

1 Introduction

The non-dominated sorting genetic algorithm II (NSGA II) was developed by Deb et al. [1] to improve the NSGA [2]. The main attraction of NSGA II is its fast non-dominated sorting algorithm that is computationally more efficient than most available non-dominated sorting techniques. In addition, a crowding distance assignment algorithm without a need for a niching parameter for maintaining the diversity among pareto-optimal solutions adds to the attraction of this algorithm. NSGA II is currently one of the most popular multiobjective evolutionary algorithms (MOEAs), with comparisons made with other well known MOEAs like SPEA [3] and PAES [4], achieving competitive superior results for many test problems.

To further improve the algorithm, this paper proposes to replace the crossover and mutation operators of the original NSGA II algorithm using a variant of differential evolution DE [5].Termed NSGAII-DE, the algorithm is tested with three selected test problems. They are based on Schaffer (SCH) [6], Kursawe (KUR) [7], and Zitzler's test problems (ZDT6) [8]. To demonstrate the ability of the proposed algorithm in real world problems, the algorithm is tested on a mass rapid transit scheduling problem improvised from [9] and the optimization of inspection frequencies for substations modeled in [10].

2 Development of Proposed Algorithm

2.1 Background and Related Works

Kwan and Chang proposed a DE-based heuristic in [9] for solution to a simplified train scheduling problem. For that problem, the variant for generating mutant vector V for the $(G+1)^{th}$ generation of the ith candidate solution X_i is:

M. Randall, H.A. Abbass, and J. Wiles (Eds.): ACAL 2007, LNAI 4828, pp. 345–356, 2007.
© Springer-Verlag Berlin Heidelberg 2007

$$V_{i,G+1} = \lambda(X_{i,G} - X_{r1,G}) + F(X_{r2,G} - X_{r3,G}) \tag{1}$$

$$r1 \neq r2 \neq r3 \neq i$$

where $r1$, $r2$, $r3 \in [1, N]$ are randomly selected solutions from the population of size N. F and λ are amplification factors for the bracketed terms. Motivated by initial success, attempts are made to further apply the proposed algorithm to other types of problems. However, subsequent tests revealed that this variant does not converge fast enough for many applications. This motivates us to explore a better variant to suit a wider range of real world applications.

Recently, multi-objective DE based techniques are also reported in works like [11] and [12]. Notably, a similar attempt to replace the crossover and mutation rates by a rotationally invariant DE variant was noted in [13], where the authors reported better performance of their NSDE than the NSGA II for a class of rotated problems. These further support the notion that the common mutation and crossover operators may not be effective in handling certain problems.

2.2 Proposed Variant

The proposed variant presented in equation (2) replaces the term $X_{r1,G}$ in equation (1) with $X_{rBest,G}$ and omit the $X_{i,G}$ term. Thus, instead of generating each mutant vector with respect to $X_{i,G}$, $V_{i,G+1}$ is generated with respect to the term $X_{rBest,G}$.

$$V_{i,G+1} = \lambda X_{rBest,G} + F(X_{r2,G} - X_{r3,G}) \tag{2}$$

$$r2 \neq r3$$

where $r2$ and $r3 \in [1, N]$ are randomly chosen solutions from the population of size N. In the single objective case, $X_{rBest,G}$ is merely the best solution for the Gth generation. In the multiobjective case, however, the notion of 'best' is no longer a single optimum term, but is chosen from a set of non-dominated or Pareto-optimal solutions. In this work, $X_{rBest,G}$ is randomly chosen from the top 30% of set of solutions of the Gth generation. The NSGA II algorithm ranks each candidate solution from '1' onwards depending on how many solutions it is dominated by. Solutions with ranking of '1' denote the non-dominated solutions in the current population. The higher the rank value of a solution, the more it is dominated by other solutions. Besides that, the crowding-distance-assignment assigns each solution based on density estimation, with a higher value representing lesser crowding of other solutions around its vicinity. Solutions at the boundary points are assigned ∞. $rBest$ is selected first based on the non-dominated ranking followed by the crowding-distance-assignment. E.g. if more than 30% of solutions are ranked '1', a crowding-distance-assignment of higher value (with the same non-dominated ranking) would be preferred to those of lower values in consideration of whether a solution should be included as one of the possible choices in the set where $X_{rBest,G}$ is chosen from.

The 'crossover' operator of DE generates a trial vector $U_{i,G+1}$ as presented in equation (3):

$$U_{ji,G+1} = \begin{cases} V_{ji,G+1}, & \text{if } r(j) \leq CR \text{ or } j = rn(i) \\ X_{ji,G}, & \text{if } r(j) > CR \text{ or } j \neq rn(i) \end{cases} \tag{3}$$

where *CR* is the crossover rate. *j* denotes the jth decision variable of the ith candidate solution. *r(j)* is a randomly chosen number in [0,1]. Thus, if the randomly generated *r(j)* of the jth decision variable is smaller or equal to the crossover rate *CR*, $U_{ji,G+1}$ will take the value of the mutated vector $V_{ji,G+1}$, else it will take the value of the original candidate solution $X_{ji,G}$. In addition, to prevent degeneration, the term *rn(i)* is a randomly chosen decision variable *j* in the ith candidate solution which will be chosen to be replaced by the mutant vector.

Unlike in the single objective version of DE proposed in [5], $U_{i,G+1}$ does not replace the current $X_{i,G}$ if it is better, but it is treated as a member of the child population candidate as provided in the NSGA II structure, and is involved in non-dominated ranking and crowding distance assignment.

The proposed variant is tested in this paper using theoretical test problems first and subsequently on two real world problems.

3 Comparison with Test Problems

In order to test the effectiveness of our proposed algorithm, we test the algorithm with the three test problems SCH, KUR and ZDT6 detailed in Table 1. All approaches are run for a maximum of 25000 function evaluations (translated to 250 generations with population size of 100).

The real-coded version of the NSGA II is used, with associated parameters, crossover and mutation distribution index (η_c and η_m), both set to 20 as recommended. The NSGAII-DE has crossover settings *CR* and *F* set to 0.8 and λ set to 1. Codes are implemented in Matlab. Ten runs are performed for each algorithm.

The three test problems are selected because of the following reasons – SCH is selected to test the convergence for a large range of variable(s). KUR is selected to test convergence to a nonconvex and discontinuous objective space, and ZDT6 is used to test convergence to a nonconvex and nonuniformly spaced objective space. Such features are common in the real-world problems, which we will be applying the algorithm to ultimately. Pareto fronts for all the cases are presented in Fig.1. In Fig.1(a) and (b), the two Pareto fronts from NSGA II and NSGA II-DE coincide with each other. Whereas in Fig.1(c), it is obvious that NSGA II-DE performs better than NSGA II.

In addition, two performance measures proposed in [1] correspond to two goals in a multiobjective optimization, i.e. convergence and diversity: The convergence metric γ measures the extent of convergence of the solutions generated by the algorithms to a known set of pareto-optimal solutions by computing the Euclidean distance between the solutions. The second diversity metric Δ (see equation (1) of [1]) provides a measure comparing the Euclidean distance between each pareto-optimal solution against all the others. Table 2 shows the mean (first rows) and variance (second rows) of the convergence metric γ. A better convergence is obtained by the NSGA II-DE for both KUR and ZDT6, but it performs slightly worse for SCH. Table 3 shows the mean and variance of the diversity metric Δ. Better diversity measures are noted for NSGA II-DE in all three test problems.

With this initial testing on theoretical problems, we established the comparable performance of the DE variant with the original NSGA II, and proceed to applying the algorithms on the two real-world problems.

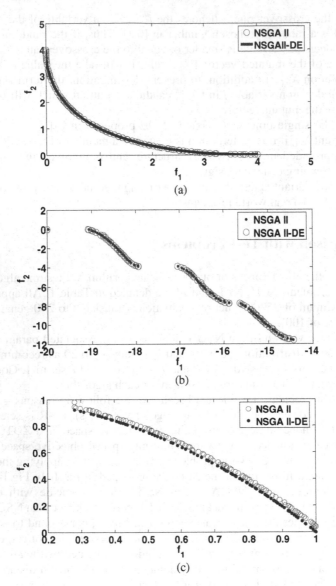

Fig. 1. Pareto fronts of test problems (a). SCH, (b). KUR, and (c). ZDT6

4   Mass Rapid Transit Schedule Optimization

4.1   Background and Formulation

Waiting time and traveling time are important service quality objectives considered in the optimal generation of a mass rapid transit schedule. In this paper, each service

Table 1. Theoretical test problems used in this study

Prob-lems	n	Variable Bounds	Objective functions	Optimall solutions	Com-ments
SCH	1	$[-10^3,10^3]$	$f_1(x) = x^2$, $f_2(x) = (x-2)^2$	$x \in [0,2]$	convex
KUR	3	$[-5,5]$	$f_1(x) = \sum_{i=1}^{n-1}(-10\exp(-0.2\sqrt{x_i^2 + x_{i+1}^2}))$ $f_2(x) = \sum_{i=1}^{n}(\mid x_i \mid^{0.8} +5\sin x_i^3)$	(refer to [16])	Non-convex
ZDT6	10	$[0,1]$	$f_1(x) = 1 - \exp(-4x_1)\sin^6(6\pi x_1)$ $f_2(x) = g(x)[1 - (f_1(x)/g(x))^2]$ $g(x) = 1 + 9[(\sum_{i=2}^{n} x_i)/(n-1)]^{0.25}$	$x_1 \in [0,1]$, $x_i=0$, $i=2, \cdots n$	Non-convex Non-uniformly spaced

Table 2. Mean (first row) and variance (second row) of the convergence metric γ

Algorithm	SCH	KUR	ZDT6
NSGA II	0.003098	0.025761	0.86599
	0	0.000564	0
NSGA II-DE	0.003262	0.015507	0.86425
	0	0.000005	0.000028

Table 3. Mean (first row) and variance (second row) of the diversity metric Δ

Algorithm	SCH	KUR	ZDT6
NSGA II	0.47102	0.53958	0.77723
	0.004493	0.006099	0.15121
NSGA II-DE	0.4554	0.40955	0.58166
	0.000527	0.000464	0.07022

quality is investigated with respect to the economic objective of operating cost. The work is extended from [9] by increasing the number of variables from 7 to 108 taking into account for instantaneous variables relating to dwell time at each passenger station and run time/ coast level profiles between all track sections. Previously, all the service qualities are combined into a weighted sum that is minimized against the operating cost. This paper investigates the effect that each of the service-related attribute has on operating cost instead.

An intuitive way to incorporate the objective(s) is to minimize the absolute waiting time and traveling time of the whole system. However, [14] has pointed out that the 'expectation' concept of passengers is a more accurate reflection of passengers' attitude towards waiting time and traveling time than the objective(s) above. Adopting that idea, a waiting time dissatisfaction index ($DI_{WT}(t)$) was derived in equation (4) and an actual to shortest journey time ratio (ASJR) defined in equation (5).

$$DI_{WT}(t) = \begin{cases} 0 & \text{for } t \leq E_{WT} \\ m_1 t + C_1 & \text{for } E_{WT} < t \leq T_{first} \\ \alpha(t)^2 & \text{for } t > T_{first} \text{ and } t > E_{WT} \end{cases} \qquad (4)$$

$DI_{WT}(t)$ denotes the average waiting time dissatisfaction incurred by passengers on a particular passenger platform, where E_{WT} denotes the amount of time that a group of passengers are willing to wait before any 'dissatisfaction' is incurred, taken as 60 seconds in this paper. T_{first} is the time interval to the arrival of the 1st train. A linear dissatisfaction measure represents the dissatisfaction of passengers up to the point where the first train arrives (under a periodic planning timetable without real time operational adjustment, this is taken as the dispatch interval). If passengers cannot get onto the first train that arrives due to congestion, the waiting time dissatisfaction index will increase in a quadratic manner (implied by the $\alpha(t)^2$ term).

To reflect the total traveling time 'dissatisfaction', we propose an actual to shortest journey time ratio. The shortest journey time is calculated based on the summation of minimum run times and the minimum dwell time allowable at each station respectively. The *ASJR* definition is:

$$ASJR = \frac{\text{Actual Total Travel Time}}{\text{Shortest Total Travel Time}} \qquad (5)$$

where the shortest total travel time can be deemed the 'expected' total traveling time of the passengers.

Under certain operating range, each of the two service qualities is found to be contradictory with the operating cost. Shortening the run time between each station to shorten the traveling time, for example, leads to an increase in electrical energy consumption [15], which in turn increases the operating cost. Reducing the waiting time by increasing the dispatch intervals increases the number of trains in the system and increases the operating cost as well. The aim of applying NSGAII-DE is to discover the range where each service quality is found to have a conflicting relationship with operating cost. This will facilitate the decision making process for implementation considerations.

Detailed definition of operating cost as well as the passenger flow model equations can be found in [9]. Note that the cost coefficients have to be changed to protect the privacy of the study system and the cost values should by no means be interpreted to represent the true values.

4.2 Decision Variables, Constraints and Other Settings

The decision variables to be optimized are dispatch frequencies, dwell times (stopping times) at each station as well as coast levels between each section track relating running time to the electrical energy consumption. Some of the constants relating to maximum train capacity and allowable passenger build-up on each station are summarized in Table 4. Besides the constant parameters, constraints on the bounds on each decision variable and the safety distances allowable between trains apply. Interested readers are encouraged to read [9] for a more detailed description of the problem.

The optimization parameters are presented in Table 5. The population size of 100 and the maximum iterations are determined after trials are conducted. The mutation, crossover distribution index and other parameters are unchanged as in the test case studies.

Table 4. Constant parameters used in the scheduling model

Parameters	Values
Expected waiting time E_{WT}	60 seconds
Maximum train capacity T_{MAX}	1500 passengers
Shortest possible total traveling time (dir1)	3163 seconds
Shortest possible total traveling time (dir2)	3154 seconds
Maximum allowable passenger build-up on each station (P_{MAX})	1000 passengers

Table 5. Optimization parameters

Algorithm	η_m	η_c	F	λ	Coding	Population Size	Max Iteration
NSGA II	20	20	-	-	Real	100	2500
NSGA II-DE	20	20	0.8	1	Real	100	2500

4.3 Pareto-Fronts Generated

Multiple runs are conducted for each algorithm to determine its consistency. The average and best results for each algorithm are presented. The average results are plotted to provide insight in terms of the consistency of the Pareto-optimal solutions.

Two sets of Pareto-fronts generated in Figs.2 and 3 demonstrate the superiority of the proposed NSGA II-DE technique over NSGA II. NSGA II-DE is seen to clearly dominate NSGA II solutions for both situations. Moreover, slight deviations were noted for the average results and best results for NSGA II-DE as compared to NSGA II for both Figs.2 and 3.

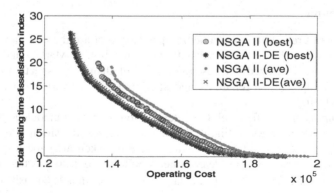

Fig. 2. Pareto plots of operating cost vs. total waiting time dissatisfaction

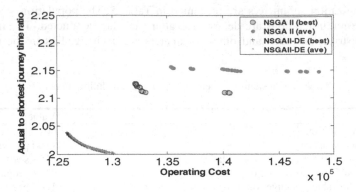

Fig. 3. Pareto plots of operating cost vs. actual to shortest journey time ratio

4.4 Summary of Results and Physical Implications

The two figures display the 'trade-off' relationships between each service quality and the corresponding range of operating cost. Operating cost trading off with the total waiting dissatisfaction index range from approximately \$129000 to \$190000, while that for actual to shortest journey time ratio is a much shorter range from \$126000 to \$130000. For the purpose of illustration, this paper has optimized each service quality with operating cost as two objective problems. The establishment of NSGAII-DE as the most appropriate algorithm in this work allows us to extend it to more objectives in the future.

Furthermore, the identification of the trade-off regions for each service quality with respect to operating costs provides greater insight into the problem itself for discovering domain knowledge which can be subsequently employed in more objectives to shorten the computational time.

5 Inspection Frequencies Optimization for Substations

5.1 Background

Improving overall reliability and reducing operating cost are the two most important but often conflicting objectives for substation optimization. A Markov and system reliability model are developed to assess the impact of changing inspection frequencies of individual component on reliability and operating cost for various substation configurations in [10].

At the component-specific level, the deterioration process is modeled with a three-state Markov process shown in Fig. 4. As time progresses, transitions from one state to the next are made. Should proper maintenance be taken after the inspection, the component can be restored from deteriorated condition back to a better one. In this model, $\lambda_{i,i+1}$ is the transition rate from state i to i+1, $\lambda_{i,f}$ denotes transition rate from state i to failure state, and $\mu_{i,j}$ represents the transition rate from state i back to j (j<i).

Generally speaking, with more frequent inspections and subsequent maintenance actions carried out, a component can be restored faster from a more deteriorated state to a better state.

For component n, the expected maintenance cost, $EC_{m,n}$, and the expected repair cost, $EC_{r,n}$, are calculated using the equations (6) and (7):

$$EC_{m,n} = \sum_{i=1}^{N} (EquivalentI_i \times \sum_{k=1}^{3} EquivalentC_{mk}) , \qquad (6)$$

$$EC_{r,n} = C_r \times Failure\,Frequency \qquad (7)$$

where $EqivalentI_i$ is a product of the state probability and inspection frequency in that state, representing the equivalent inspection frequency in state i, and $EqivalentC_{mk}$ is the equivalent cost of maintenance type k (k=1, 2, 3). C_r is the failure cost each time.

At the system-specific level, the adopted system reliability model was developed to assess the composite reliability of power generation and distribution [16], which views substation configurations as being connected in series or parallel or a combination of both.

The overall cost containing the capital and operating costs and the Loss of Expected Energy (LOEE) are the two criteria to evaluate the performance of substation configurations.

The overall cost (C) in one substation can be easily calculated by:

$$C = \sum_{n=1}^{M} (EC_{m,n} + EC_{r,n}) + CapC \times Rate \qquad (8)$$

where CapC is the capital cost, and Rate is the interest and depression rate. M is the number of components.

LOEE is the reliability objective which measures the reliability worth associated with the cost of the customers due to the failure, which is expressed as:

$$LOEE = \sum_{p=1}^{m} Pf_p \times L_p \times Du_p \qquad (9)$$

where m is the number of load points in one substation, Pf_p is the probability of failure at load point p. L_p is the loss of load (MW) due to the failure at load point p, and Du_p is the duration of failure at load point p.

5.2 Case Studies

Two basic substation configurations analyzed in this paper are shown in Fig. 5. The capital cost is calculated based on the typical data about the length of bus, number of breakers, transformers and other system equipment. Other parameters are set as: Rate = 12%, N = 3. The optimization parameters are laid out in Table 6.

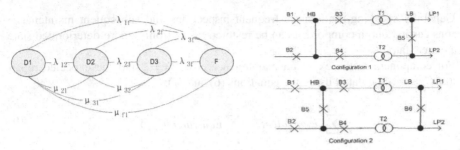

Fig. 4. Three state Markov model   **Fig. 5.** Typical substation configurations

Table 6. Optimization parameters (c1—configuraiton 1; c2—congfiguration 2)

Algorithm	η_m	η_c	F	λ	Coding	Population Size	Max Iteration
NSGA II	20	20	-	-	Real	110(c1) 120(c2)	65 (c1) 85(c2)
NSGA IIDE	20	20	0.8	1	Real	110(c1) 120(c2)	65 (c1) 85(c2)

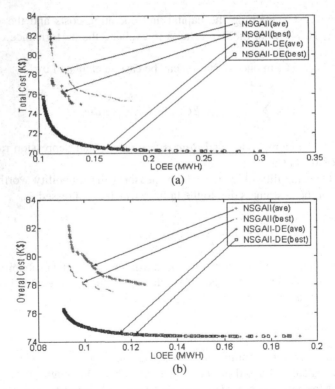

Fig. 6. Pareto plots of LOEE vs. overall cost for (a) configuration 1, and (2) configuration 2

5.3 Optimization Results

The minimization of overall cost (equation (9)) against *LOEE* (equation (10)) is presented for the two configurations depicted in Fig 6. The same model with four configurations has been studied in [10]. The two sets of Pareto-fronts generated in Fig.6 demonstrate the superiority of the proposed NSGA II-DE technique over NSGA II. NSGA II-DE is seen to clearly dominate NSGA II solutions for both situations.

6 Conclusion

This paper has identified a differential evolution (DE) variant of the NSGA II that is well suited to real world applications. The algorithm, termed NSGAII-DE, was tested against three test problems and found to outperform NSGA II in terms of both convergence for KUR and ZDT6 and diversity for all three test problems. The extension of the problem to two real world problems, an optimal train scheduling problem and a substation optimization problem, demonstrate the effectiveness of this algorithm when compared with NSGA II.

References

1. Deb, K., Agrawal, S., Pratab, A., Meyarivan, T.: A fast elitist non-dominated sorting genetic algorithm for multi-objective optimization: NSGA-II. IEEE Transactions on Evolutionary Computation 6, 182–197 (2002)
2. Srinivas, N., Deb, K.: Multiobjective function optimization using nondominated sorting genetic algorithms. Evol. Comput. 2, 221–248 (1995)
3. Zitzler, E., Thiele, L.: Multiobjective evolutionary algorithms: A comparative case study and the Strength Pareto Approach. IEEE Transactions on Evolutionary Computation 3, 257–271 (1999)
4. Knowles, J., Corne, D.: The Pareto archived evolution strategy: A new baseline algorithm for multiobjective optimization. In: Proceedings of the 1999 Congress of Evolutionary Computation, pp. 98–105. IEEE Press, Piscataway, NJ (1999)
5. Storn, R., Price, K.: Differential Evolution-A simple and efficient adaptive scheme for global optimization over continuous space. Technical Report TR-95-012, ICSI
6. Schaffer, J.D.: Multiple objective optimisation with vector evaluated genetic algorithms. In: Grefensttete, J.J. (ed.) Proceedings of the First International Conference on Genetic Algorithms, pp. 93–100. Lawrence Erlbaum, Hillsdale, NJ (1987)
7. Kursawe, F.: A variant of evolution strategies for vector optimization. In: Schwefel, H.-P., Männer, R. (eds.) Parallel Problem Solving from Nature - PPSN I. LNCS, vol. 496, pp. 193–197. Springer, Heidelberg (1991)
8. Zitzler, E.: Evolutionary algorithms for multiobjective optimization: Methods and applications. Doctoral dissertations ETH 13398, Swiss Federal Institute of Technology (ETH), Zurich, Switzerland (1999)
9. Kwan, C.M., Chang, C.S.: Application of Evolutionary Algorithm on a Transportation Scheduling Problem - The Mass Rapid Transit. In: IEEE Congress on Evolutionary Computation 2005, Edinburgh, vol. 2, pp. 987–994 (September 2005)

10. Chang, C.S., Yang, F.: Evolutionary Multi-objective optimization of Inspection Frequencies for Substation Condition-based Maintenance. In: Proceedings of 11th Naval Platform Technology Seminar, Singapore (accepted for publication, 2007)
11. Abbass, H.A., Sharker, R.: The Pareto Differential Evolution Algorithm. International Journal on Artificial Intelligence Tools 11, 531–552 (2002)
12. Xue, F.: Multi-objective Differential Evolution and its Application to Enterprise Planning. In: Proceedings of the 2003 IEEE International Conference on Robotics and Automation (ICRA 2003), vol. 3, pp. 3535–3541. IEEE Press, Los Alamitos (2003)
13. Iorio, Li, X.: Solving Rotated Multi-objective Optimization Problems Using Differential Evolution. In: Webb, G.I., Yu, X. (eds.) AI 2004. LNCS (LNAI), vol. 3339, pp. 861–872. Springer, Heidelberg (2004)
14. Murata, S., Goodman, C.J.: An optimal traffic regulation method for metro type railways based on passenger orientated traffic evaluation. In: Proceedings of COMPRAIL 1998, pp. 573–584 (1998)
15. Chang, C.S., Sim, S.S.: Optimizing train movements through coast control using genetic algorithms. In: IEE Proceedings of Electr. Power Appl. vol. 144 (1997)
16. Propst, J.E.: Calculating Electrical Risk and Reliability. IEEE Trans. Industry Applications 31, 1197–1205 (1995)

Investigating a Hybrid Metaheuristic for Job Shop Rescheduling

Salwani Abdullah[2], Uwe Aickelin[1], Edmund Burke[1], Aniza Mohamed Din[1], and Rong Qu[1]

[1] School of Computer Science
University of Nottingham
Nottingham, UK
{uxa,ekb,amd,rxq}@cs.nott.ac.uk
[2] Faculty of Information Science and Technology
Universiti Kebangsaan Malaysia
Bangi, Selangor, Malaysia
salwani@ftsm.ukm.my

Abstract. Previous research has shown that artificial immune systems can be used to produce robust schedules in a manufacturing environment. The main goal is to develop building blocks (antibodies) of partial schedules that can be used to construct backup solutions (antigens) when disturbances occur during production. The building blocks are created based upon underpinning ideas from artificial immune systems and evolved using a genetic algorithm (Phase I). Each partial schedule (antibody) is assigned a fitness value and the best partial schedules are selected to be converted into complete schedules (antigens). We further investigate whether simulated annealing and the great deluge algorithm can improve the results when hybridised with our artificial immune system (Phase II). We use ten fixed solutions as our target and measure how well we cover these specific scenarios.

Keywords: Artificial immune systems, simulated annealing, great deluge algorithm, job shop scheduling.

1 Introduction

Job shop scheduling problems are concerned with tackling the problem of assigning n jobs to m machines and are very well studied. The problem has been addressed using several local search techniques such as tabu search, genetic algorithms and simulated annealing as observed by Jain and Meeran in [20] who analysed some of the techniques used and made comparisons between them. In this paper, we are specifically trying to tackle the problem of changing job shop environments. Such changes include the unexpected arrival dates of jobs into the factory. If jobs arrive too early, it could lead to them being stored for long periods of time and if they arrive late, it could cause delays in processing other jobs [12,21]. An efficient method of rescheduling is needed to manage the problem.

M. Randall, H.A. Abbass, and J. Wiles (Eds.): ACAL 2007, LNAI 4828, pp. 357–368, 2007.
© Springer-Verlag Berlin Heidelberg 2007

2 Problem Description

Job shop schedules require constant revision as problems could happen during production and delays could cost money and time. For a detailed discussion of job shop problems, see [4] and [24]. Rescheduling is important to ensure the production can maintain its flow and minimize interruption. A quick solution to such a problem is usually very much preferred compared to starting from scratch. This observation represents the motivation for this paper. It addresses the goal of being able to generate a diverse range of partial schedules that could be used as a replacement in the event of changes in a job shop environment. These partial schedules should enable us to generate a new complete schedule in order to keep the manufacturing process flowing smoothly with a low level of interruption. In this paper, we will employ an artificial immune system algorithm to build these partial schedules. We use previous, complete schedules (later known as the antigen universe) to build a collection of partial schedules. This data stems from [18]: the number of jobs used is 15, assigned to five machines. We employ precedence constraints to the jobs when building the partial schedules. These partial schedules are then evolved using a genetic algorithm. These processes will be explained in Section 3.1. In Section 3.2, we hybridise the newly developed artificial immune system with local search to see whether there is improvement to the results.

3 A Hybrid Metaheuristic Model

Artificial immune systems (AIS) are motivated by immunology. The biological immune system defends the body from antigens. It generates antibodies that can attack specific antigen. An overview of artificial immune systems research can be seen in [2] and [7].

Previous research on AIS for scheduling has shown that an AIS model can be used in a job shop setting. Different scheduling problems have been addressed including the job shop scheduling problem [5,6,13], the hybrid flow shop scheduling problem [11] and the job shop rescheduling problem [16,17,18], which is the main concern of this paper. Hart and Ross [17], in their research, tackled this problem by building a block of partial schedules. There are many definitions given to the antibody and the antigen for the problem, which are used to build the partial schedules. We are employing the definition given by Hart and Ross in [17]. The key definitions used in this research are outlined below:

- An **antigen** is defined as *"the sequence of jobs on a particular machine given a particular scenario"* [17], which represents a full schedule for the problem. The antigens are represented by a sequence of numbers of length 15 for the problem tested here.
- An **antibody** is defined as *"a short sequence of jobs that is common to more than one schedule"* [17], which is also known as a partial schedule. The antibodies are represented by sequences of numbers of length 5, where the length of an antibody is less than the length of an antigen.

- An **antigen universe** is considered to be a collection of antigens (sequence of jobs) to be matched with the antibodies (partial schedules). An antigen universe has to be prepared before we can build an antibody population.
- An **antibody population** is a collection of partial schedules constructed from gene libraries.
- **Gene libraries** consist of genotypes [19,23]. The gene libraries in this research are constructed from all the antigens in the antigen universe.
- A **final population** consists of a collection of best antibodies. When we hybridise our AIS model with a local search method, the final population from our AIS model will be the initial solutions to the local search method.
- **Fitness** represents the value assigned to each antibody in the antibody population to evaluate the coverage of an antibody over the antigens. The higher the fitness, the better an antibody will be.

We have divided our work into two phases. In the first phase, an AIS model is used to generate the antibody population, with $l = 5$ where l is the length of an antibody. A genetic algorithm is then used to evolve the antibodies. The idea is that only the antibodies with the highest fitness (i.e. the best antibodies) that have most of the jobs matched with the antigens will be kept in the final population. It is important to note here that we used a genetic algorithm to evolve the antibody population as used in [17,18]. We modified the algorithm in [17,18] with the aim of improving the results. The simulated annealing and great deluge algorithms are then applied respectively in the second phase by using the best antibodies selected in the first phase (final population) as initial solutions. In this research, we are using the parameters adapted from [3]. The aim is to investigate if we can improve the fitness of the antibodies developed in the final population in Phase I as both local search methods have been known to produce good results for other scheduling problems such as examination timetabling (e.g. [3,8]).

3.1 Phase I: The Artificial Immune System Model

Before we generate the antibody populations, we need to have an antigen universe. The antigen universe for this research is the same as that used by Hart and Ross [18], which is based on a benchmark problem by Morton and Pentico [22]. The number of jobs used in this problem is 15 and the jobs have to be assigned to five machines. Hart and Ross created ten test scenarios by mutating the arrival dates of the jobs to a random date between $0 - 300$ with a probability of 0.2. The arrival dates must not be less than p_t days before the due date, where p_t is the processing time of the job. A genetic algorithm developed by Fang et al [12] is used to generate five schedules for each of these test-scenarios. This resulted in five sets of ten schedules; one for each machine, and these schedules became the antigen universe for this research. This research uses the antigen universe generated from one of the machines with the assumption that all machines have a similar pattern of jobs.

Generating the Antibody Population. The first step in this model is to generate an antibody population (a collection of antibodies) from gene libraries [6,17,18,26]. The gene libraries in this research are constructed from all the antigens in the antigen universe. The antigens are divided into five libraries, where each library consists of ten partial schedules of size 3 also known as components. An antibody for this research is constructed based on a modular design method [14,19,23,25] where the length of each antibody is 1/3 the length of each antigen. We are using a small size problem because we are interested in evolving the antibodies using the AIS algorithm and hybridising the model with a local search method to see if we can improve the results. In our future work, a larger size of problem will be used.

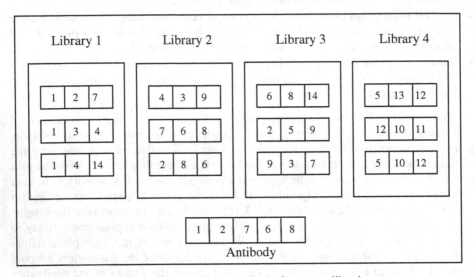

Fig. 1. Constructing an antibody from gene libraries

In the example in Figure 1, the gene libraries consist of four libraries and each library contains three components. Three jobs are allocated in each component. Following the modular design method, there are several ways to combine the components from gene libraries to produce an antibody. In Figure 1, we select the first component from Library 1 and combine it with the second component from Library 2 to produce an antibody. Since the length of an antibody is 5 jobs, a possible combination between components in Library 1and Library 2

$$C\binom{n_1+n_2}{r_1+r_2} = \frac{(n_1+n_2)!}{(n_1+n_2-r_1-r_2)!(r_1+r_2)!}. \tag{1}$$

can be constructed from this example, where n_1 and n_2 represent the number of jobs in the components from the first and second library, respectively, and r_1 and r_2 represent the number of jobs to be selected from the components ($r_1 + r_2 = 5$). In Figure 1, we can see a combination of three jobs from the first component and two jobs from the second component. Therefore, jobs 1, 2 and 7 from the first component in Library 1 are combined with jobs 6 and 8 selected from the second component in Library 2. We can get other combinations from these two components using (1) above to generate an antibody population. This process is repeated until all the components in Library 1 have been combined with all the components in Library 2 as well as all the other libraries.

We also have to ensure that there will be no duplicate jobs in the antibody. We compare each antibody generated in the population and eliminate the antibodies with duplicated jobs. The process will go on until a population of antibodies is generated. By doing this, we develop a level of antibody diversity.

We generate three types of antibody populations:

1. A population with antibody duplication (we can have several similar antibodies in one population) – Type A (4514 antibodies)
2. A population with no antibody duplication regardless of the source gene libraries (no similar antibodies in one population) – Type B (2416 antibodies)
3. A population with antibody duplication (only when the antibodies are constructed from different source libraries) – Type C (2839 antibodies)

We generated these versions to see whether having a large number of similar antibodies in one population would affect the coverage of the antigen universe by the antibody population. An initial antibody population of size 100 is selected randomly.

The Matching Function. A matching function is used as the evaluation function within the genetic algorithm to calculate the fitness of each antibody in the antibody population. A sample of antigens is first selected from the antigen universe. Each antibody is then matched against each of the antigens selected by aligning an antigen string with an antibody string and calculating a matching score.

Antigen	1	2	7	4	3	9	6	8	14	5	13	12	10	11	15	Match Score
	4	3	9	5	12											0
		4	3	9	5	12										0
			4	3	9	5	12									0
				4	**3**	**9**	5	12								15
Antibody					4	3	9	5	12							0
						4	3	9	5	12						0
							4	3	9	**5**	12					5
								4	3	9	5	**12**				5
									4	3	9	5	12			0
										4	3	9	5	12		0
											4	3	9	5	12	0

Fig. 2. The process of matching an antibody with an antigen by aligning the antibody at every possible alignment position

Based on the example in Figure 2, if there is an antigen string '1 2 7 4 3 9 6 8 14 5 13 12 10 11 15', and an antibody string '4 3 9 5 12', we have to align the antibody at every possible alignment position with the antigen gene by gene in order to calculate a matching score. A matching score is calculated by summing up the scores from the matches where a match of each position contributes a score of five. Therefore, based on the number of matches between both the antibody and the antigen, the matching score for the example given above is 15, which is the best possible match found (highest matching score) by this process. Since an antibody is matched with each of the antigens in the sample, if the antibody is matched against more than one antigen, a total matching score for the antibody is arrived at by summing up the highest matching scores of matching the antibody with each of the antigens in the previous process.

Hart and Ross [17] selected certain samples of antibodies from the antibody population to be matched with a sample of antigens and repeated the matching process for a certain number of iterations based on the number of antigens selected. In our algorithm, we matched all the antibodies in the population with the antigens and ran the matching process only once. We would also like to note that, for our preliminary experiments, we did not include any wildcard genes in any antibody in the antibody population as we wanted to see the exact fitness of the antibodies as we matched them with the antigens. In [17], the authors allow a wild card match between the antibody and the antigen. A wild card is used as a substitute to any job.

Crossover and Mutation. A genetic algorithm was implemented based on GENESIS [15] and this was used to evolve the antibody population. We used an order-based crossover operator, as it can ensure no job duplication in an antibody for any relationship between two parent antibodies. During crossover, we applied tournament selection to select the best antibody to be included in the next generation. We evaluated the fitness of the children produced and compared their fitness with the fitness of the parents. If the children had lower fitness than the parents, they were discarded, and the parents were selected for inclusion in the next generation. Only the best antibodies, i.e. antibodies with the highest fitness, were considered for the next generation. A mutation operator, which randomly mutates each gene with a probability of 0.2, was also applied as used in [17].

3.2   Phase II: Simulated Annealing and the Great Deluge Algorithm

In the second phase, we apply local search methods on the final population generated from the first phase to improve the fitness of the antibodies.

Simulated Annealing. The simulated annealing algorithm is well studied and an overview and description is presented in [1].

As mentioned above, the initial solution for this algorithm is provided by the final population developed using the model described in Section 3.1. We set the initial temperature T_0 to 5000 and the final temperature T_f to 0.05. The temperature will be decreased by α, where α is defined as 0.98 which has been found to be an effective value in the literature [8,9,27].

While the current temperature is greater than the final temperature, new antibodies, Ab_{new} are generated. This is done by applying two different operators, respectively in two different experiments; changing one job in Ab or swapping two jobs in Ab, where Ab represents the antibodies in the antibody population. The fitness of each antibody is then calculated using the same matching function as applied in the artificial immune system model. The new antibody will be kept if the fitness of the new antibody is better than the fitness of the current best antibody in the antibody population. Otherwise, it is accepted with a probability of $e^{-\delta T}$. Here, δ is defined as the difference between the fitness of the new antibody and the old antibody. We also record the best antibody found overall. This is included in the antibody population (final solution) if it is better than the original antibody.

The Great Deluge Algorithm. Dueck [10] introduced the Great Deluge algorithm in 1993. This algorithm is similar to Simulated Annealing but it has a different acceptance process for worse solutions. The control parameter in this algorithm is called a level or boundary. A worse solution is still acceptable as long as it is within the boundary, which, at the beginning, is set to the fitness of the initial solution. The boundary is then decreased by a fixed decay rate, β, at every iteration of the search.

The initial solution for this algorithm is also provided by the final population generated using the artificial immune system model in Phase I. Here we set the number of iterations, *iter* to 120, which is the possible number of new antibodies generated by an antibody. We also set the estimated quality of the final solution, $f(EQ)$, which is the maximum fitness value for an antibody, depending on the number of antigens selected in the matching function. If the number of antigens selected is one, the maximum fitness value for $f(EQ)$ is 25. This estimated quality represents the final estimated fitness value of an antibody. The boundary to the fitness of each antibody known as *boundary* is decreased by a decreasing rate, β [3] which is defined as follows:

$$\beta = (f(Ab) - f(EQ)) \, / \, iter \, . \tag{2}$$

While the number of iterations does not exceed *iter*, new antibodies are generated by using the same two operators used in the simulated annealing algorithm. We then calculate the fitness of each new antibody generated, $f(Ab)$, by using the same matching function as described in Phase I. A new antibody which is worse than the old one will only be accepted if its fitness is less than the boundary. This loop will also stop if there is no more improvement within a fixed number of iterations.

4 Experiments and Results

As described in Section 3.1, Hart and Ross created ten test scenarios from a base problem, *jb11*, taken from Morton and Pentico [18,22] and the schedules generated from the problem became the antigen universe for this research. We generate three types of antibody populations in order to determine whether having a large number of similar antibodies in one population would affect the coverage of the antigen universe by the antibody population. Our program was coded in C and the experiments were executed on a PC in Windows XP environment with a Pentium 4-2.4 GHz processor and 512 MB RAM.

In the first phase, an initial population of size 100 was selected randomly for each type of antibody population and these populations were evolved using a genetic algorithm for 250 generations, with a crossover rate of 0.7 as used in [17]. We used two mutation rates in the experiments. A mutation rate of 0.2 is employed as it is the same parameter used in [17] and, therefore, it is easier for us to make a comparison with those results. We then used a mutation rate of 0.001 as this gave us a steady growth of the fitness of the antibodies in the antibody population. The antibodies evolved here were the antibodies with the highest fitness value in each generation. At the end of the generation, the antibody library should consist of a collection of general and specific antibodies, which could either match many antigens or only one specific antigen.

Tables 1 and 2 show the average number of antigens that cannot be matched by any antibody for a matching threshold ranging from 2 to 5. A matching threshold, t_m, is a guideline on when we can determine whether an antibody and antigen are matched. The number of genes to bind or match must be greater or equal to the threshold value of t_m [17]. This experiment tests the coverage of the antigen universe by the antibody population. Table 1 shows the results of the experiment by Hart and Ross [17]. Table 2 shows the results of our experiments performed on final populations generated from the antibody population Type A, Type B and Type C, respectively (from Phase I) with a mutation rate of 0.2.

Table 1. Average number of antigens (out of a possible 10) not matched by any antibody as generated by Hart and Ross [17]

Match Thres-hold	Ag = 1			Ag = 4			Ag = 8		
	Ab			Ab			Ab		
	5	10	30	5	10	30	5	10	30
2	0.9	0.0	0.0	2.2	0.9	0.0	3.5	2.5	0.9
3	5.3	2.6	1.6	5.4	3.2	2.0	5.5	4.7	4.1
4	8.7	7.1	5.2	7.8	7.3	6.3	8.6	8.1	8.2
5	9.7	9.5	8.8	9.5	9.5	8.7	9.7	9.6	9.5

In Table 1, the results from Hart and Ross managed to create a trend where the average number of antigens not matched by any antibody decreases as the size of the antibody samples, s increases from 5 to 30. The results in Table 2 are in line with the trend where the average number of unmatched antigens still decreases when the whole population is matched against the antigens. However, the main difference between the results compared to Hart and Ross's was that as we increase the number of antigens, the average number of antigens that cannot be matched by any antibody decreases. While the result in [17] could be interpreted as evidence that more specific antibodies have been produced, we believe that, as we expose more antigens to the antibodies, the fitness of the antibodies would increase and therefore would result in more antigens getting matched or recognized. Therefore with our model, we can produce partial schedules that can be used as replacement to an actual schedule when disturbances occur.

We also ran experiments to see if our Phase II could improve the results. Two different sets of experiments have been carried out, where we use the final populations generated using our artificial immune system as initial solutions to the simulated annealing and the great deluge algorithms separately (Phase II). Ten different sets of antibody populations (initial solutions) were used for each sample of antigens. The final populations generated from this phase were then matched with all the existing ten antigens to illustrate the diversity of the antibodies/partial schedules created. Two different operators were tested. As the operator swapping two jobs generated similar results, we present only the results tested with the operator of changing one job in each antibody.

Table 2. Average number of antigens (out of a possible 10) not matched by any antibody (modified algorithm for AIS)

Match Thres-hold	Ab = 100								
	Type A			Type B			Type C		
	Ag			Ag			Ag		
	1	4	8	1	4	8	1	4	8
2	0.0	0.0	0.0	0.0	0.0	0.0	0.0	0.0	0.0
3	0.4	0.0	0.0	0.9	0.1	0.0	0.8	0.1	0.0
4	6.5	3.6	1.3	6.2	3.4	1.4	6.6	3.2	1.3
5	8.5	6.3	4.7	8.3	6.6	5.3	8.2	7.1	5.8

Table 3. Average number of antigens (out of a possible 10) not matched by any antibody in population Type A (for New AIS (our artificial immune system), AIS+SA (our artificial immune system hybridised with the simulated annealing algorithm) and AIS+GD (our artificial immune system hybridised with the great deluge algorithm))

Match Thres-hold	Ab = 100								
	New AIS			AIS + SA			AIS + GD		
	Ag			Ag			Ag		
	1	4	8	1	4	8	1	4	8
2	0.0	0.0	0.0	0.0	0.0	0.0	0.0	0.0	0.0
3	0.6	0.1	0.0	1.5	0.3	0.0	1.4	0.4	0.0
4	6.8	3.0	1.0	**6.6**	4.7	**0.6**	6.5	4.0	1.4
5	7.9	6.0	4.4	8.3	**5.3**	3.5	8.2	**5.1**	4.8
Fitness Diff. (%)				28.5	11.7	4.8	28.8	10.7	4.7

The results depicted in Table 3 are the average number of antigens not matched by any antibody for both hybrid models compared to the artificial immune system alone with a mutation rate of 0.001 on antibody population Type A. We also show the percentage of the fitness improvement on antibodies generated using the hybrid search algorithm compared with the fitness of the antibodies generated using our new artificial immune system algorithm in the table. It is important to note that the time taken to generate initial antibody populations is less than one minute. The time taken to get a final population (antibody population) using our artificial immune system algorithm (from Phase I) is between one to two minutes while the time taken to get a final solution (antibody population) using a hybrid with the simulated annealing and great deluge algorithms, respectively (Phase II) is one minute or less. This applies to any parameter used to evolve the final populations. We believe this is due to the cooling schedule that is used in the simulated annealing algorithm and the number of iterations set in the great deluge algorithm.

The results of the experiments show, not surprisingly, that the hybrid search algorithms do improve on the artificial immune system algorithm developed in [17] and in this research. However, the hybrid algorithm does not improve the coverage of the antigen universe compared to our artificial immune system algorithm alone except for certain combinations of the number of antigens and the matching threshold. This is probably due to the large number of general antibodies (partial schedules) produced using the artificial immune system that can be matched with most of the antigens. Both hybrid models produced more specific antibodies and, therefore, could not cover most of the antigens.

The fitness of the antibodies in the population, however, does improve, as depicted in the last row in Table 3. Here we total up the fitness of all the antibodies in all ten different sets of antibody populations for each sample of antigens for both hybrid search algorithms and our artificial immune system algorithm. The fitness of the whole antibody populations generated using the hybrid simulated annealing and the hybrid great deluge algorithm, respectively increases by more than 28% over the antibodies using the artificial immune system alone. However, the percentage drops gradually as the number of antigens selected increases.

5 Conclusion

This paper has solved a simple job shop scheduling problem. We have developed an artificial immune system model by drawing upon the research in [17,18]. Our empirical results represent an improvement upon those in [17,18]. We also investigated the use of local search methods to further improve the partial schedules developed in the antibody population. The results obtained indicated that the hybridisation of our artificial immune system approach with simulated annealing and great deluge, respectively, did not yield improvement in terms of the coverage of the antigen universe. However, they did improve the fitness of the antibodies produced in the population. This is important, as we need to provide a range of good partial schedules that can be used to replace certain jobs in the actual schedule when we have changes in the arrival dates of the jobs. We will also use the results as a platform for our future work on hyper heuristic. For the problem, the antibodies will represent a sequence of low level heuristic.

References

1. Aarts, E., Korst, J., Michiels, W.: Simulated Annealing. In: Burke, E.K., Kendall, G. (eds.) Search Methodologies: Introductory Tutorials in Optimisation and Decision Support Techniques, ch. 7, Springer, Heidelberg (2005)
2. Aickelin, U., Dasgupta, D.: Artificial Immune Systems. In: Burke, E.K., Kendall, G. (eds.) Search Methodologies: Introductory Tutorials in Optimisation and Decision Support Techniques, ch. 13, Springer, Heidelberg (2005)
3. Burke, E.K., Bykov, Y., Newall, J.P., Petrovic, S.: A Time-Predefined Local Search Approach to Exam Timetabling Problem. IIE Transactions 36(3), 509–528 (2004)
4. Brucker, P.: Scheduling Algorithms, 5th edn. Springer, Heidelberg (2007)

5. Chandrasekaran, M., Asokan, P., Kumanan, S., Balamurugan, T., Nickolas, S.: Solving Job Shop Scheduling Problems Using Artificial Immune System. The International Journal of Advanced Manufacturing Technology 31(5-6), 580–593 (2006)
6. Coello Coello, C.A., Cortés Rivera, D., Cruz Cortés, N.: Use of an Artificial Immune System for Job Shop Scheduling. In: Timmis, J., Bentley, P.J., Hart, E. (eds.) ICARIS 2003. LNCS, vol. 2787, pp. 1–10. Springer, Heidelberg (2003)
7. de Castro, L.N., Timmis, J.: Artificial Immune Systems: A New Computational Intelligence Approach. Springer, Heidelberg (2002)
8. Dowsland, K.A.: Off-the-Peg or Made-to-Measure? Timetabling and Scheduling with SA and TS. In: Burke, E.K., Carter, M. (eds.) PATAT 1997. LNCS, vol. 1408, pp. 37–52. Springer, Heidelberg (1998)
9. Dowsland, K.A.: Simulated Annealing. In: Reeves, C.R. (ed.) Modern Heuristic Techniques For Combinatorial Problems, ch. 2, McGraw Hill, New York (1995)
10. Dueck, G.: New Optimization Heuristics: The Great Deluge Algorithm and the Record-to-Record Travel. Journal of Computational Physics 104, 86–92 (1993)
11. Engin, O., Doyen, A.: A New Approach to Solve Hybrid Flow Shop Scheduling Problems by Artificial Immune System. Future Generation Computer Systems 20, 1083–1095 (2004)
12. Fang, H.-L., Ross, P., Corne, D.: A Promising Genetic Algorithm Approach to Job-Shop Scheduling, Rescheduling and Open-Shop Scheduling Problems. In: Forrest, S. (ed.) The Fifth International Conference on Genetic Algorithms, pp. 375–382. Morgan Kaufmann, San Francisco (1993)
13. Ge, H.W., Sun, L., Liang, Y.C.: Solving Job-Shop Scheduling Problems by a Novel Artificial Immune Systems. In: Zhang, S., Jarvis, R. (eds.) AI 2005. LNCS (LNAI), vol. 3809, pp. 839–842. Springer, Heidelberg (2005)
14. Goldsby, R.A., Kindt, T.J., Osbourne, B.A.: Kuby Immunology, 4th edn. W.H. Freeman, New York (2000)
15. Grefenstette, J.: Genesis: A System For Using Genetic Search Procedures.In: Conference on Intelligent Systems and Machines, pp. 161–165 (1984)
16. Hart, E., Ross, P., Nelson, J.: Producing Robust Schedules Via An Artificial Immune System. In: ICEC 1998, pp. 464–469. IEEE Press, Los Alamitos (1998)
17. Hart, E., Ross, P.: An Immune System Approach to Scheduling in Changing Environments. In: Banzhaf, W., Daida, J., Eiben, A.E., Garzon, M.H., Honavar, V., Jakiela, M., Smith, R.E. (eds.) GECCO 1999, pp. 1559–1565. Morgan Kaufmann, San Francisco (1999)
18. Hart, E., Ross, P.: The Evolution and Analysis of a Potential Antibody Library for Job-Shop Scheduling. In: Corne, D., Dorigo, M., Glover, F. (eds.) New Ideas in Optimisation, pp. 185–202. McGraw-Hill, London (1999)
19. Hightower, R.R., Forrest, S., Perelson, A.S.: The Evolution of Emergent Organization in Immune System Gene Libraries. In: Eshelman, L. (ed.) The Sixth Annual Conference on Genetic Algorithms, pp. 344–350. Morgan Kaufmann, San Francisco (1995)
20. Jain, A.S., Meeran, S.: A State-of-the-Art Review of Job-Shop Scheduling Techniques, Technical Report, Department of Applied Physics, Electronics and Mechanical Engineering, University of Dundee, Scotland (1998)
21. Jensen, M.T., Hansen, T.K.: Robust Solutions to Job Shop Problems. In: The 1999 Congress on Evolutionary Computing (CEC 1999), pp. 1138–1144 (1999)
22. Morton, T.E., Pentico, D.W.: Heuristic Scheduling Systems. John Wiley, Chichester (1993)
23. Oprea, M., Forrest, F.: Simulated Evolution of Antibody Gene Libraries Under Pathogen Selection. In: The 1998 IEEE International Conference on Systems, Man and Cybernetics (1998)
24. Pinedo, M.: Scheduling: theory, algorithms, and systems, 2nd edn. Prentice-Hall, Englewood Cliffs (2002)

25. Sompayrac, L.: How the Immune System Works, 2nd edn. Blackwell Publishing, Oxford (2003)
26. Spellward, P., Kovacs, T.: On the Contribution of Gene Libraries to Artificial Immune Systems. In: The 2005 Conference on Genetic and Evolutionary Computation (GECCO 2005), pp. 313–319. ACM Press, New York (2005)
27. Thompson, J., Dowsland, K.A.: General Cooling Schedules for a Simulated Annealing Based Timetabling System. In: Burke, E.K., Ross, P. (eds.) Practice and Theory of Automated Timetabling. LNCS, vol. 1153, pp. 345–363. Springer, Heidelberg (1996)

Enhancements to Extremal Optimisation for Generalised Assignment

Marcus Randall

School of Information Technology
Bond University, QLD 4229, Australia
Tel.: +61 7 55953361
mrandall@bond.edu.au

Abstract. Extremal optimisation (EO) is a relatively new meta-heuristic technique that is based on the principles of self organising criticality. It allows for a poorly performing solution component to be removed at each iteration of the algorithm and be replaced by a random one. Over time, improvements emerge and the system is driven towards good quality solutions. There has been very little literature concerning EO and combinatorial optimisation and relatively few computational results have been reported. In this paper, an enhanced model of EO, which allows the traversal feasible and infeasible spaces, is presented. This improved version is able to operate on single solutions as well as populations of solutions. In addition to local search, a simple partial feasibility restoration heuristic is introduced. The computational results for the generalised assignment problem indicate that it provides significantly better quality solutions over a sophisticated ant colony optimisation implementation.

Keywords: extremal optimisation, generalised assignment problem.

1 Introduction

The idea of self-organising criticality (SOC) has been an important part of biology and evolutionary theory for a number of years. It is only recently that these concepts have been applied to solving optimisation problems [1]. One of the main expressions of these ideas has been by Boettcher and Percus [1, 3] in the form of Extremal Optimisation (EO). The following is a representative sample of the work in this area.

In terms of benchmark problems, Boettcher and Percus [1, 3] have described and carried out only limited experimentation on the travelling salesman problem (TSP). More successful application has been in graph (bi)partitioning [1] and the MAX-CUT problems [3]. For these at least, EO can locate optimal and near optimal solutions and is comparable to other meta-heuristics.

Beyond the original applications, some work has been done to adapt the standard EO algorithm to dynamic combinatorial optimisation. As an example, Moser and Hendtlass [11, 12] apply EO to a dynamic version of the composition

M. Randall, H.A. Abbass, and J. Wiles (Eds.): ACAL 2007, LNAI 4828, pp. 369–380, 2007.

problem. During the course of solving the problem with EO, the problem may undergo a variety of transformations to its structure and/or data. Despite the EO solver not being made aware of specific changes, it is able to adapt to them more readily than a standard ACS solver. This, however, was the reverse for the static version of the problem.

Randall and Lewis [15] present an extended form of EO. A controlling heuristic is used to manage a number of EO solutions. This is referred to as evolutionary population dynamics (EPD). Experiments on multi-dimensional knapsack problem instances showed that EO with EPD achieved equal or better results than EO alone on almost all tests cases (in one case it was slightly worse, with a difference of only 1%). On all test cases EO with EPD also achieved its results in less time. On average, EO required 80% more time to obtain generally poorer results. The results using EO with EPD were also compared with tests using a standard ant colony optimisation (ACO) solver. EO with EPD was been found to deliver near-optimal results faster than the existing ACO algorithm, and the speed of the algorithm could be expected to improve for larger problem sizes.

For EO to be widely adopted for combinatorial problems, it needs to be shown that it is capable of being competitive with other meta-heuristics. The enhancements to EO outlined in this paper demonstrate this. The remainder is organised as follows. Section 2 gives an overview of EO. Section 3 describes some enhancements to EO. In simple terms, these are the ability to move between feasible and infeasible space (while being driven towards the former) and a population approach. The results across a set of generalised assignment problems (GAPs) are discussed in Section 4. Finally, the conclusions and future research directions are given in Section 5.

2   Extremal Optimisation

Extremal optimisation is one of a number of emerging biologically inspired metaphors for solving combinatorial optimisation problems. As it is relatively new in the field, compared to more established techniques such as ant colony optimisation [6], genetic algorithms [7] and particle swarm optimisation [8], there exists wide scope to apply and to extend this meta-heuristic.

Boettcher and Percus [1, 3, 4] describe the general tenets of EO. Nature can be seen as an optimising system in which the aim is to allow competitive species to populate environments. In the course of the evolutionary process, successful species will have the ability to adapt to changing conditions while the less successful will suffer and may become extinct. This notion extends to the genetic level as well. Poorly performing genes are replaced using random mutation. Over time, if the species does not become extinct, its overall fitness will increase.

For combinatorial optimisation problems, the genes represent solution components. To illustrate the operation of EO, the TSP is used. This is because it is a well-known benchmark problem; has been studied extensively in the literature and is one of the few problems that has been solved by EO. The application to the target problem, the GAP, is dealt with in detail in the next section.

In the case of the TSP, solution components are the cities that make up a tour. In the original EO algorithm, the worst component would be replaced by one generated at random. The worst component is the city which is connected to its two furthest neighbours (given by ranks). However, as the performance of this version of the algorithm was unacceptable, τ-EO was introduced [3]. In this form a poor component is chosen probabilistically. This is achieved using a parameter τ and Equation 1.

$$P_i = i^{-\tau} \qquad 1 \le i \le n \tag{1}$$

Where:

i is the rank of the component,
P_i is the probability ($P_i = [0, 1]$) that component i is chosen and
n is the number of components.

Ranking components from worst to best (i.e., rank one is given to the worst component), roulette wheel selection is used to choose the component to replace. For permutation problems such as the TSP, the way to replace the chosen node is not immediately obvious. Choosing a random node to swap this one with is likely to produce a worse result. To increase the chance of a better solution being produced, the worst edge connecting this node is first chosen and deleted. The node is then reconnected to its probabilistically chosen best neighbour. The third edge from this node (i.e., the one needed to be removed to ensure a valid permutation) is removed. The two remaining nodes that only have one edge each are then connected together, ensuring a valid permutation.

The entire procedure is repeated for a user-specified number of iterations, or until the search obtains a particular solution quality.

3   Some EO Enhancements

The enhancements to EO will be discussed in terms of the test application problem, the GAP. Their applicability, however, is much wider than just this problem. In particular, the emphasis is on constrained problems - other examples of which are bin packing, graph colouring, knapsack, scheduling and timetabling. The generalised assignment problem [10] is a problem in which jobs are assigned to agents for these agents to perform subject to capacity constraints. Each job may be performed by one agent only. The aim is to minimise the total cost of assigning the jobs to the set of agents. Its mathematical formulation is given in Equations 2-5.

$$\text{Minimise} \sum_{i=1}^{N} \sum_{j=1}^{M} c_{ij} x_{ij} \tag{2}$$

s.t.

$$\sum_{i=1}^{N} a_{ij}x_{ij} \leq b_j \quad \forall j \quad 1 \leq j \leq M \tag{3}$$

$$\sum_{j=1}^{M} x_{ij} = 1 \quad \forall i \quad 1 \leq i \leq N \tag{4}$$

$$x_{ij} \in \{0,1\} \quad \forall i \quad 1 \leq i \leq N \quad \forall j \quad 1 \leq j \leq M \tag{5}$$

Where:

c_{ij} is the cost of assigning job i to agent j,
a_{ij} is the resource required by agent j to perform job i,
x_{ij} is 1 if job i is assigned to agent j, 0 otherwise,
b_j is the capacity of agent j,
M is the number of agents and
N is the number of jobs.

3.1   Moving Between Feasible and Infeasible Space

EO's initial applications to combinatorial optimisation problems (such as the TSP) were not promising, with results being not competitive with other heuristics. Problems for which the solution is a permutation, such as the TSP and quadratic assignment problem, are particularly problematic. These types of problems often require omplex sets of movements in search space to ensure that another permutation can be produced. This somewhat strict approach of forcing solutions to be structurally feasible often results in a decrease in overall solution quality.

For problems that do not possess such structural properties, but have capacity type constraints, EO is a natural fit. As EO only makes a small change at each iteration - allowing many transitions to be made in computationally reasonable time - it may not matter that the solution is feasible at all times. Overall, many feasible solutions will be still produced. At each iteration of the algorithm, the component to be changed will depend on whether the solution is currently feasible or infeasible:

- *A feasible solution* - In essence, a probabilistic greedy move is performed in order to optimise the objective function. This is done by changing a high cost job-to-agent assignment to a random agent. As a result of this transition, it is possible that an infeasible solution has been produced.
- *An infeasible solution* - The focus changes to moving the solution back to a feasible state. As such, a component to change is chosen according to the amount of infeasibility it contributes to the solution. This is the amount of resource that the job-agent pairing requires - i.e., the higher this value, the more likely it is to be chosen. Initial experimentation revealed that the search process spent a great number of iterations in infeasible space. Therefore, a simple, non-degenerative, parameter-free heuristic was devised (see

Algorithm 1) that would reduce the amount of infeasibility of a solution. This is only a partial feasibility restoration algorithm and has a computational complexity of only $O(MN)$. It does not need to guarantee feasibility because of EO's own ability to move back to feasible space. It's use is akin to local search for feasible solutions.

Algorithm 1. The partial feasibility restoration algorithm for the GAP.

---

for all agents **do**
    if this agent is overloaded (infeasible) **then**
        Determine the job whose resource requirement most closely matches that of this
        agent's amount of infeasibility
        Find a new agent that can take this job without becoming infeasible
        if such an agent exists **then**
            Update the solution and it's cost
        else
            Do nothing
        end if
    end if
end for

3.2 Local Search

Local search is solely driven toward optimising the objective function, moving only in feasible space, and making purely greedy moves. As such, it is in contrast to the balanced EO strategy described previously. However, it is useful for obtaining locally optimal solutions which cannot be guaranteed by EO.

For the GAP, in which the overall costs of assignment of jobs to agents, two possible operations exist. "Move" moves an item from one group to another. The job and agent are chosen such that the (negative) change in the objective function is the greatest. This is a variable length search stopping when an improving move cannot be found. "Swap" works in a similar way except that at each iteration, two items are chosen such that their swap will lead to the most improvement in the objective function. The combination of the two operators hass been shown to be more effective than when either is used alone (see for example Randall [13, 14]).

3.3 Population of EO Solutions

According to the principles of the Bak-Sneppen model of evolution, the weakest member of a population and its closest neighbours die and are replaced by new members. A simple computational model of this behaviour can be used to extend EO to being a population based approach. It is implemented in the following way:

1. Create a population of EO solutions (rather than just the normal single solution).
2. Allow all of the solutions to be changed according to EO principles.

3. At every pre-determined iteration count for which each population member has completed N EO operations, select the member with the worst solution cost. Also choose its two closest neighbours (in terms of number of common solution components).
4. Replace these three solutions with ones generated at random.
5. Resume the search process.

This approach represents a simplification of the EPSOC (Evolutionary Programming using Self Organising Criticality) algorithm [9, 15]. The only extra parameter that the new population EO requires is the population size.

4   Computational Experiments

The computing platform used to perform the experiments is a 3GHz Pentium 4 based PC. Each problem instance is run across ten random seeds. The experimental programs are coded in the C language and compiled with `gcc`. The only two EO parameters are τ and the number of iterations it is to run for. τ is set as 1.5 (a value consistent with Boettcher and Percus [2]), while the latter is 500,000. This value is the same as used by Randall and Lewis [15]. Additionally it is an underestimate of the equivalent steps used by Randall's [14] ACO implementations. These used 3000 iterative colonies having ten ants each and performing N constructive steps. Given the lower value of $N = 100$, this equates to 3,000,000 EO transitions. It must be borne in mind, however, that EO accesses local search more frequently than ACO.

The test suite of problems is the large-sized set of Chu and Beasley [5]. The definitions are as follows and are reproduced from this work. Generally, the type

Table 1. The problem instances used in this study. Note that the first number in the name represents the number of agents while the second is the number of jobs. For instance, 'A5-100' is of Type A with 5 agents and 100 jobs.

Instance	Best known cost	Instance	Best known cost
A5-100	1698	C5-100	1931
A5-200	3235	C5-200	3458
A10-100	1360	C10-100	1403
A10-200	2623	C10-200	2814
A20-100	1158	C20-100	1244
A20-200	2339	C20-200	2397
B5-100	1843	D5-100	6373
B5-200	3553	D5-200	12796
B10-100	1407	D10-100	6379
B10-200	2831	D10-200	12601
B20-100	1166	D20-100	6269
B20-200	2340	D20-200	12452

A problems are the easiest to solve while the D problems are the hardest (as they are the most constrained). The characteristics of each of the problem instances are given in Table 1.

Type A: a_{ij} are integers from $U(5, 25)$, c_{ij} are integers from $U(10, 50)$ and $b_i = 0.6(\frac{n}{m})15 + 0.4R$ where $R = \max_{i \in I} \sum_{j \in J, I_j = i} a_{ij}$ and $I_j = \min[i \mid c_{ij} \leq c_{kj}, \forall k \in I]$.

Type B: a_{ij} and c_{ij} are the same as Type A and b_i is set to 70% of the value given in Type A.

Type C: a_{ij} and c_{ij} are the same as Type A and $b_i = 0.8 \sum_{j \in J} \frac{a_{ij}}{m}$.

Type D: a_{ij} are integers from $U(1, 100)$, $c_{ij} = 111 - a_{ij} + e$ where e are integers from $U(-10, 10)$ and $b_i = 0.8 \sum_{j \in J} \frac{a_{ij}}{m}$.

Table 2. The cost results of running the four variants of EO. These are for the entire problem set across ten random seeds. The results are represented as relative percentage differences (RPDs) from the optimal cost. RPD is given as $\frac{a-b}{b} \times 100$ where a is the best cost achieved in a run and b is the optimal cost of the problem instance. Note that 'Min', 'Med' and 'Max' denote minimum, median and maximum respectively. 'ls' refers to the local search heuristic and 'pfr' is the partial feasibility restoration algorithm. Bolded items indicate the best result for combination and measure.

Problem	ls off, pfr off Min	Med	Max	ls off, pfr on Min	Med	Max	ls on, pfr off Min	Med	Max	ls on, pfr on Min	Med	Max
A5-100	4.42	4.98	5.65	7.01	8.69	11.43	0	0	0	0	0	0
A5-200	4.05	4.85	5.22	5.5	6.12	7.11	0	0	0	0	0	0
A10-100	5.96	7.17	7.5	5.44	6.14	6.69	0	0	0	0	0	0
A10-200	5.26	5.66	5.91	5.34	5.87	6.14	0	0	0	0	0	0
A20-100	11.83	13.17	13.56	13.82	14.98	16.41	0	0	0	0	0	0
A20-200	7.65	8.61	9.32	9.36	10.99	11.37	0	0	0	0	0	0
B5-100	16.33	17.39	19.37	15.63	17.77	18.29	1.36	1.71	1.95	**0.71**	**1.03**	**1.41**
B5-200	11.54	12.86	14.07	18.18	19.69	20.8	1.91	2.14	2.45	**0.42**	**0.48**	**0.56**
B10-100	17.77	20.97	21.96	22.89	24.02	25.66	0	0.07	0.14	**0**	**0**	**0**
B10-200	19.64	21.03	21.62	29.28	33.36	34.23	1.06	1.8	1.98	**0.6**	**0.76**	**1.02**
B20-100	27.7	30.66	31.56	29.33	32.72	33.36	0.17	0.39	0.51	**0.09**	**0.17**	**0.26**
B20-200	19.23	20.51	21.71	40.21	41.86	43.68	0.3	0.43	0.51	**0.17**	**0.21**	**0.26**
C5-100	12.12	13.15	14.66	14.45	16.24	17.45	0.57	1.04	1.24	**0.36**	**0.6**	**0.78**
C5-200	12.9	13.69	13.91	20.71	22.33	22.99	1.3	1.74	2.34	**0.29**	**0.52**	**0.67**
C10-100	28.08	31.93	33.43	23.31	27.33	29.72	1.35	2.07	2.21	**0.71**	**1.1**	**1.28**
C10-200	15.81	16.68	17.87	28.71	30.69	31.17	1.53	1.71	2.03	**0.46**	**0.82**	**1.03**
C20-100	45.74	53.05	55.71	31.35	34.28	36.82	1.61	2.17	2.73	**0.72**	**1.05**	**1.13**
C20-200	28.79	30.14	32.17	41.97	45.62	46.93	1.96	2.09	2.29	**0.92**	**1.17**	**1.29**
D5-100	6.5	7.45	8.39	5.37	5.64	5.87	2.43	2.93	3.17	**1.37**	**1.54**	**1.65**
D5-200	7.46	7.72	8.07	6.46	6.7	6.89	2.88	3.1	3.33	**1.52**	**1.63**	**1.71**
D10-100	13.54	14.59	15.6	8.29	8.58	8.87	2.82	3.29	3.54	**2.15**	**2.56**	**2.76**
D10-200	12.07	12.6	13.15	8.49	8.65	8.97	1.84	2.07	2.39	**1.24**	**1.54**	**1.6**
D20-100	23.26	24.53	24.76	8.17	8.53	8.82	2.81	3.19	3.46	**2.14**	**2.47**	**2.58**
D20-200	18.89	19.81	20.56	8.95	9.12	9.48	2.51	2.85	2.92	**1.55**	**1.69**	**1.81**

The computational experiments are naturally divided into two parts. The first part is to test the effect of the combinations of the local search and partial feasibility restoration heuristics. Given that these may be turned on and off, this leads to four groups of results. The Kruskal-Wallis statistical procedure will be used to detect if there is any significant difference between the groups. If one is present, post-hoc testing will be used to determine the nature of the difference. These results will be compared to those of Randall's [14] competitive ant colony optimisation implementation. In the second part of the experiments, the population approach will be trialed. The population size parameter will be tested using the four values of {5, 10, 15, 20} on the D class problem instances. Again, Kruskal-Wallis will be used to help to determine the most appropriate population size. These results will be compared against those of the first part of the experiments.

Table 3. Table 1 as reproduced from Randall [14]. Three different ACO implementations were used. The first two (i.e., "Compete 1" and "Compete 2') allowed ants to compete for solution components, while "Control" was a standard ACO implementation. For details of the three solvers, please refer to this paper. Note that bolded items indicate the best result for the combination of problem and measure.

Problem	Optimal Cost	Compete 1			Compete 2			Control		
		Min	Med	Max	Min	Med	Max	Min	Med	Max
A5-100	1698	0	0	0	0	0	0	0	0	0
A5-200	3235	0	0	0	0	0	0	0	0	0
A10-100	1360	0	0	0	0	0	0	0	0	0
A10-200	2623	0	0	0	0	0	0	0	0	0
A20-100	1158	0	0	0	0	0	0	0	0	0
A20-200	2339	0.04	0.04	**0.04**	0	0	0.04	0	0	0.04
B5-100	1843	**0.43**	**0.71**	**0.92**	0.65	1	1.14	3.36	4.72	6.02
B5-200	3553	**0.51**	**0.65**	**0.79**	0.53	0.66	0.87	0.73	0.89	1.04
B10-100	1407	0.07	0.28	0.43	**0**	**0**	**0.07**	0.07	0.14	0.36
B10-200	2831	1.31	**1.45**	1.66	**1.27**	1.54	**1.62**	1.38	1.7	2.08
B20-100	1166	0.69	1.03	1.11	0.34	**0.6**	**0.77**	**0.26**	0.69	0.94
B20-200	2340	0.64	0.75	0.81	**0.43**	**0.56**	**0.64**	0.51	0.68	0.81
C5-100	1931	**0.31**	**0.57**	**0.73**	0.67	0.85	1.04	0.62	1.09	1.35
C5-200	3458	0.52	0.67	0.81	**0.35**	**0.64**	**0.75**	0.49	0.78	1.01
C10-100	1403	1.14	1.43	**1.57**	**0.93**	1.18	**1.57**	1.5	1.75	2.07
C10-200	2814	**0.92**	**1.33**	**1.46**	1.39	1.58	1.78	1.46	1.74	1.92
C20-100	1244	1.69	1.89	2.49	**0.88**	**1.17**	**1.53**	1.53	1.65	2.25
C20-200	2397	1.63	1.98	2.13	**1.34**	**1.81**	**2**	1.88	2.07	2.17
D5-100	6373	**1.99**	**2.38**	**2.51**	2.97	3.21	3.4	3.36	3.88	4.14
D5-200	12796	**2.31**	**2.5**	**2.79**	3.17	3.26	3.39	3.4	3.56	3.74
D10-100	6379	**3.45**	**3.66**	**3.87**	4.04	4.37	4.61	4.33	4.45	4.73
D10-200	12601	**2.77**	**3.02**	**3.18**	3.56	3.84	3.95	3.4	3.92	4.03
D20-100	6269	**3.51**	**3.55**	**3.96**	3.53	3.88	4.15	3.73	4.07	4.4
D20-200	12452	**3.17**	**3.48**	**3.7**	4.09	4.26	4.41	4.14	4.33	4.49

Table 2 shows the results of part 1 of the experiments. For reference, the results of competitive ACO [14] are reproduced in Table 3.

The patterns that can be observed in Table 2 are as follows. EO requires local search in order to be effective. In terms of the first two result columns, it is interesting to note that using the partial feasibility restoration heuristic is, on the whole, worse than not using it. However, the reverse is very much the case when local search is switched on. Improved solutions are produced by using the two heuristics in combination, rather than just using local search alone. The extra computational effort of the partial feasibility restoration algorithm was negligable (typically less than one percent more CPU time). Running the Kruskal-Wallis statitical procedure confirms this. Unsurprisingly, a significant difference is recorded between the four groups of results. This is significant at the 0.05 level. Post-hoc testing also revealed that indeed the combination of the two heuristics was significantly better than local search on its own.

Further comparison of these results against those in Table 3 showed that this variant of EO is significantly better than the competitive ACO implementations. The only problem on which it does not better or equal its performance is B5-100 (though the difference is negligible). It might be argued that ACO would benefit from the use of the partial feasibility restoration heuristic. However, as the ACO implementations are constructive, and discard the colonys's solutions after each iteration of the algorithm, the heuristic in its present form is incompatible with it. Another interesting point of comaparison is that, unlike the ACO implementations, EO often locates improved solutions late in the run. This suggest that it is less prone to premature convergence. Figure 1 presents a graph of a typical run of EO as visual affirmation of this.

Initial investigation of different population sizes for part two of the experiments showed that there was no statistically significant difference between them.

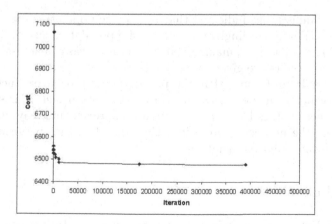

Fig. 1. A typical run for EO on the problem D5-100. The markers indicate the iterations at which EO finds new best solutions. Notice that EO is able to quickly come to a good solution, and spends the remaining time making slight improvements.

Table 4. The RPD cost results of running the population version of EO. Note that a bolded itme indicates that the particular result is strictly better than the corresponding result of the fourth group of results in Table 2.

Problem	Min	Med	Max
A5-100	0	0	0
A5-200	0	0	0
A10-100	0	0	0
A10-200	0	0	0
A20-100	0	0	0
A20-200	0	0	0
B5-100	**0.27**	**0.73**	**0.81**
B5-200	**0.34**	**0.37**	**0.45**
B10-100	0	0	0
B10-200	**0.39**	**0.44**	**0.53**
B20-100	0	**0.09**	**0.09**
B20-200	**0.09**	**0.13**	**0.13**
C5-100	**0.05**	**0.34**	**0.47**
C5-200	**0.23**	**0.35**	**0.43**
C10-100	**0.29**	**0.5**	**0.71**
C10-200	**0.28**	**0.41**	**0.57**
C20-100	**0.24**	**0.32**	**0.4**
C20-200	**0.5**	**0.56**	**0.71**
D5-100	**1**	**1.21**	**1.43**
D5-200	**1.04**	**1.17**	**1.32**
D10-100	**1.83**	**2.12**	**2.27**
D10-200	**0.87**	**0.92**	**1.02**
D20-100	**1.8**	**1.97**	**2.11**
D20-200	**1.15**	**1.24**	**1.4**

The Kruskal-Wallis ranks indicated that larger size populations tended to produce better results. Accordingly, the setting of a population size of 20 members across 25,000 iterations (i.e., making 500,000 in total) was run across all problem istances. These results are given in Table 4.

It is evident from Table 4 that the population approach produces superior results, particularly on the harder problem instances. In fact, it is statistically significantly better. It is believed that further improvements may be gained by experimenting the number of iterations that the algorithm performs before it replaces the worst solution members.

5   Conclusions

Even though EO is, on the surface, a relatively simple meta-heuristic, it is capable of delivering very good quality solutions. However, as demonstrated herein for the generalised assignment problem, it does require supporting algorithms to achieve this success. One of EO's strength is that can easy be adapted so that it can move between feasible and infeasible space. When in feasible space, it is able to use local search. When it is not, it can use a partial feasibility restoration heuristic that will drive it toward feasible space. The combination of these two support mechanisms produces a powerful search effect that is able to outperform a sophisticated ACO implementation. Ongoing work, using the bin packing and graph colouring problems, suggests that this pattern of results is a consistent one.

It is believed that further improvement can be achieved by considering other meta-heuristic devices and the manipulation of the τ parameter. These devices include such techniques as intensification/ diversification, a tabu memory and candidate set strategies. Of particular interest will be the development of the population approach sketched out in this paper. Some aspects of investigation include the addition of population learning mechanisms (such as ACO's pheromone) and different methods of determining the extinction of solution members.

References

[1] Boettcher, S., Percus, A.: Extremal optimization: Methods derived from Co-evolution. In: Banzhaf, W., Daida, J., Eiben, A., Garzon, M., Honavar, V., Jakiela, M., Smith, R. (eds.) Proceedings of the Genetic and Evolutionary Computation Conference, pp. 825–832 (1999)

[2] Boettcher, S., Percus, A.: Combining local search with co-evolution in a remarkably simple way. In: Proceedings of the Congress on Evolutionary Computation, Piscataway, NJ, pp. 1578–1584. IEEE Service Center (2000)

[3] Boettcher, S., Percus, A.: Nature's way of optimizing. Artificial Intelligence 119, 275–286 (2000)

[4] Boettcher, S., Percus, A.: Extremal Optimization: An evolutionary local search algorithm. In: Proceedings of the 8th INFORMS Computer Society Conference, Norwell, MA. Interfaces in Computer Science and Operations Research, vol. 21, pp. 61–78. Kluwer Academic Publishers, Dordrecht (2003)

[5] Chu, P., Beasley, J.: A genetic algorithm for the generalised assignment problem. Computers and Operations Research 24, 17–23 (1997)

[6] Dorigo, M., Di Caro, G.: The ant colony optimization meta-heuristic. In: Corne, D., Dorigo, M., Glover, F. (eds.) New Ideas in Optimization, McGraw-Hill, London (1999)

[7] Goldberg, D.: Genetic Algorithms in Search, Optimization and Machine Learning. Addison Wesley, Reading MA (1989)

[8] Kennedy, J., Eberhart, R.: The particle swam: Social adaptation in social information-processing systems. In: New Ideas in Optimization, pp. 379–387. McGraw-Hill, London (1999)

[9] Lewis, A., Abramson, D., Peachey, T.: An evolutionary programming algorithm for automatic engineering design. In: Wyrzykowski, R., Dongarra, J.J., Paprzycki, M., Waśniewski, J. (eds.) PPAM 2003. LNCS, vol. 3019, pp. 586–594. Springer, Heidelberg (2004)

[10] Martello, S., Toth, P.: An algorithm for the generalised assignment problem. In: Proceedings of the 9th IFORS Conference, Hamburg, Germany (1981)

[11] Moser, I., Hendtlass, T.: On the behaviour of extremal optimisation when solving problems with hidden dynamics. In: Ali, M., Dapoigny, R. (eds.) IEA/AIE 2006. LNCS (LNAI), vol. 4031, Springer, Heidelberg (2006)

[12] Moser, I., Hendtlass, T.: Solving problems with hidden dynamics - comparison of extremal optimisation and ant colony system. In: Congress on Evolutionary Computing, pp. 1248–1255 (2006)

[13] Randall, M.: Heuristics for ant colony optimisation using the generalised assignment problem. In: Proceedings of the Congress on Evolutionary Computing 2004, Portland, Oregon, pp. 1916–1923 (2004)

[14] Randall, M.: Competitive ant colony optimisation. In: Okuno, H., Ali, M. (eds.) IEA/AIE 2007. LNCS (LNAI), vol. 4570, pp. 974–983. Springer, Heidelberg (2007)

[15] Randall, M., Lewis, A.: An extended extremal optimisation model for parallel architectures. In: 2nd IEEE International e-Science and Grid Computing Conference. Workshop on Biologically-inspired Optimisation Methods for Parallel and Distributed Architectures: Algorithms, Systems and Applications, IEEE Computer Society, Los Alamitos (2006)

Identification of Marker Genes Discriminating the Pathological Stages in Ovarian Carcinoma by Using Support Vector Machine and Systems Biology

Meng-Hsiun Tsai[1], Jun-Dong Chang[2], Sheng-Hsiung Chiu[3], and Ching-Hao Lai[2,*]

[1] Department of Management Information Systems, National Chung Hsing University,
Taiwan, R.O.C.
[2] Department of Computer Science and Engineering, National Chung Hsing University,
Taiwan, R.O.C.
[3] Troilus Bio-technology Co. LTD., Taiwan, R.O.C.
phd9514@cs.nchu.edu.tw

Abstract. Ovarian cancer is a primary gynecological cancer which pathological stages include benign, borderline and invasive stages cause death in many countries. In this paper, linear regression, analysis of variance (ANOVA) and support vector machine (SVM) are used to identify the gene markers of ovarian cancer for an authentic cDNA expression datasets among 8 normal ovarian tumors, 6 borderline of cancers, 7 ovarian cancer at stage I and 9 ovarian cancer at stage III samples. First, the linear regression analysis obtains 200 useful genes with largest residuals. Further select 14 genes by ANOVA and Scheffe when P-value is less than 0.000005. Then, we use support vector machine to classify the pathological stages by gene expressions. Five experiments are performed with clustering conditions. In the first clustering experiment, the cluster 1 includes BOT, and other pathological stages are in cluster 2. They have significant differences at BOT stage and can get average accuracy about 95.686% in cross-validation. It is quite precise for classifying pathological stages by gene expressions. The average accuracy of all clustering experiments is about 88.541% in cross-validation. Besides, we also develop a statistical analysis system including linear regression and ANOVA function for gene expression analysis. The experimental results and our analysis system can assist biologists and doctors to research and diagnose ovarian cancer by gene expressions.

Keywords: Microarray, systems biology, ovarian cancer, ANOVA, support vector machine (SVM), gene expression.

1 Introduction

Ovarian cancer is one of the primary gynecological cancers to cause death in many countries [1], especially in the United States, Canada and Europe [2]. In the comparisons of 97% ovarian tumor cases, ovarian cancer can be classified into three

---

* Corresponding author .

M. Randall, H.A. Abbass, and J. Wiles (Eds.): ACAL 2007, LNAI 4828, pp. 381–389, 2007.
© Springer-Verlag Berlin Heidelberg 2007

major pathological stages: benign, borderline and invasive stages. It is more difficult to diagnose the symptom before the benign tumors becoming malignancies than other gynecological cancers, and the overall 5-year survival rate is about 46% [3]. Hence, how to diagnose the ovarian cancer before entering the invasive stages is very important. In recent years, many researchers have discovered many oncogenes which are related to ovarian cancer and have different gene expressions for different pathological stages in biological experiments. This paper uses support vector machine and statistical methods to analyze the expressions of a great deal of genes to find the high related genes to help cancer researchers find the more useful gene markers and also to assist doctors diagnose the ovarian cancer more precisely.

In many research papers about ovarian cancer, the relationship between gene expressions and ovarian cancer has been reported [7-12]. These papers present important and useful information for some ovarian cancer and more specific markers have been discovered from many gene expression analysis and identification. The researchers discovered the ovarian cancer specific markers by distinguishing the difference of gene expressions at different pathogenic stages of ovarian cancer. Analyzing the difference of gene expressions at each pathogenic stage is also helpful to describe the sequential states from benign to invasive tissue.

Microarray is a popular and useful tool to analyze the gene expressions. This technology helps cancer researchers to distinguish the difference of gene expressions profile between normal and malignant tissues [6]. Furthermore, microarray datasets also are useful for classifying tumors by their expression profiles and provide useful biological, prognostic and diagnostic information for mechanistic research and contrivance [3]. Most microarray databases usually include tens of thousands of genes extracted from many anamnesis data. Identifying the cancer specific gene markers from a large number of data by manpower is imprecise, expensive and time consuming. Hence, we can perform this work by computer science techniques to get more performance, and further to develop therapy software systems for assisting diagnosis and therapy.

In this paper, the ovarian cDNA expression database is obtained from ovarian tissues of patients and collected in years 2001-2003 under collaborative efforts of surgical and pathological units at China Medical University Hospital in Taiwan. This database includes ovarian cDNA expression datasets of 30 patients, and each sample includes 9,600 genes. The pathogenic stages of ovarian cancer include benign ovarian tumor (OVT), borderline tumor (BOT), ovarian cancer at stage I (OVCA-I), and ovarian cancer at stage III (OVCA-III). The number of samples for OVT, BOT, OVCA-I and OVCA-III is 8, 6, 7 and 9, respectively. In statistics, when the number of samples is more than thirty, the samples are enough to represent the population. Hence, the size of our ovarian cDNA expression database is large enough to be analyzed in statistical methods and can provide reliable analysis results. The analysis methods include two stages: the first stage is preprocessing stage for culling the most of nonspecific genes, and the second stage uses classification method to identify the oncogenes which might be the ovarian cancer specific markers. Because of our database includes 9,600 genes, it is too much for analyzing. At the first stage, the linear regression analysis is used to select the genes that their expressions are appearing different from others. The analysis of variable (ANOVA) method is used to test and select the gene expression values are appearing different in various pathogenic stages. At the second stage, a supervised classification method, support vector machine (SVM) is used to classify each pathogenic stage by the expressions of the selected genes. When one or some genes can provide well classification results, it means that these genes can used to identify the tissue of a patient is belonged to which pathogenic stages.

The analysis procedure and methods are presented in Section 2. Some analysis data and results are shown in Section 3. Section 4 presents some discussions and future works of this paper.

2 Related Analysis Methods

2.1 Linear Regression

In statistics, a classic problem is how to define the relationship between two sets of random variables X and Y. Linear regression is used to obtain a straight line to represent a fit relationship between two set of data [25]. The equation of linear regression line is defined as Equation (1), and the variables X and Y can be classified into an explanatory and a dependent variable. Least squares method is the most common one to fit a regression line of linear regression. It finds the best fitting line for the data with the least sum of the squares of the vertical deviations between each data point and the line, and the vertical deviations also are called residuals.

$$Y = a + bX + e \tag{1}$$

where e is a residual and is a random variable with mean zero; a and b are the coefficients which can get the smallest sum of the square residuals.

2.2 Analysis of Variance (ANOVA)

Analysis of variance (ANOVA) is a statistical method which used to test the difference exists in the means of more than two populations. Another similar method is t-test which can be used to test the difference of means between two populations. ANOVA and t-test both are based on the well-known hypothesis testing. Hypothesis testing is used to test the means of more than one population are the same or not. Before testing, there is a hypothesis that the means of all populations are the same, and this hypothesis is called null hypothesis. Another hypothesis is an alternative hypothesis which supposes the null hypothesis is not true. In actual cases, when the means of different data are not equal in certain probability, the null hypothesis must be rejected and the alternative hypothesis is accepted. The probability are called P-value, when the P-value is less then the lower bound α, it is more difficult to reject null hypothesis. In ANOVA and t-test, the difference of means is estimated by the standard error. The hypothesis testing has two well-known kinds of errors. The first kind of errors is occurring when the null hypothesis actually is true, but the statistical decision is false, this error is called "Type I error". The second kind of errors is occurring when the null hypothesis is actually false, but the statistical decision is true, and this error is called "Type II error". In practice, ANOVA uses pairwise t-tests to test the differences in more than two populations.

In this paper, ANOVA is used to test and distinguish the difference of gene expressions at OVT, BOT, OVCA-I and OVCA-III stages. When a gene which have specific different between each pathological stage, it can be used to classify each pathological stage more easily. Hence, ANOVA is an important and useful preprocessing method for selecting these suspect disease linked genes, and let the following classification procedure perform effectively to get the oncogenes to be ovarian cancer specific marks.

2.3 Support Vector Machine (SVM)

Support vector machine (SVM) is a superior learning algorithm based on statistical theory proposed by Vapnik and collaborators [13, 14]. In recent years, SVM is generally applied in bioinformatics, pattern recognition, feature classification and function approximation. SVM is based on the idea of structured risk minimization on a nested set structure of separating hyperplanes. SVM maps the nonlinear feature space to the linear feature space, the linear feature space is in high dimensional space where the classification becomes efficiently and precisely.

In general, when SVM in binary classification condition, the sample data set $S = \{(x_i, y_i) \mid i = 1, 2, ..., l\}$, $x_i \in R^n$, $y_i \in R$, that x_i denotes the ith input pattern with n tuples, and $y_i \in \{1, -1\}$, is the ith output result. According to above factors, the trained SVM should minimize the generalization error, or at least minimizes an upper bound on it. It is shown that the hyperplane with this property is the one that leaves the maximum margin between the two classes. Given a new data z to classify, the optimal function can be described as follow:

$$f(z) = \text{sign}\left(\sum_{i=1}^{l} \alpha_i y_i \langle x_i, z \rangle + b \right). \tag{2}$$

3   Methods and Analyzed Results

Microarray chips include many fluorescence spot ellipses to measure the expression intensities of different genes. Before hybridizing, the researches should determine which genes they want to measure and locate each gene on the microarray chip. Generally speaking, the microarray datasets must include the samples of different pathological stages for distinguishing the gene expression of each stage. The most common method of estimating the gene expression is determined by the mean fluorescence intensities of the all pixels within the spot ellipse. The gene expressions which are extracted from the normal and cancerous cases are defined as *NEI* and *CEI*. In order to let the gene expressions have consistent base, the background value of normal and cancerous cases are defined as *NEB* and *CEB*. The normalized gene expressions *NNEI* and *NCEI* can be obtained by Equation (3). In our ovarian cDNA expression database, the gene expressions are calculated by Equation (4) and are stored in decimal fraction form.

$$NNEI = NEI - NEB,$$
$$NCEI = CEI - CEB. \tag{3}$$

$$NE = \log_2(NNEI),$$
$$NC = \log_2(NCEI). \tag{4}$$

In this paper, the research topic is about ovarian cancer specific markers identification. The database is obtained from the ovarian tissues of the patients, was collected in years 2001-2003 under collaborative efforts of surgical and pathological units at China Medical University Hospital in Taiwan. This database includes ovarian cDNA expression datasets of 30 patients, and each sample includes 9,600 genes. The pathogenic stages of ovarian cancer are defined as benign ovarian tumor (OVT), borderline tumor (BOT), ovarian cancer at stage I (OVCA-I), and ovarian cancer at stage III (OVCA-III). The number of samples at OVT, BOT, OVCA-I and OVCA-III is 8, 6, 7 and 9. In statistics, when the number of samples is more than thirty, the samples are enough to represent the statistical population. Hence, our database is large enough to be analyzed in statistical methods and can provide reliable analysis results.

Because of our cDNA expression datasets of ovarian cancer include 9,600 genes for each sample. Expression of many genes may not have significant difference between each pathological stage. In our analysis system, each gene expression can be shown as a curve line graph and further analyze each gene expression using medians, means and distribution with different pathological stages. The linear regression analysis can be used to reduce the number of genes by genetic expression analysis. In our linear regression, variables X are the means of each gene in the samples of OVT, and variables Y are the means of each gene in samples of BOT, OVCA-I and OVCA-III. All genes are sorted by their residuals, and the genes which have 200 largest residuals are selected to be the output of linear regression analysis. The data point with largest residuals have significant differences between the values of X and Y, it also means the gene expressions may have significant differences between normal and cancerous tissues.

The ovarian cancer gene markers are genes that have a significant difference between each pathological stage. Analysis of variance (ANOVA) and Scheffe are used to test and identify the differences of gene expressions. The genes with significant differences may have correlation with ovarian cancer and can assist the biologists to identify the oncogenes more effectively. This paper performs ANOVA analysis for each gene respectively. When the testing result of a gene rejects H_0, it presents the gene has significant differences between each pathological stage and can be used to get better results of clustering some stages. After ANOVA, the Scheffe is used to perform some multiple comparisons for each gene between each pair of all pathological stages. Theoretically, when the P-value of hypothesis testing is less than $\alpha\ (= 0.05)$, the data has significant difference. In our experiment, when the α-value is set to 0.000005, we still can get 14 genes which have significant differences. It means the expressions of 14 genes have very significant differences at each pathetical stage, They include: EGR1 (early growth response 1), MMP2 (matrix metalloproteinase 2), C1S (complement component 1, s subcomponent), SPARCL1

(SPARC like 1), FN1 (fibronectin 1), C19orf7 (chromosome 19 open reading frame 7), C9orf40 (chromosome 9 open reading frame 40), WEDC2 (WAP four disulfide core domain 2), CDK5R1 (Cyclin-ependent kinase 5), DHFR (dihydrofolate reductase), KIAA0576 (KIAA0576 protein), MYST2 (MYST histone acetyltransferase 2), PGK1 (phosphoglycerate kinase 1) and MDFI (MyoD family inhibitor).

According to the 14 genes by ANOVA analysis, the procedure in this section is to classify the datasets of these 14 genes into 5 conditions by BOT, respectively. The gene expression datasets of 14 genes analyzed by ANOVA are going to train SVM. After training SVM, the optimal decision function $f(x)$ also can be obtained. Consequently, the testing datasets can be classified by the optimal decision function $f(x)$ with 5 classified conditions by BOT.

Considering the classified conditions, there are 5 classified conditions which are:

Classifier 1 $\begin{cases} C_1: BOT \\ C_2: OVT, OVCA\text{-}I, OVCA\text{-}III \end{cases}$, Classifier 2 $\begin{cases} C_1: OVT, BOT, OVCA\text{-}I \\ C_2: OVCA\text{-}III \end{cases}$,

Classifier 3 $\begin{cases} C_1: OVT, BOT \\ C_2: OVCA\text{-}I, OVCA\text{-}III \end{cases}$, Classifier 4 $\begin{cases} C_1: BOT, OVCA\text{-}I \\ C_2: OVT, OVCA\text{-}III \end{cases}$,

Classifier 5 $\begin{cases} C_1: OVT \\ C_2: BOT, OVCA\text{-}I, OVCA\text{-}III \end{cases}$.

In the classification experiment, LIBSVM [26] is applied in classification where the kernel type is radial basis function (RBF). The gene expression datasets of 14 genes are classified by the trained SVM for 5 classification conditions, respectively. The classification results of these 5 conditions are shown in Table 1 to Table 5, respectively. In the classification results, we can discover the expression in different classification conditions for 14 genes by BOT. In classifier 1, there are 4 notable genes including early growth response 1, dihydrofolate reductase, KIAA0576 protein and MyoD family inhibitor. The classifications for classifier 2 to classifier 4 also have high accuracy of classification; the genes are notable by SVM classifications. In classifier 5, the average accuracy of classification is 79.269%. However, fibronectin 1 has 80.488% accuracy to distinguish whether the tumor is cancerous or not in our experiments.

From the classification results, we can discover the notable expression of 14 genes at BOT stage. Therefore, the proposed method can clearly analyze human ovarian cancer by these 14 genes at BOT stage before turning to ovarian cancer stage.

Table 1. Classification Results of Classifier 1

Gene Name	Accuracy (%)
early growth response 1	92.5
dihydrofolate reductase	95.122
KIAA0576 protein	100
MyoD family inhibitor	95.122
Average	95.686

Table 2. Classification Results of Classifier 2

Gene Name	Accuracy (%)
chromosome 19 open reading frame 7	90.244
cyclin-dependent kinase 5	92.683
MYST histone acetyltransferase 2	90.244
Average	91.057

Table 3. Classification Results of Classifier 3

Gene Name	Accuracy (%)
matrix metalloproteinase 2	85.366
SPARC-like 1	87.543
Average	86.455

Table 4. Classification Results of Classifier 4

Gene Name	Accuracy (%)
WAP four-disulfide core domain 2	90.239
phosphoglycerate kinase 1	85.366
chromosome 9 open reading frame 40	95.122
Average	90.242

Table 5. Classification Results of Classifier 5

Gene Name	Accuracy (%)
complement component 1, s subcomponent	78.049
fibronectin 1	80.488
Average	79.269

4   Discussions

In this paper, we use some statistical methods and support vector machine to analyze and identify the ovarian cancer gene markers for an authentic ovarian cDNA expression database at various stages of ovarian tumors (among 8 normal ovarian tumors, 6 borderline of cancers, 7 ovarian cancer at stage I and 9 ovarian cancer at stage III). The total number of samples is 30, hence the analysis results are authentic in statistical theory. First, linear regression analysis can be used to obtain 200 useful genes with largest residuals which in order to perform statistical testing and multiple comparisons for these genes by ANOVA and Scheffe. It can select 14 genes which have significant differences for each pathological stage when the α-value is 0.000005. Most of these genes are the oncogenes of cancers presented in many biological experiments and research papers [15-24]. Hence, it means that our ANOVA analysis can obtain the genes wanted for us correctly and effectively.

Then, we use SVM to classify the pathological stages by gene expressions. In the first classification experiment, the cluster 1 includes BOT, and other pathological stages are in cluster 2. We find the expressions of 4 genes have significant differences at BOT stage and can get average accuracy about 95.686% in cross-validation. It is quite precise for classifying pathological stages by gene expressions. This classification is very useful and can be applied to assist biologists or doctors to identify or diagnose whether the patients are passing the borderline of tumor stage into malignant tumors. The second classification experiment includes 3 genes, and its average accuracy is about 91.057% in cross-validation. This classification can be used to identify the patients who are entering ovarian cancer. The other kinds of classification experiments can get accuracy about 86.455%, 90.242% and 79.269% in cross-validation, respectively. They also can be used to identify pathological stages of ovarian cancer in different requirements and applications, which can keep stable accuracy rate.

In average expressions of the discovered genes at four stages, we can discover which gene is remarkable and probably transform to ovarian cancer. Finally, this paper uses statistical methods and support vector machine for identifying various stages of ovarian cancer by gene expressions. The analyzed results show that the proposed method can get the average accuracy is about 88.541%, and it can provide assistances for biologists and doctors. The proposed method can be extensively applied to microarray analysis and bioinformatics in different kinds of cancer for gene expression analysis.

References

1. Jemal, A., Thomas, A., Murray, T., Thun, M.: Cancer statistics 2002. CA Cancer J. Clin. 52, 23–47 (2002)
2. Website: http://www.healthandenvironment.org/ovarian_cancer
3. Huang, G.-S., Hung, Y.-C., Chen, A., Hong, M.-Y.: Microarray analysis of ovarian cancer. In: 2005 IEEE International Conference on Systems, Man and Cybernetics, vol. 2, pp. 1036–1041 (2005)
4. Schena, M., Shalon, D., Davis, R.W., Brown, P.O.: Quantitative monitoring of gene expression patterns with a complementary DNA microarray. Science 270, 467–470 (1995)
5. Alizadeh, A.A., Eisen, M.B., Davis, R.E., et al.: Distinct types of diffuse large B-cell lymphoma identified by gene expression profiling. Nature 403, 503–511 (2000)
6. DeRisi, J., Penland, L., Brown, P.O., Bittner, M.L., Meltzer, P.S., Ray, M., Chen, Y., Su, Y.A., Trent, J.M.: Use of a cDNA microarray to analyze gene expression patterns in human cancer. Nat. Genet. 14, 457–460 (1996)
7. Schummer, M., Ng, W.V., Bumgarner, R.E., Nelson, P.S., Schummer, B., Ednarski, D.W., Hassell, L., Baldwin, R.L., Karlan, B.Y., Hood, L.: Comparative hybridization of an array of 21,500 ovarian cDNAs for the discovery of genes overexpressed in ovarian carcinomas. Gene 238, 375–385 (1999)
8. Wang, K., Gan, L., Jeffery, E., Gayle, M., Gown, A.M., Skelly, M., Nelson, P.S., Ng, W.V., Schummer, M., Hood, L., Mulligan, J.: Monitoring gene expression profile changes in ovarian carcinomas using cDNA microarray. Gene 229, 101–108 (1999)

9. Ismail, R.S., Baldwin, R.L., Fang, J., Browning, D., Karlan, B.Y., Gasson, J.C., Chang, D.D.: Differential gene expression between normal and tumor-derived ovarian epithelial cells. Cancer Res. 60, 6744–6749 (2000)
10. Martoglio, A.M., Tom, B.D., Starkey, M., Corps, A.N., Charnock-Jones, D.S., Smith, S.K.: Changes in tumorigenesis- and angiogenesis-related gene transcript abundance profiles in ovarian cancer detected by tailored high density cDNA arrays. Mol. Med. 6, 750–765 (2000)
11. Ono, K., Tanaka, T., Tsunoda, T., Kitahara, O., Kihara, C., Okamoto, A., Ochiai, K.: Identification by cDNA microarray of genes involved in ovarian carcinogenesis. Cancer Res. 60, 5007–5011 (2000)
12. Welsh, J.B., Zarrinkar, P.P., Sapinoso, L.M., Kern, S.G., Behling, C.A., Monk, B.J., Lockhart, D.J., Burger, R.A., Hampton, G.M.: Analysis of gene expression profiles in normal and neoplastic ovarian tissue samples identifies candidate molecular markers of epithelial ovarian cancer. Proc. Natl. Acad. Sci. USA 98, 1176–1181 (2001)
13. Vapnik, V.: The Nature of Statistical Learning Theory. Springer, New York (1995)
14. Vapnik, V.: Statistical Learning Theory. John Wiley, New York (1998)
15. Yang, S.Z., Eltoum, I.A., Abdulkadir, S.A.: Enhanced EGR1 activity promotes the growth of prostate cancer cells in an androgen-depleted environment. J. Cell Biochem. 97, 1292–1299 (2006)
16. Jeng, J.-T., Lee, T.-T., Lee, Y.-C.: Classification of ovarian cancer based on intelligent systems with microarray data. IEEE International Conference on Systems, Man and Cybernetics 2, 1053–1058 (2005)
17. Yemelyanov, A., Czwornog, J., Chebotaev, D., Karseladze, A., Kulevitch, E., Yang, X., Budunova, I.: Tumor suppressor activity of glucocorticoid receptor in the prostate. Oncogene, pp. 1885–1896 (2006)
18. Ustach, C.V., Kim, H.R.: Platelet-derived growth factor D is activated by urokinase plasminogen activator in prostate carcinoma cells. Mol. Cell Biol. 24, 6279–6288 (2005)
19. Watermann, I., Gerspach, J., Lehne, M., Seufert, J., Schneider, B., Pfizenmaier, K., Wajant, H.: Activation of CD95L fusion protein prodrugs by tumor-associated proteases. Cell Death Differ 14, 765–774 (2007)
20. Vanharanta, S., Wortham, N.C., Arola, J., Tomlinson, I.P., Karhu, A., Arango, D., Aaltonen, L.A.: 7q deletion mapping and expression profiling in uterine fibroids. Oncogene, 6545–6554 (2005)
21. Hanauske, A.R.: Translational research with pemetrexed in breast cancer. Oncology 18, 66–69 (2004)
22. Morales, C., Ribas, M., Aiza, G., Peinado, M.A.: Genetic determinants of methotrexate responsiveness and resistance in colon cancer cells. Oncogene, 6842–6847 (2005)
23. Yu, J., Baron, V., Mercola, D., Mustelin, T., Adamson, E.D.: A network of p73, p53 and Egr1 is required for efficient apoptosis in tumor cells. Cell Death Differ, 436–446 (2007)
24. Selvamurugan, N., Kwok, S., Partridge, N.C.: Smad3 interacts with JunB and Cbfa1/Runx2 for transforming growth factor-beta1-stimulated collagenase-3 expression in human breast cancer cells. J. Biol. Chem. 279, 27764–27773 (2004)
25. Edwards, A.L.: An Introduction to Linear Regression and Correlation. W. H. Freeman, San Francisco, CA (1976)
26. Chang, C.-C., Lin, C.-J.: LIBSVM: a library for support vector machines (2001), software available at http://www.csie.ntu.edu.tw/ cjlin/libsvm

Ancestral DNA Sequence Reconstruction Using Recursive Genetic Algorithms

Mauricio Martínez[1], Edgar E. Vallejo[1], and Enrique Morett[2]

[1] ITESM-CEM, Computer Science Dept.
Atizapán de Zaragoza, Edo. de México, 52926, México
{A00964166,vallejo}@itesm.mx
[2] IBT UNAM, Dept. of Cellular Engineering and Biocatalysis
Cuernavaca, Morelos, 62210, México
emorett@ibt.unam.mx

Abstract. This paper explores the capabilities of genetic algorithms for reconstructing ancestral DNA sequences. We conducted a series of experiments on reconstructing ancestral states from a given collection of taxa and their phylogenetic relationships. We tested the proposed model using simulated phylogenies obtained from actual DNA sequences by applying realistic mutation rates. Experimental results demonstrated that the recursive application of genetic algorithms to smaller instances of the problem allows us to reconstruct ancestral DNA states accurately.

1 Introduction

Nowadays, an increasing number of complete genomes from diverse organisms is available to enable the understanding of life at the molecular level. The elucidation of accurate evolutionary relationships among these organisms (and perhaps others, including extinct species) from their molecular data remains a challenge for molecular evolution research.

Molecular evolution investigations incorporating ancestral sequences hold the potential for providing a larger picture of the evolutionary process. In effect, these studies would help us to understand fundamental mechanisms of evolutionary change that operate at the molecular level, such as gene duplication and horizontal gene transfer. For instance, ancestral sequence reconstruction algorithms have been used for predicting protein function, and for discovering potential gene homologues, among other applications (Edwards and Shields, 2005; Collins, et al, 2003, Blanchette, et al, 2004).

Ancestral sequence reconstruction algorithms work by searching among possible states for those that exhibit desirable properties according to some specified optimality criterion (*e. g.* maximum parsimony, maximum likelihood) (Felsenstein, 2004). However, it is currently impossible for even the faster computers to exhaustively search all possible states for more than a moderate number of taxa, since they can be associated to huge numbers of different sequences, but only one of which is presumably the correct representation of the actual evolutionary history.

M. Randall, H.A. Abbass, and J. Wiles (Eds.): ACAL 2007, LNAI 4828, pp. 390–400, 2007.
© Springer-Verlag Berlin Heidelberg 2007

There have been many attempts to develop robust and fast ancestral sequence reconstruction algorithms (Hall, 2006; Koshi and Goldstein, 1996; Pupko, et al, 2000; Pupko, et al, 2002; Yang, et al, 1995). However, there is no general, sound and complete method capable of producing the optimal ancestral sequences from an arbitrary collection of taxa. For example, maximum parsimony are capable of producing the correct states only when there are no backward and no parallel substitutions at each nucleotide site. In practice, however, nucleotide sequences are often subject to these changes (Avise, 2004).

There is an increasing interest in applying evolutionary algorithms to computationally intensive bioinformatics problems (Fogel and Corne, 2003). For example, previous work on phylogenetic reconstruction using genetic algorithms has yielded competitive results (Lewis, 1998; Matsuda, 1996). We would like to posit here that artificial evolution (*i. e.* genetic algorithms) can be used to simulate natural evolution backwards to reconstruct ancestral states accurately.

This paper describe a series of experiments on the reconstruction of ancestral DNA sequences using simulated phylogenies. Preliminary results indicate that the proposed method is able to reconstruct ancestral DNA sequences accurately. In effect, recursive genetic algorithms slightly outperform maximum parsimony and maximum likelihood methods in our experiments. In addition, we found that the recursive application of the GAs is a promising heuristic for approximating hierarchically structured problems.

2   Methods

2.1   Data Set

A common complication in ancestral sequence reconstruction experiments is that true ancestral states are seldom known with absolute certainty. In effect, there is commonly a lack of direct evidence to evaluate the performance of ancestral inference. The creation of artificial phylogenies using simulation (Hillis, 1995) is becoming a widely used method for the validation of phylogenetic reconstruction algorithms. In the experiments reported here, we use a collection of artificial phylogenies generated from fragments of actual bacterial genomes to test the performance of the proposed method.

Particularly, we created a data set of fixed length nucleotide ancestral sequences by simulating the mutation process recursively. Using Kimura's two-parameter model, we obtained a collection of 4-level trees by the progressive application of the mutation rate (0.1) to each node of the tree (beginning at the ancestral node) to obtain 0, 1 or 2 child nodes. According to the underlying substitution model, mutations were further divided into transitions (70%) and transversions (30%) (Kimura, 1980). We tested the proposed method using both symmetric and asymmetric trees as shown in Figure 1 and Figure 2, respectively.

The existence of simulated phylogenies including ancestral sequences allows us to validate the accuracy of the proposed method objectively. Particularly, the performance of the genetic algorithm will be assessed by comparing the obtained ancestral sequences to those of the simulated phylogenies.

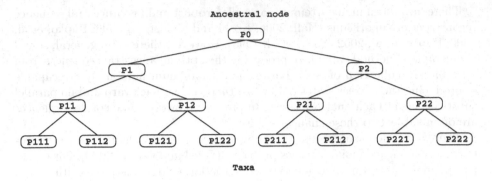

Fig. 1. Symetric tree

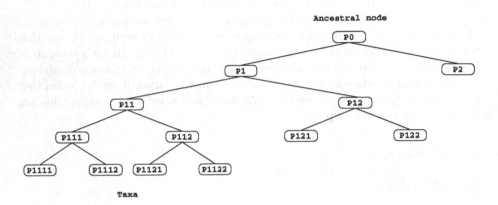

Fig. 2. Asymetric tree

2.2 Genetic Algorithms

We designed a genetic algorithm to evolve a population of ancestral sequences. In general, the design of a genetic algorithm involves the definition of the genome representation, genetic operators, fitness function, and parameters for the simulations (Mitchell, 1996).

Genome representation. Each individual of the population consists of a fixed-lenght binary string representing the concatenation of the ancestral sequences intended to be reconstructed. Figure 3 shows the chromosome that represents a candidate solution for the tree of Figure 1, assuming the objective of reconstructing all of the ancestral sequences of the tree.

Fig. 3. GA chromosomes consist of binary coded ancestral sequences

Fitness function. Ancestral sequences along with evolutionary change should be able to explain the observed taxa; the converse should also be true. In effect, mutation progressively alters the genomes of organisms over generations. However, mutation rates are often small so as to allow natural selection to preserve characteristics that contribute to the adaptation of organisms. As a consequence, we expect taxa sequences to be highly similar to those of their immediate ancestor. However, these similarities are expected to be progressively diluted as we approach the origin of the underlying phylogenetic tree.

Following these observations, we formulated the fitness function as the pairwise comparisons of the evolving ancestral sequences to those of their immediate descendants. Specifically, matchings between an ancestral sequence and a descendant sequence at corresponding positions are added to overall fitness (+1), mismatchings at corresponding positions do not contribute to the fitness of the candidate solution.

Note that the fitness function does not carry information on the underlying substitution model (i.e. transition-tranversition rate). We hypothesize that the proposed fitness function is sufficiently informative for the accurate reconstruction of ancestral states.

Genetic operators. We used fitness proportional selection with linear scaling, one-point crossover and point mutation that operate on fixed length binary chromosomes. This genetic operators combination is often used in the practice of genetic algorithms (Mitchell, 1996).

Parameters for the simulations. Simulations were conducted using different combinations of parameter values: generations (1000–10000), population size (1000–6000), chromosome length (100–1600), crossover probability (0.6-0.7) and mutation probability (0.001-0.01). This parameters were obtained from empirical observations of preliminary experiments.

3   Experiments and Results

3.1   Canonical GAs: Symmetric Trees

In this experiment, we explore the capabilities of simple GAs for simultaneously reconstructing all the ancestral sequences of a symmetric 4-level tree. Particularly, ancestral sequences P_i were reconstructed from taxa T_j (see Figure 4).

Figure 5 presents the results of this experiment. In can be appreciated that fitness increased rapidly in early stages of the simulation. However, the algorithm showed a poor performance when comparing the evolved sequences with the original ancestral sequences belonging to the simulated phylogenies.

Therefore, we concluded that reconstructing ancestral sequences simultaneously would be extremely hard in practice as errors at a particular ancestral sequence would be easily propagated in the chromosome. This is an indication of the existence of a high degree of epistasis (*i. e.* interactions among genes) in the representation. To overcome this limitation, we believe the problem should

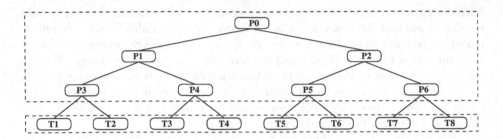

Fig. 4. Simultaneous reconstruction of ancestral sequences

be approached using a divide-and-conquer strategy. As a consequence, we arrived to the formulation of recursive genetic algorithms (RGAs).

3.2   Recursive GAs: Symmetric Trees

A recursive GA consists of the progressive application of the GA to approximate a hierarchically structured problem. The proposed GA uses the results produced in previous applications of the algorithm in solving smaller nested instances of the problem.

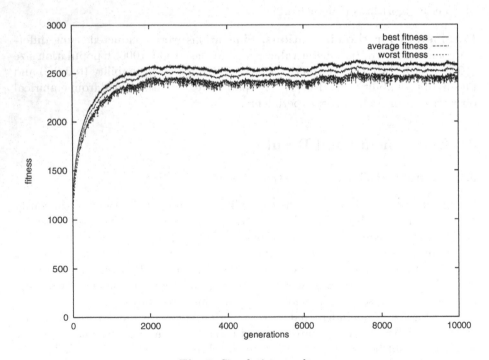

Fig. 5. Simulation results

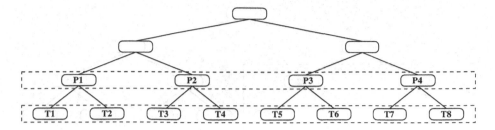

Fig. 6. Reconstruction of ancestral sequences by level

```
proc reconstruct (node)
    for each child i of node
        if (i ≠ leaf)
            reconstruct(i)
    end for
    run GA to reconstruct node from all child i
    return (node)
end proc
```

Fig. 7. reconstruct procedure

Particularly, in reconstructing ancestral sequences, the recursive GA may be viewed as a bottom-up procedure. The ancestral sequences produced by the application of the GA to the taxa are used to progressively reconstruct the ancestral sequences of the nodes belonging to higher levels of the phylogenetic tree. This procedure may be organized as the top-down nesting of calls to the GA and bottom-up reconstruction of each level of the tree (see Figure 6).

The pseudocode of the recursive procedure is delineated in figure 7. The procedure is called recursively from the root of the tree until the leaves are reached. Then the reconstruction of ancestral sequences is achieved by the application of the GA during the bottom-up transversal of the tree.

Figure 8 presents the results of this experiment. Each graph shows the convergence of the algorithm for each of the three reconstructed levels of the tree. Note there are differences in the degree of fitness of different levels as there

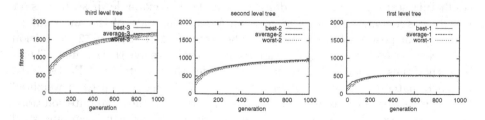

Fig. 8. Results by level

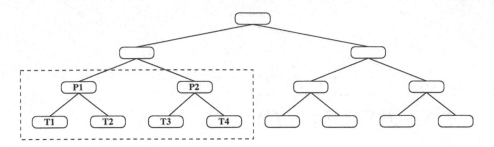

Fig. 9. 4×2 strategy

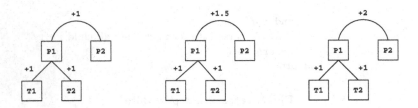

Fig. 10. Weighted similarity to the co-evolving sibling

are fewer sequences at the upper levels of the tree. However, convergence was achieved consistently. Particularly, the algorithm showed a better performance than the previous experiment when comparing the reconstructed sequences with the original ancestral sequences of the simulated phylogenies.

We also conducted a series of experiments consisting of the application of the recursive GA to small instances of the ancestral sequence reconstruction problem. Particularly, we inferred two ancestral sequences from four taxa in one application of the GA. We called this approach the 4×2 strategy. Figure 9 shows and example in which sequences P1 and P2 are simultaneously reconstructed from sequences T1, T2, T3 and T4.

In this strategy two ancestral sequences will be evolving simultaneously. The existence of a co-evolving sibling in the chromosome can be useful to solve conflicting states among descendants of a node. For example, if sequences T1 and T2 has a conflicting state at position k then the state of the ancestral sequence P1 at position k could be either the state of T1 or state of T2 at position k. However, the state of P2 at position k can be used to solve this conflict. Therefore, the fitness function is modified to consider these matchings. Particularly, for each matching at corresponding positions of sibling sequences, a weighted similarity coefficient w is added to the overall fitness of the candidate solution.

We considered different values for the weighted similarity w of an ancestral sequence to his co-evolving sibling as shown in Figure 10. The value of $w = 2.0$ seemed to be the most appropriate in our preliminary experiments.

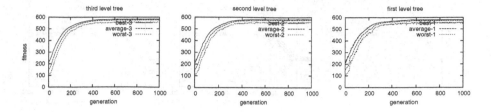

Fig. 11. Results by level 4 × 2 strategy

Figure 11 presents the results of this experiment. Each graph shows the convergence of the algorithm for each of the three levels of the tree. Note there are similarities in the degree of fitness at different instances of the algorithm as there are always two evolving ancestral sequences. The algorithm showed a better performance than both previous experiments when comparing the reconstructed sequences to the original ancestral sequences of the simulated phylogenies.

3.3   Recursive GAs: Asymmetric Trees

In this experiment, we were unable to use the 4×2 strategy as an asymmetric tree would not possess this structure in general. Instead, we used a 2 × 1 strategy, in which the recursive GA is used to infer one ancestral sequence at each application of the GA (see Figure 12).

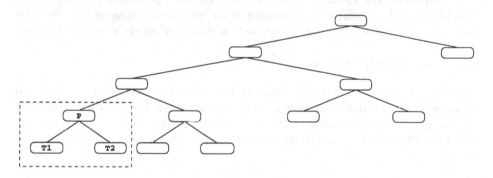

Fig. 12. 2 × 1 strategy

Figure 13 presents the results of this experiment. Each graph shows the convergence of the algorithm for each of the three levels of the tree. This experiment produced results that are comparable to those produced by the 4 × 2 strategy with respect to the original ancestral sequences.

3.4   Overall Results

Table 1 presents the average accuracy of each of the previous experiments with respect to the original sequences over 10 simulations. Results indicate that the

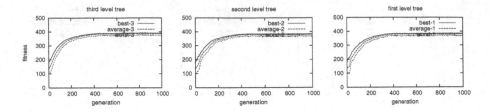

Fig. 13. Results by level 2 × 1 strategy

application of the recursive GA to smaller instances of the problem produces better results. Particularly, the 4 × 2 strategy produced very competitive results, comparable to those produced by traditional methods (Zhang and Nei, 1997).

Table 1. Overall results

Level of the tree	Level-by-level	4 × 2 strategy	2 × 1 strategy
Level 0	71%	94%	90%
Level 1	88%	91%	87%
Level 2	93%	91%	91%

It is important to point out that asymmetric trees are more common in practice than symmetric trees. Therefore, the fact that the performance of the recursive GA scaled gracefully from symmetric to asymmetric trees shows that the proposed method is a promising approach to ancestral sequence reconstruction.

3.5 Comparative Study

In addition, we conducted a preliminary comparative study with conventional approaches. Particularly, we contrasted the obtained results with those produced by the maximum parsimony (MP) and maximum likelihood (ML) algorithms using the PHYLIP program (Retief, 2000). The comparisons are shown in figures 16 and 17, respectively.

Table 2. Comparative study: MP vs. RGA

Method	Symmetric trees	Asymmetric trees
Maximum Parsimony (MP)	95%	94%
Recursive GA (RGA)	96%	97%

Both MP and ML algorithms were unsuccessful in reconstructing the topology of the underlying tree. Therefore, comparisons were conducted exclusively with respect to the root node of the tree. Experiments using simulated phylogenies allowed us to uncover this limitations of conventional phylogenetic reconstruction algorithms.

Table 3. Comparative study: ML vs. RGA

Method	Symmetric trees	Asymmetric trees
Maximum Likelihood (ML)	96%	96%
Recursive GA (RGA)	96%	97%

4   Discussion

The preliminary results reported here indicate that genetic algorithms are capable of reconstructing ancestral DNA sequences accurately. However, due to the potential propagation of errors produced by a high degree of epistasis in the representation, it is necessary to apply this search procedure using a divide-and-conquer strategy. We propose here to perform genetic search recursively, by the subsequent application of the GA to the results produced by the previous application of the GA. In this way, we arrived to the formulation of the recursive genetic algorithm search procedure.

Overall, we believe that GAs provide an appropriate methodology for searching the space of DNA ancestral sequences. In effect, phylogenetic studies are often confronted to the fact that most phylogenetic inference algorithms for reconstructing ancestral sequences are intractable for realistic applications. On the contrary, GAs are able to produce competitive results using a moderate amount of computational resources.

The construction of simulated phylogenetic trees allowed us to validate the performance of the proposed method objectively. The results were consistent even when either a moderate or a large number of artificially created ancestral sequences were considered.

The focus of this study has been on the reconstruction of DNA sequences. An immediate extension of this work would be the consideration of protein sequences (Cai, et al, 2004). In addition, careful statistical validations are required to attest the accuracy of the GA approach to ancestral sequence reconstruction in general.

Once reliable ancestral sequences from extinct organisms reconstructed *in silico* are available, it would be interesting to formulate and test new hypothesis on molecular evolution. These studies would provide a larger framework for understanding the properties of evolution.

Acknowledgments

This work was supported by the Consejo Nacional de Ciencia y Tecnología (CONACYT) under SEP-CONACYT award No. SEP-2004-C01-47434 . Any opinions, findings and conclusions or recommendations expressed in this publication are those of the authors and do not necessarily reflect the views of the sponsoring agency.

References

Avise, J.C.: Molecular Markers, Natural History and Evolution, 2nd edn. Sinauer Associates Inc. (2004)

Blanchette, M., Green, E.D., Miller, W., Haussler, D.: Reconstructing large regions of an ancestral mammalian genome in silico. Genome Res. 14, 2412–2423 (2004)

Cai, W., Pei, J., Grishin, N.V.: Reconstruction of ancestral protein sequences and its applications. BMC Evol. Biol. 4, 33 (2004)

Collins, L.J., Poole, A.M., Penny, D.: Using ancestral sequences to uncover potential gene homologues. Appl. Bioinformatics 2(Suppl. 3), S85–95 (2003)

Edwards, R.J., Shields, D.C.: BADASP: predicting functional specificity in protein families using ancestral sequences. Bioinformatics 21, 4190–4191 (2005)

Felsenstein, J.: Inferring Phylogenies. Sinauer, Associates (2004)

Fogel, G.B., Corne, D.W.: Evolutionary Computation in Bioinformatics. MIT Press, Cambridge (2003)

Hall, B.G.: Simple and accurate estimation of ancestral protein sequences. Proc. Natl. Acad. Sci. USA 103, 5431–5436 (2006)

Hillis, D.M.: Approaches for assesing phylogenetic accuracy. Syst. Biol. 44, 3–16 (1995)

Kimura, M.: A simple method for estimating evolutionary rate of base substitution through comparative studies of nucleotide sequences. J. Mol. Evol. 16, 111–120 (1980)

Koshi, J.M., Goldstein, R.A.: Probabilistic reconstruction of ancestral protein sequences. J. Mol. Evol. 42, 313–320 (1996)

Lewis, P.O.: A genetic algorithm for maximum-likelihood phylogeny inference using nucleotide sequence data. Mol. Biol. Evol. 15, 277–283 (1998)

Matsuda, H.: Protein phylogenetic inference using maximum-likelihood and genetic algorithms. In: Proceedings of the Pacific Symposium on Biocomputing 1996, World Scientific (1996)

Mitchell, M.: An Introduction to Genetic Algorithms. MIT Press, Cambridge (1996)

Pupko, T., Peer, I., Hasegawa, M., Graur, D., Friedman, N.: A branch-and-bound algorithm for the inference of ancestral amino-acid sequences when the replacement rate varies among sites: Application to the evolution of five gene families. Bioinformatics 18, 1116–1123 (2002)

Pupko, T., Peer, I., Shamir, R., Graur, D.: A fast algorithm for joint reconstruction of ancestral amino acid sequences. Mol. Biol. Evol. 17, 890–896 (2000)

Retief, J.D.: Phylogenetic analysis using PHYLIP. Methods Mol. Biol. 132, 243–258 (2000)

Yang, Z., Kumar, S., Nei, M.: A new method of inference of ancestral nucleotide and amino acid sequences. Genetics 141, 1641–1650 (1995)

Zhang, J., Nei, M.: Accuracies of ancestral amino acid sequences inferred by the parsimony, likelihood, and distance methods. J. Mol. Evol. 44(Suppl. 1), S139–146 (1997)

Author Index

Lecture Notes in Artificial Intelligence (LNAI)